THE EXPANDING WORLD OF CHEMICAL ENGINEERING

THE EXPANDING WORLD OF CHEMICAL ENGINEERING

SECOND EDITION

EDITED BY

SHINTARO FURUSAKI
DEPARTMENT OF CHEMICAL SYSTEMS AND ENGINEERING
KYUSHU UNIVERSITY
FUKUOKA, JAPAN

JOHN GARSIDE
DEPARTMENT OF CHEMICAL ENGINEERING
UNIVERSITY OF MANCHESTER INSTITUTE OF SCIENCE AND TECHNOLOGY (UMIST)
MANCHESTER, UK

AND

L.S. FAN
DEPARTMENT OF CHEMICAL ENGINEERING
OHIO STATE UNIVERSITY
COLUMBUS OHIO, USA

TAYLOR & FRANCIS
NEW YORK • LONDON

Published in 2002 by
Taylor & Francis
29 West 35th Street
New York, NY 10001

Published in Great Britain by
Taylor & Francis
11 New Fetter Lane
London EC4P 4EE

Printed in the United States of America on acid-free paper.

Library of Congress Cataloging-in-Publication Data

Kemikaru enjiniaringu no susume. English.
The expanding world of chemical engineering.--2nd ed. / edited by
Shintaro Furusaki, John Garside, and L.S. Fan.
p. cm.
Includes index.
ISBN 1-56032-917-3 (alk. paper)
1. Chemical engineering. I. Furusaki, S. (Shintaro), 1938- II. Garside, John.
II. Fan, Liang-Shih. IV. Title.

TP155 .K4613 2001
660--dc21 2001043408

CONTENTS

PREFACE TO SECOND EDITION

The aim of this second edition is to expand the book's content to include new chemical engineering fields and technologies expected to be vital in the 21st century, thus providing an understanding of chemical engineering's varied and growing role. The contributions of several additional North American authors have also made this edition more suitable for use as a textbook of technical English in non-English-speaking countries.

The newly added chapters are explained as follows:

Chapter 1. Computer chemistry is a useful technique for designing chemical materials for reactions and separation media. This chapter discusses the application of computers in chemical technology.

Chapter 5. Polymers are important materials for the construction of buildings and plants. They are used in many structural materials for a wide variety of equipment, and as functional materials for chemical reactions and separation of materials. This chapter covers the production and use of polymers with new functions.

Chapter 6. Catalysts are used for many reactions to increase the reaction rate. This chapter introduces new catalysts which will become important in the 21st century, such as chelating catalysts.

Chapter 7. Supercritical fluids are studied intensively for the separation of thermally unstable materials. They are also expected to be applicable for reactions with mass transfer. This new area of chemical engineering is described in this chapter.

Chapter 8. Colloid and interfacial phenomena are important topics to study in chemical engineering for interfacial reactions and separations. This chapter covers recent developments in this field.

Chapters 9–11. Bubble columns and fluidized beds are important apparatuses for chemical reactions, and particle technology is a basic technology not only for fluidization but also for many chemical processes using particles. New scopes in these fields are described in these chapters.

Chapter 12. Genetic engineering is the key technology in biotechnology. Chemical engineering approaches in this field and applications to biochemical processes are introduced in this chapter.

Chapter 15. Membranes are now used in many separation processes. Characteristics of biomembranes are investigated for application to the development of functional membranes. Membranes are used for environmental treatment and for medical technology as well. Many aspects of the development of membranes and such processes are presented in this chapter.

Chapter 18. Robots and micro-machines are interesting tools for chemical and biochemical industry. They can be used in precise control systems in chemical processes. This chapter describes recent trends in the application of robots and micro-machines in chemical and biochemical processes.

Chapter 19. Solar energy is one of the promising substitutes for the fossil fuel energy supply. The contributions of chemical engineering in this field are described in this chapter.

Other chapters from those described above were included in the first edition, and have now been revised to present recent developments in the research field of each chapter concerned.

The editors would like to thank Dr Norman Lloyd for correcting the English of the manuscripts by Japanese contributors. We also would like to thank Karen Gonser for refining the English in the chapters from North America.

Shintaro Furusaki
John Garside
L. S. Fan

January 2000

PREFACE TO FIRST EDITION

As will be clear from the preface to the original edition, this book has its origins in a Japanese language text published in 1987. By bringing together a number of Japanese chemical engineers working in the newer areas of the discipline this aimed to convey the challenges of recent developments in chemical engineering and the excitement experienced by those working in these fields. The present English language edition retains this objective, preserving many of the original chapters but also introducing some new material.

The book has been prepared with a number of different audiences in mind. By highlighting the expanding world of chemical engineering we hope it will encourage those in the later stages of their secondary school career to consider chemical engineering as a subject for study at university. It should also prove useful to those in their first year of chemical engineering courses by helping them to appreciate the enormous range of issues to which their chosen discipline can contribute. We hope that it may also be of interest to professional chemical engineers by bringing home to them the recent achievements and the future potential of chemical engineering. As an added benefit, its primarily Japanese authorship will also allow an international audience to appreciate some of the current concerns of Japanese chemical engineering.

It is a pleasure to record the help and enthusiasm of the individual authors in developing their material, to acknowledge the support offered by the publishers for this project, and to thank Debbie Walker and Sylvia Petherick for their painstaking typing of the manuscript.

John Garside
Shintaro Furusaki August 1993

EXTRACT FROM THE PREFACE TO THE ORIGINAL JAPANESE EDITION

Since its birth in the early part of the 20th century chemical engineering has developed to its present status as a result of many pioneering research activities and professional achievements. Over this time there have been a number of important milestones such as the concept of unit operations in the 1920s, the systematization of transport phenomena and reaction engineering in the 1950s and 1960s, and the development of a systems engineering approach in the 1970s. In addition to this foundation of a solid chemical engineering science base, chemical engineers have broadened their areas of study, particularly in the 1970s and 1980s, to embrace fields such as the environment, energy, biotechnology and materials science. It has thus established itself as both a fundamental and an applied discipline.

It is now necessary to expand the range of interest of chemical engineering still further and to make it more exact. This new chemical engineering must interact strongly with those fields of science and engineering that exist around its boundaries in order to enable it to make a useful and effective contribution.

This book is written by chemical engineering experts who have carried out research in different aspects of advanced technology related to chemical engineering. Its main objective is to attract young people in colleges and high schools to an interest in chemical engineering so that they might consider studying chemical engineering in the future. It will also introduce some novel aspects of chemical engineering to new students who have already joined chemical engineering departments. We have included chapters on functional materials, computer applications, biotechnology, food and medical technology, environmental engineering, energy and natural resources. As a result it is hoped that the reader will come to appreciate the breadth of chemical engineering.

Shintaro Furusaki July 1987

CONTRIBUTORS

Tadafumi Adschiri
Department of Chemical Engineering
Tohoku University
Sendai, Japan

Hajime Asama
Advanced Technology Center
The Institute of Physical and Chemical Research (RIKEN)
Wako, Japan

Isao Endos
Biochemical Systems Laboratory
The Institute of Physical and Chemical Research (RIKEN)
Wako, Japan

Larry E. Erickson
Department of Chemical Engineering
Kansas State University
Manhattan, Kansas, USA

L.-S. Fan
Department of Chemical Engineering
The Ohio State University
Columbus, Ohio, USA

Teruo Fujii
Institute of Industrial Science
University of Tokyo
Tokyo, Japan

Shintaro Furusaki
Department of Chemical Systems and Engineering
Kyushu University
Fukuoka, Japan

John R. Grace
Chemical Engineering
University of British Columbia
Vancouver, British Columbia, Canada

Manabu Ihara
Institute for Chemical Reaction Science
Tohoku University
Sendai, Japan

Shinji Iijima
Department of Biotechnology
Nagoya University
Nagoya, Japan

Hayato Kaetsu
Biochemical Systems Lab.
The Institute of Physical and Chemical Research (RIKEN)
Wako, Japan

Shigeo Katoh
Department of Chemical Science and Engineering
Kobe University
Kobe, Japan

Kurt Koelling
Department of Chemical Engineering
The Ohio State University
Columbus, Ohio, USA

Toshinori Kojima
Department of Industrial Chemistry
Seikei University
Tokyo, Japan

Hiroshi Komiyama
Department of Chemical System Engineering
University of Tokyo
Tokyo, Japan

D.J. Lee
Department of Chemical Engineering
The Ohio State University
Columbus, Ohio, USA

Masakuni Matsuoka
Department of Chemical Engineering
Tokyo University of Agriculture and Technology
Tokyo, Japan

Takeshi Matsuura
Industrial Membrane Research Institute
Department of Chemical Engineering
University of Ottawa
Ottawa, Ontario, Canada

Shuzo Ohe
Department of Industrial Engineering
Science University of Tokyo
Tokyo, Japan

Umit S. Ozkan
Department of Chemical Engineering
The Ohio State University
Columbus, Ohio, USA

Raj Rajagopalan
Department of Chemical Engineering
University of Florida
Gainesville, Florida, USA

Anthony D. Rosato
Mechanical Engineering Department
New Jersey Institute of Technology
Newark, New Jersey, USA

Kyoichi Saito
Department of Materials Technology
Chiba University
Chiba, Japan

Kiyotaka Sakai
Department of Chemical Engineering
Waseda University
Tokyo, Japan

Yoshito Sakino
Osaka Prefectural Police
Osaka, Japan

Hayatoshi Sayama
Process Management Institute
Okayama, Japan

Takeshi Sekine
Ajinomoto Co., Inc.
Kanagawa, Japan

Gavin W. Sinclair
Computer Integrated Process Operations Consortium
School of Chemical Engineering
Purdue University
West Lafayette, Illinois, USA

Jennifer L. Sinclair
School of Chemical Engineering
Purdue University
West Lafayette, Illinois, USA

Kazuhiko Suzuki
Department of Systems Engineering
Okayama University
Okayama, Japan

Tsuyoshi Suzuki
Advanced Technology Center
The Institute of Physical and Chemical Research (RIKEN)
Wako, Japan

Rick B. Watson
Department of Chemical Engineering
The Ohio State University
Columbus, Ohio, USA

1. COMPUTATION CHEMISTRY AND ENGINEERING

JENNIFER SINCLAIR and GAVIN SINCLAIR
School of Chemical Engineering, Purdue University, West Lafayette IN 47907-1283, USA

INTRODUCTION

Computers have fundamentally changed the field of chemical engineering. In the old days, engineers only did hands-on experiments and engineering calculations by hand. However, the need to solve highly complex engineering problems requires that the best scientific tools be employed. Hence, today, with the presence of the computer, the chemical engineer also conducts computer simulations. Whether a chemical engineer is designing a new process or operating an existing process, computers are an essential part of the job.

In this chapter, we will discuss six key technical areas relating to computation chemistry and engineering. In the first section, *Computation Chemistry*, we will look at how scientists and engineers design molecules and actually predict the performance of these molecules without ever *making* the molecule. This is especially important in the pharmaceutical industry, where a research group can screen millions of possible molecules to find the best products to synthesize in the physical laboratory, focusing their efforts on the most likely candidates for laboratory and human trials.

The second key field that we will review is *Computational Fluid Dynamics* or *CFD*. CFD describes the flow of fluids, with or without dispersed particles or droplets, which is involved in many chemical engineering applications. When there is an avalanche on a mountain, how do people know if the ski lodge will be buried in snow? This is a flow problem. The snow particles flow down the mountain based on the fundamental laws of physics, and this flow can be modeled and predicted using computer simulation. In the chemical industry, what if a chemical engineer is trying to understand the mixing patterns of species that are reacting in a chemical reactor (the reactants) in order to maximize the conversion of these reactants to the desired products? These problems can be very complicated, but recent advances in computational fluid dynamics are helping engineers design and operate plants more efficiently.

The third key field that we will consider is *Process Control*. Process control is how chemical plants automatically control processing variables like temperature and pressure. A simple example of process control is the thermostat on your furnace. The thermostat controls the temperature of the house within a certain preset range. The same principle applies to a chemical plant on a much larger and more complex scale. Process control is especially important to operate a chemical plant safely.

The fourth key area is *Process Simulation*. In this section, we will describe how engineers can simulate an entire chemical plant and actually test different operating strategies using the computer. The most successful strategies are then implemented in the real world. Process simulation is also essential for new plant design. Different types of equipment can be tested in a process simulation to determine what is the best way to design the plant.

In the fifth section, we will look at *Process Scheduling and Optimization*. This is a special application of Process Simulation. Imagine you are making 100 different products in the chemical plant, and you can make each product by three different production paths. If you examine all the different options that an engineer has to produce these 100 products, there are literally trillions of combinations. Special computer software for Process Simulation and Optimization helps an engineer decide what is the most efficient plan for scheduling the plant.

In the sixth section, we will look at *Artificial Intelligence*. Artificial intelligence is the process of taking human reasoning and programming a computer algorithm based on the human thought process.

In the final section, we will look at a non-technical area, that is, how computers can help engineers make *Business Decisions* in chemical companies, especially decisions that involve uncertainty.

All of these areas of chemical engineering have been growing and changing rapidly as computers have become more powerful and more widely available. Ten years ago, most of these fields were in their infancy. Today, due to the computer, these fields are revolutionizing the way chemical engineering is done.

COMPUTATIONAL CHEMISTRY

Have you ever done an experiment in chemistry lab? You carefully weighed out the reagents, mixed them together, and perhaps filtered to get the final product. You then carefully weighed the final product to determine the conversion. Perhaps you performed some tests on the final product to measure the purity or physical characteristics. How long did all of this take? 30 minutes? One hour?

Imagine you are now in the laboratory of a pharmaceutical company. It might take hours or days to synthesize a drug of potential interest. It may take weeks or months to characterize all the physical properties of this new compound, and perhaps months or years to gather biological data. If the first drug does not end up exhibiting the desired properties, you start the whole process over. But don't worry; there are only several trillion possibilities left to try.

Fortunately, today most screening of new compounds or molecules is done with computational chemistry. Instead of actually making and testing molecules in the lab, scientists and engineers design molecules on the computer. By going back to basic information about a molecule such as the size of the atoms and bond lengths, an engineer or scientist can build molecules on a computer and then have these molecules "react" with other molecules to predict the physical and biologi-

Figure 1.1 The active site of HIV protease, shown in space filling balls, is visualized via a structure-focusing, computational chemistry program developed by Molecular Simulations Inc. (Chemical & Engineering News, April 27, 1998).

cal characteristics of each molecule. There are many different molecular designs that can be tried, which is where *combinatorial chemistry* comes into the picture. Combinatorial chemistry is a way to generate different combinations of atoms and arrangements of atoms to generate all the different molecules possible. Since there are so many combinations possible, combinatorial chemistry usually employs some rules or algorithms to focus on the molecules that are most likely to be effective.

For example, let's say you want a drug to inhibit a certain enzyme involved in a disease, such as the HIV protease involved in AIDS. If the enzyme isn't well characterized yet, such as the HIV protease, it is difficult to find a drug molecule that will interact with the drug target receptor. A structuring-focusing program can be used to analyze the active site of the HIV protease and then to search virtual molecular libraries for candidate drugs that might bind to the active site. Hence, computational chemistry, which allows for high throughput screening, is an integral part of drug discovery.

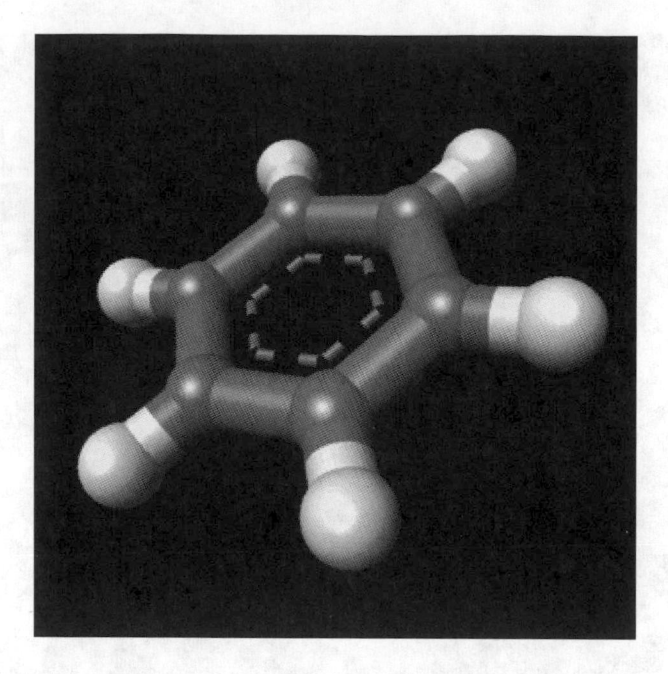

Figure 1.2 Benzene image using Molecular Simulations Inc. software. <http://www.msi.com/life/gallery/lsmolecules/descriptions/molecules3.html>.

The best candidates from the computational screening process are then synthesized and evaluated for their physical and biological properties, and ultimately, in the case of a pharmaceutical product, animal and human trials are conducted. The computer has not replaced these steps, but it has greatly focused the search for good molecules.

There are many companies that write software for computation chemistry. Using the software, a molecule can be built very quickly and bond vibrations animated, selecting individual bonds to see how a molecule moves. Three-dimensional animations are even possible.

Some of these programs generate molecules based on combinatorics, a mathematical field that generates millions of different combinations of molecules in different arrangements. (We will see another application of combinatorics in the *Plant Scheduling and Optimization* section later on in this chapter.) Some software is written based on bioinformatics, which are algorithms that predict the biological properties of molecules designed on the computer. Other programs are written to predict the physical or chemical characteristics of computer-generated molecules. For example, computational chemistry can provide extremely useful insight into how a catalyst works. Various types of surface interactions of the reacting molecule(s) with the proposed, computer-generated catalyst can be determined and visualized.

Figure 1.3 Students in the Molecular Modeling Laboratory at Middlebury College viewing 3-D animations using Spartan software by Wavefunction. (Chemical & Engineering News, May 26, 1997).

Figure 1.4 Interaction of ethylene with a proposed catalyst generated by Molecular Simulations Inc. software. Surfaces illustrate charge density. <http://www.msi.com/materials/gallery/catalysis.html>.

By understanding these interactions, the chemist or chemical engineer can design improved catalysts leading to potentially enormous impacts on a chemical process. For example, even a factor of two in improvement in catalyst efficiency may determine the economic feasibility of a chemical process.

Computation chemistry is a rapidly growing field and has now evolved to the stage where chemical process engineers who know nothing about computational chemistry calculations can apply its results. Chemical companies routinely use computers to compute physical properties of molecules. Also, if an engineer furnishes a computational chemistry program with a 3-D geometry estimate for the molecules involved in a chemical reaction, the computer program can predict the energy change associated with the reaction. These types of predictions from the computer-generated results are getting much closer to reality. By the time you graduate from college, computation chemistry may be more important than chemical synthesis.

COMPUTATIONAL FLUID DYNAMICS

Computational fluid dynamics, abbreviated CFD, is the name given to the numerical solution of the equations that describe fluid motion, namely, mass and momentum balances applied to the fluid phase. In general, however, the term CFD also includes the numerical solution of equations that describe the transport of energy within the fluid phase, or the transport of different reacting or non-reacting species within the fluid phase. These types of simultaneous processes, heat transfer and fluid flow or chemical reaction and fluid flow, pervade the chemical industry, so a computer tool needs to be able to simulate the fluid flow patterns under these conditions. If the simulation tool can accurately predict the fluid flow patterns, engineers can then use the computer to design and optimize fluid flow patterns in various types of devices such as reactors, mixers or pipelines. In chemical engineering practice today, these types of simulations *are* possible and are being conducted extensively within the chemical industry. Chemical engineers not only use the computer and CFD to design and optimize chemical processes, but they also use flow simulations as a way to "see inside" various devices that are not working efficiently, or not working at all, in order to figure out the problem(s) and analyze potential solutions. Chemical engineers can reliably scale-up, that is, take the data from a small-scale device and predict the behavior in a larger-scale device, using CFD. Chemical engineers also apply CFD to reduce experimental testing of trial devices or trial operations. The computer simulates the trial device or trial run much more quickly and at a much lower cost than performing the actual hands-on experimental runs. Continual increases in computer speed and power only serve to continuously increase the cost and time savings of computer or numerical experiments over actual physical experiments. The virtual laboratory is rapidly becoming a reality. However, best design practice will always dictate a combined approach that involves complementary numerical experiments and actual physical experiments.

Figure 1.5 Actual fitting. <http://www.fluent.com/software/aboutcfd.htm> (©Fluent Inc. Courtesy of Fluent Inc., Lebanon, NH).

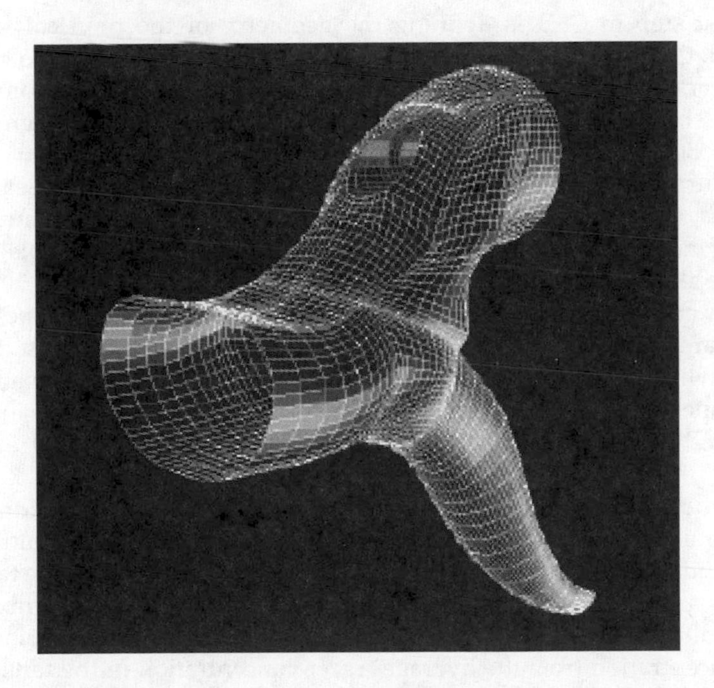

Figure 1.6 Computational mesh. <http://www.fluent.com/software/aboutcfd.htm> (©Fluent Inc. Courtesy of Fluent Inc., Lebanon, NH).

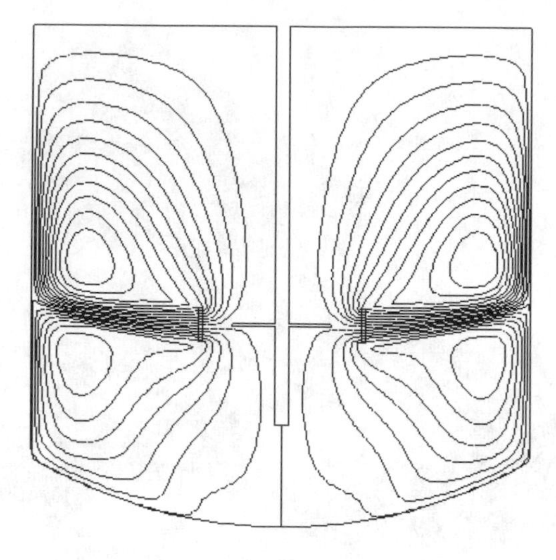

Figure 1.7 The flow pattern created by a Rushton turbine is shown. The impeller pumps fluid radially outward, giving rise to recirculation zones above and below the impeller. (CACHE News, Fall 1999. ©Fluent Inc. Courtesy of Fluent Inc., Lebanon, NH).

The first step in CFD is defining the geometry of the physical setup to be simulated. Current CFD simulation capability allows the engineer to easily build a computer model of a very highly complex geometry using a computational mesh. As shown below, in order to simulate the fluid flow through the actual fitting, a computational mesh or grid is generated.

The accuracy of the fluid flow predictions is dependent on the preset coarseness or fineness of the mesh. Current CFD simulation capability can also automatically refine the mesh in specific regions of the flow as needed, for example, in regions where the fluid velocity is changing very rapidly.

As an example of the design capabilities of CFD, two different types of mixers are compared below. Obviously, the chemical engineer wants to use the mixing impeller which results in the shortest blending time for a given tank size and impeller speed. In Figures 1.7 and 1.8, the simulated mixing patterns in a Rushton turbine and a pitched-blade turbine are visualized using flow streamlines or contours.

The blending time in each mixer can be determined by first adding a small amount of a tracer species whose properties are identical to those of the fluid in the tank and then simulating the mixing of that tracer throughout the tank. Figure 1.9 shows a layer of tracer species being added to the top of the mixing vessel with the pitched-blade turbine. By computing the ratio of the deviation in the tracer concentration from the average tracer concentration in the tank as a function of time, the blending time to achieve the desired concentration uniformity

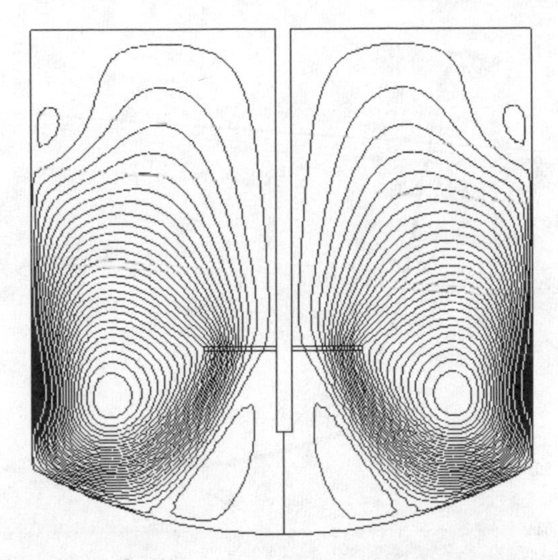

Figure 1.8 The flow produced by a pitched blade turbine is axial, as shown. When operating at low speeds, however, the radial component of the flow becomes comparable to the axial component, altering the flow pattern considerably. (CACHE News, Fall 1999. ©Fluent Inc. Courtesy of Fluent Inc., Lebanon, NH).

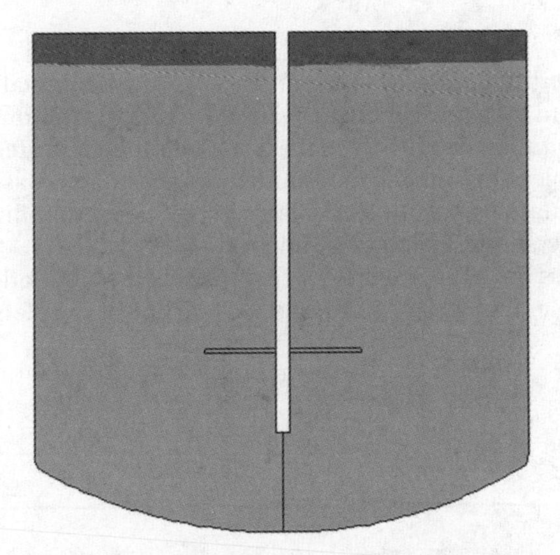

Figure 1.9 For blending time studies, a tracer species can be introduced to the vessel as a pure layer on top, as shown. The concentration of the tracer can then be monitored as a function of time at several locations within the tank to determine the time required for the mixture to reach uniformity. (CACHE News, Fall 1999. ©Fluent Inc. Courtesy of Fluent Inc., Lebanon, NH).

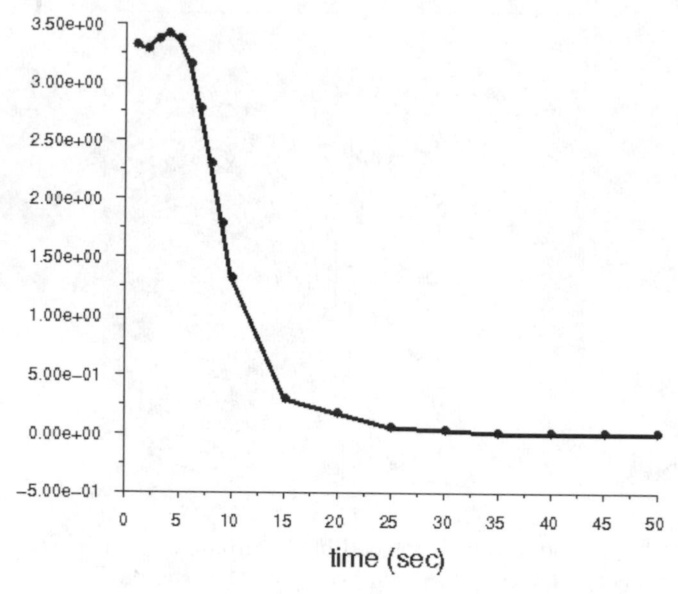

Figure 1.10 Deviation in species concentration as a function of time. (CACHE News, Fall 1999. ©Fluent Inc. Courtesy of Fluent Inc., Lebanon, NH).

can be determined. Figure 1.10 shows the decay in the deviation in the tracer concentration from the average concentration as a function of time for the pitched-blade turbine. A simulation with the tracer species is then repeated for the Rushton turbine and the blending time is determined for this mixer. Based on the results of these two simulations, the mixer design can be finalized, utilizing the impeller that produces the shortest blending time.

Recent advances in CFD software now also enable the simulation of multiphase flows, for example, the flow of solid particles or liquid droplets in a gas, or the

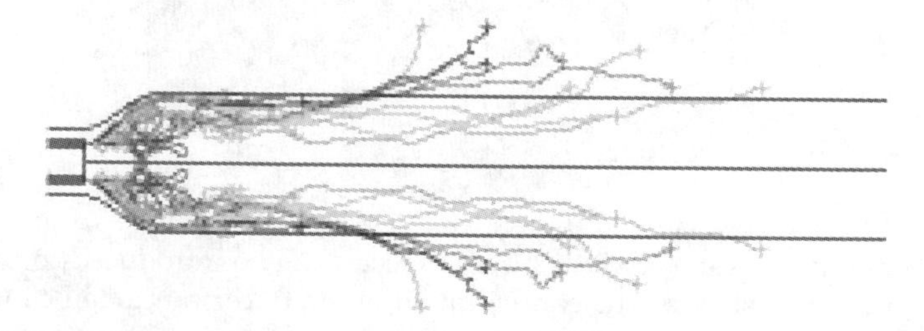

Figure 1.11 Coal particle trajectories in the coal furnace. (Fluent Newsletter, Winter 1995. ©Fluent Inc. Courtesy of Fluent Inc., Lebanon, NH).

flow of two immiscible fluids. A very important multiphase flow process in the chemical and energy industries is coal combustion. In coal combustion, coal particles are burned in air to produce a large quantity of useful heat. However, in this process, undesirable reaction by-products are produced, such as NO which is an environmental hazard. It is important to properly design the burner, which feeds the coal and air into the reactor furnace, not only because this design influences the stabilization of the flame and the amount of energy produced, but also because the exact flow paths of the coal particles in the reacting and flowing gaseous environment influences the quantity of NO produced. Shown below is a CFD simulation of coal particle trajectories issuing from a conventional burner in which the central nozzle pipe containing air and coal particles is surrounded by a secondary annular pipe in which the air is swirled as it is fed into the furnace.

These coal particle trajectories are combined with kinetic equations that describe the many reactions occurring during the combustion process, including the reaction which converts the nitrogen bound to the coal fuel to the undesirable NO by-product. CFD simulations then predict the concentration of NO everywhere in the coal furnace as shown in Figure 1.12 below.

Actual measurements of NO concentration in the furnace can be compared to the simulated concentrations and can be used to validate the assumed reaction model. Furthermore, once the reaction model is validated, CFD can be used to explore new burner designs and changes in operating conditions such as air velocity, solids concentration, and particle size in order to achieve NO reduction and meet environmental standards.

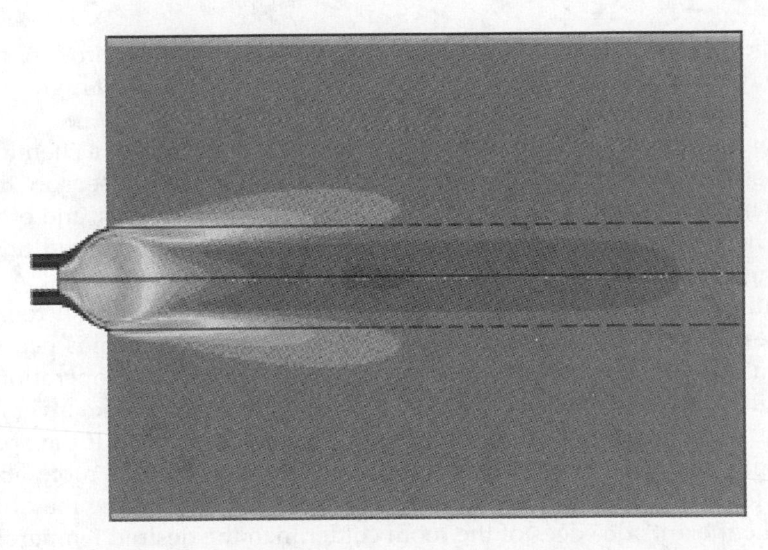

Figure 1.12 Predicted NO concentration in the coal furnace. (Fluent Newsletter, Winter 1995. ©Fluent Inc. Courtesy of Fluent Inc., Lebanon, NH).

Figure 1.13 Typical control room with computer control. (Essentials of Process Control, Luyben, McGraw-Hill. Courtesy of Honeywell).

PROCESS CONTROL

A chemical plant is very complicated and involves many hazards. A chemical plant may involve hazardous chemicals, high temperatures, and high pressures. Safety is the foremost consideration.

One of the most important features ensuring safe operation of a chemical plant is process control. Throughout the chemical process, there are sensors that continually measure temperatures, pressures, species concentrations, and other variables. These sensors are connected to computers that monitor and adjust the operating conditions of the plant according to predetermined ranges. All of this information is sent to a central area of the plant called the control room.

Besides the primary objective of safety, careful process control is necessary to maintain product quality, production rate, and the economic operation of the plant. The control schemes can become very complicated. To take a very simple example, imagine controlling the temperature in an apartment. If the apartment is too cold, the furnace will turn on, and it will tend to heat the room above the desired temperature. Near the windows of the apartment, there may be drafts that make the outside edges of the room colder than the desired temperature. As the outside temperature changes, the furnace must adjust to the new outside conditions. The apartment on one side might be kept at a very high temperature, and the apartment on the other side might be kept at a very low temperature. All of these additional factors make controlling the temperature in an apartment more difficult. It is the same way in a chemical plant involving many different processes, except the acceptable temperature ranges are usually very tight and the interrelationships between the different processes can have a much bigger effect.

Typically, the response of even a single process in a chemical plant to some type of change or disturbance is very complex and can be described using linear or non-linear differential equations. When a control scheme is added to the process, the mathematical description of the response of the process becomes even more complex. The basic operation needed to simulate control systems is integration of this set of differential equations. Fifteen years ago, this task was impossible. Today, simulation software packages can be used to determine the response of the process that is controlled. Also, due to the microprocessor, today the entire chemical plant can be computer controlled. The control instruments in the various process loops communicate with each other through networks; this is known as distributed control. These internal computer connections are called softwiring because the connections between the various processes are made through computer software. In both of these ways, the computer plays a critical role in process control.

PROCESS SIMULATION

A core application for computers in chemical engineering is process simulation. Process simulation is an area in which all of the essential elements of a chemical plant are simulated together on a computer.

A chemical process is constructed using building blocks called *unit operations*. One example of a unit operation is chemical reaction. Another unit operation is distillation, a process where a mixture of chemicals is heated and separated based on their difference in volatility (ease with which they form vapor). There are many other unit operations such as filtering, centrifuging, blending, and condensing. Each of these unit operations can be simulated on a computer. These unit operations are combined together to generate a process simulation flowsheet.

The following figures (1.14 and 1.15) show how the various unit operations are combined with a computer software program to build a process simulation flowsheet.

Once the process flowsheet has been developed, it becomes a valuable tool for engineers. One important use of a process simulation flowsheet is for plant design. A new plant can be created and tested on a computer before it is actually built. Another important use of a process simulation flowsheet is to test different operating scenarios. As we will see in the following section on *Process Scheduling and Optimization*, a process simulation flowsheet can also be used to help determine an efficient production schedule to maximize profits.

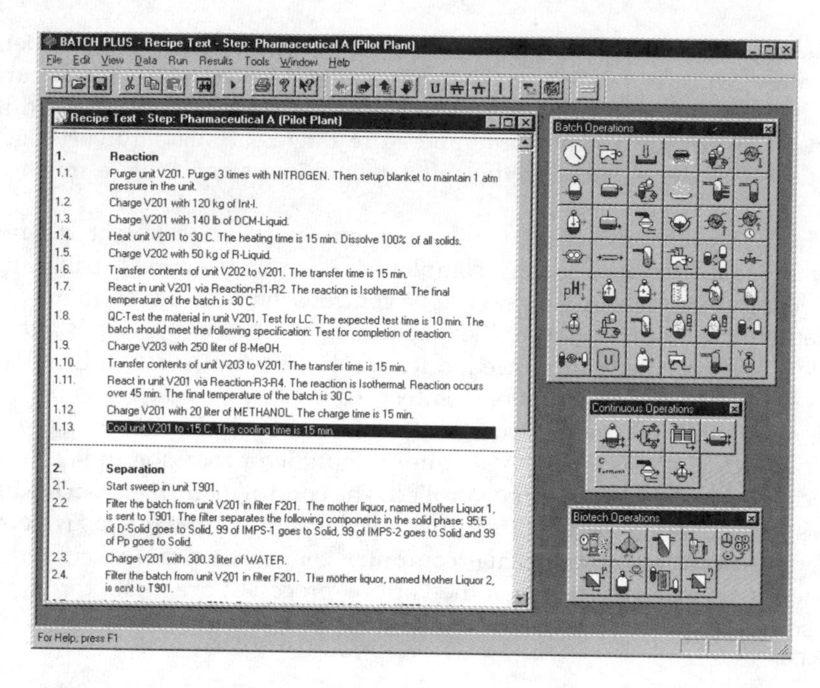

Figure 1.14 Process recipe generated by BATCH PLUS for an example step. <http://www.aspentech.com> (Ajay Modi and Reiner Muster. Courtesy of Aspen Technology).

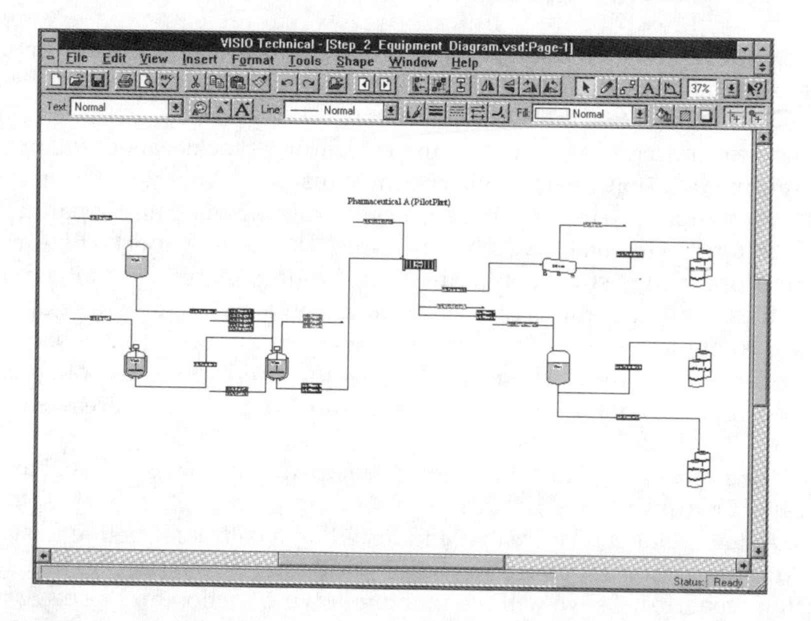

Figure 1.15 Process flowsheet generated by BATCH PLUS. <http://www.aspentech.com> (Ajay Modi and Reiner Muster. Courtesy of Aspen Technology).

PROCESS SCHEDULING AND OPTIMIZATION

Efficiency and safety are key goals for all production facilities. Production scheduling is critical to both of these issues. If a chemical plant makes a single product along a single production path, establishing a production schedule is very simple. Most chemical plants, however, manufacture numerous products along many possible production paths. If a company needs to schedule a plant that makes 100 different products by 20 different production paths, there are literally millions of combinations that must be considered to determine an optimal plant schedule, both in terms of efficiency and safety. This is called a combinatorial problem because engineers need to find an optimal plant schedule from many different combinations of products and production paths.

Plant design is very difficult in these combinatorial situations. When designing a new plant that involves sharing equipment among different products, the optimal plant design is dependent on how the plant is scheduled. The plant schedule, of course, is dependent on the plant design, leading to a seemingly endless loop that is usually resolved by over-designing and spending more on capital (equipment) than is really needed.

Until recently, even computers were not much help in these combinatorial scheduling and design situations. Linear programming models, where an engineer is trying to optimize a certain variable such as total profit or production volume subject to a series of constraints, work well for continuous operations like oil refining. However, if the operation includes discrete (yes-no) decisions such as "Should I make Product A in Reactor 1?" you have gone beyond linear programming to something called a Mixed Integer Linear Program (MILP) problem. A MILP problem includes continuous variables as well as discrete variables. Discrete decisions lead to a large number of combinations in the following way. If a person needs to make a single yes-no decision, there are $2^1 = 2$ combinations.

Table 1.1 The number of combinations grows exponentially with the number of discrete decisions.

Number of discrete decisions	*Number of combinations*
1	2
2	4
3	8
4	16
5	32
10	1 024
20	1 048 576
30	1.1×10^{9}
40	1.1×10^{12}
50	1.1×10^{15}
100	1.3×10^{30}

If a person needs to make two yes-no decisions, there are $2^2 = 4$ combinations. As the following chart shows, even a comparatively small number of discrete decisions leads to a large number of combinations that need to be considered.

Engineers have developed mathematical and software technology to solve very large MILPs of the type that arise in scheduling and planning applications. A large number of constraints, the increasing scope of operations, time and efficiency pressures, and the combinatorial complexity implied by discrete decisions make scheduling and planning problems difficult to solve.

The key to solving these problems is based on the fact that the computer algorithms do not search every possible combination. Instead, algorithms are developed that look for the most likely region of a good solution. To help understand this, imagine that you are trying to find a gas station in an unfamiliar town. You would probably eliminate residential areas from your search and instead look for crowded commercial areas, especially the intersections of major streets. In the same way, production scheduling software specifically looks for the "crowded intersections" of a problem. The software can find an "optimized" solution (not necessarily THE optimal one) without searching the millions of combinations.

To set up a production scheduling problem for the plant, the engineer would first start with a process flowsheet like the one shown in the last section. This flowsheet would include product recipes that describe how much of the different raw materials are needed to produce each final product, how long it takes during each step of the flowsheet, what are the by-products of the reaction, and then complete information on the economic aspects of the product including the cost of all the raw materials, the processing cost, the price that each final product can be sold for, the quantity demanded of all the end products, and so forth. The typical "objective function" for the simulation problem is to maximize profit. The computer program will take all of the information into account and then generate a production schedule that maximizes profit. The result of running the optimization software is a "Gantt Chart" like the one in Figure 1.16, which shows what products should be made in which equipment at what time in order maximize profit.

The economic gains of using the computer and optimization software are tremendous. Chemical plants can generally increase their production capacity by 30% when plant scheduling software is installed. The capital requirements for a new plant can often be reduced by a comparable percentage by designing plants that can share equipment for different products.

ARTIFICIAL INTELLIGENCE

Another exciting and growing application of computers in chemical engineering is artificial intelligence. Artificial intelligence, generally speaking, is training a computer to make decisions or think through problems that would normally be

Figure 1.16 A computer-generated Gantt chart. <http://www.aspentech.com> (Ajay Modi and Reiner Muster. Courtesy of Aspen Technology).

handled by people. Say, for example, a person needs to recommend a catalyst for a particular process. The choice of the catalyst will depend on the specific end product being manufactured, the purity of the end product desired, and also the operating conditions such as temperature and pressure. Instead of a chemical engineer making a recommendation each time a situation arises, the thought process of the chemical engineer can be captured in a software algorithm. Many artificial intelligence programs take the form of directed questions. The first question may be "What is the product being produced?" The answer to that question will cause the computer program to branch off into a different set of questions. The second question might be "What raw material sources are being used?" This question-and-answer process would continue until the final recommendation is made. It takes a good deal of time initially for the engineer to go through all the possible answers to the questions, but once the thought process has been distilled into the software, the computer can handle most routine requests.

Another growing area for chemical engineering is hazard analysis. Every part of a chemical plant must be evaluated for process safety, and this is best done by a standardized procedure that takes an exhaustive look at all the possible hazards in all of the processes. Artificial intelligence software has been designed not only to evaluate every area of the plant through a comprehensive search algorithm, but also to generate the standard reports and procedures needed by the hazard assessment procedure. In this application, the computer software is usually superior to the human evaluation process because the computer does not get tired of asking the same questions over and over again.

BUSINESS DECISIONS

Business decisions in a chemical company are becoming increasing dependent on computers. A major driving force is risk and uncertainty. Often people use the terms "uncertainty" and "risk" interchangeably. They are actually two different concepts, and specific definitions might be helpful. *Uncertainty* is the randomness of the external environment. This randomness cannot be changed, only managed. *Risk* comprises the potential adverse economic consequences of a company's exposure to uncertainty.

Business decisions always involve a great deal of uncertainty. There may be less demand than expected for a pesticide because of the weather conditions. Crude oil prices may spike up unexpectedly, changing the economics of products manufactured from petroleum. There is nothing that can be done about this randomness of the external environment.

An agricultural company's product line could be balanced so that some products do better when there is a lot of rain, and other products do well in dry conditions. Some companies may decide to backward integrate, meaning they produce the raw material as well as the downstream products. If raw material prices go up, the company is protected because they also produce and sell the raw material. Both of these are examples of the portfolio approach to managing risk, which is not unlike how stock market investors manage the risk of their investment portfolios. In fact, Wall Street has hired many engineers and scientists over the last decade to write computer codes for portfolio management software.

In most major business decisions impacting a chemical company, a financial model is built to simulate the expected financial results of a period of five or ten years. There are major uncertainties in these extrapolations. What will happen to prices over the next ten years? What will happen to raw material costs? What will happen to product demand?

Chemical companies use simulation software to try to assess their risk in making these major business decisions. Instead of forecasting a single selling price over the next decade, these software programs will ask for a probability distribution that represents selling prices. The software will also ask for probability distributions for the other critical business variables. These distributions are programmed into the computer, and then Monte Carlo simulation is usually employed to derive a probability distribution of the financial results. The end result, for example, may be a probability distribution of the rate of return for the project that indicates there is a 70% chance of exceeding the minimum financial standards established by the corporation. The software will also generate sensitivity analysis, showing how sensitive the financial results are on particular variables. If the result is especially sensitive to the projection for raw material costs, for example, the company may decide to spend more time and resources getting a better estimate of the future trend in raw material costs.

Although this section may seem far from what is normally considered "chemical engineering", ultimately many of these types of decisions are often made by engineers, and these decisions are driven by economics. Today, computers are used to assist chemical engineers in making these business decisions.

CONCLUSIONS

Dramatic improvements in computers have occurred over the past 20 years. In this chapter, we have described several key areas in which chemical engineers are involved that have been impacted by the change in computer power, speed, cost and availability. If you are a student looking for a career that involves computers in a very applied and significant way, then one of these areas may be for you.

2. APPLICATION OF FUZZY SYSTEMS AND NEURAL NETS TO CHEMICAL ENGINEERING PROBLEMS

KAZUHIKO SUZUKI[1], HAYATOSHI SAYAMA[2] and SHUZO OHE[3]

[1]*Department of Systems Engineering, Okayama University, Okayama, Japan*
[2]*Process Management Institute, Okayama, Japan*
[3]*Department of Industrial Engineering, Science University of Tokyo, Tokyo, Japan*

INTRODUCTION

Neural networks and fuzzy systems share the common ability to deal with difficulties arising from uncertainty, impression, and noise in this natural environment. Recent successful applications of both systems and their techniques to various fields have spurred growing research interests in both systems and their integration. In this chapter, we discuss the way in which fuzzy systems and neural net technology can be applied to process modeling and control and to fault detection and diagnosis in chemical process engineering. Fuzzy logic evolved from the need to model the type of vague or ill-defined systems that are difficult to handle using conventional binary value logic, but the methodology itself is grounded in mathematical theory. We will introduce the mathematical notation required to describe fuzzy systems and review existing application and research area. The use of neural nets, which are sometimes called Artificial Neural Networks (ANNs), in data analysis, pattern recognition and decision making has increased rapidly and their use is expected to have a significant impact in many technological and business areas. We will here outline why neural nets are so advantageous for representation of information in chemical process engineering and explain the characteristics of neural nets focusing on the neuron, the configurations and the learning algorithm. We will also introduce some simple case studies to illustrate the concept of neural nets.

CRISP SETS AND FUZZY SETS

The set theory with which we are most familiar can be described as binary-valued since, for any given universe of objects, X, a binary-valued characteristic function $\mu_A(x)$ can be used to represent whether the object x ($x \in X$) belongs to the set A or not. μ_A is described as follows:

$$\mu_A : X \to \{0,1\}$$
$$\mu_A(x) = \begin{cases} 1 \text{ when } x \in A \\ 0 \text{ otherwise} \end{cases} \tag{1}$$

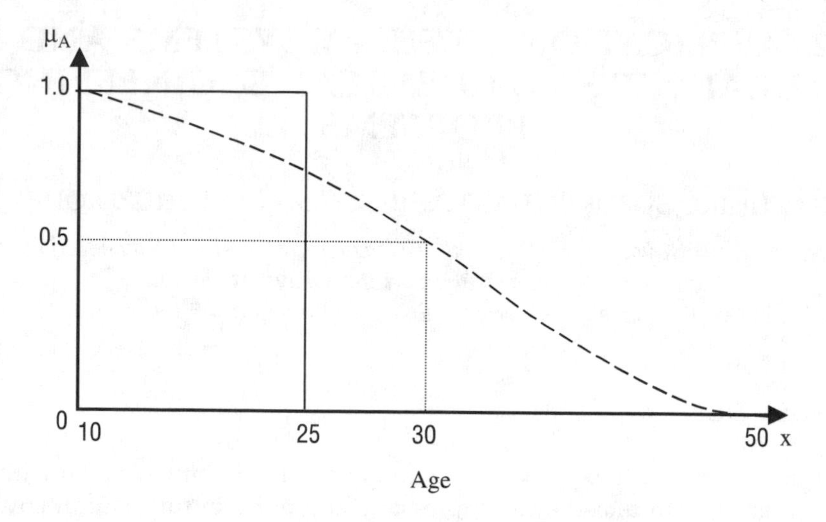

Figure 2.1 Membership function of fuzzy set of "young people".

Such a set is often described as "crisp" since there is a distinct point at which set membership starts and ends.

Consider the concept of "YOUNG" applied to persons. Given a particular value for an individual's age, what values are considered young? In classical set theory, we are forced to choose an arbitrary cutoff point, say 25 years old. Since the boundaries between what is in a set and what is outside a set are very sharp, these types of constructs are called crisp sets. A characteristic function for such a set appears as follows:

$$\mu_{\mathrm{YOUNG}} = \{\mathrm{age} \leqq 25\} \tag{2}$$

Thus, anyone under 25 years old is young. The membership graph for this set appears as solid line in Figure 2.1. The discriminant or characteristic function for this set reflects its Boolean nature. As we move along the domain, the membership of youth in the set young remains true (one) until we reach exactly 25 years old when it jumps immediately to false (zero). Note that in the membership transition graph for crisp sets the line connecting nonmembership and membership is dimensionless. All classical or crisp sets have this kind of membership function.

Unfortunately, many terms used in human reasoning, such as "the set of young persons", "the sets of tall persons" and " the sets of beautiful women" are less well defined and are not easily represented by binary-valued set theory. To treat a vague set not clear of boundaries quantitatively, Zadeh (1965) proposed the concept of the Fuzzy set and has developed a theory.

The idea of "YOUNG", illustrated as dotted line in Figure 2.1, is the classical example of a fuzzy set and illustrates the intrinsic properties of fuzzy spaces. The domain of this set, indicated along the horizontal axis, is the range of youth between 10 years old and 50 years old. The degree of membership or truth function is indicated on the vertical axis to the far left. In general, the membership goes from zero (no membership) to one (complete membership.) The membership function and the domain are connected, in this case, by a simple curve. Now, given a value for age, we can determine its degree of membership in the fuzzy set.

Thus a 30 years old has 0.5 degree of membership. The interpretation of this value corresponds to the truth of the proposition, " 30 years old is YOUNG". If the value for age is less than 10, its membership is one. If the age is greater than or equal to 50, then its membership value is zero. This membership function, as we will see, is interpreted as a measure of the compatibility between a value from the domain and the idea underlying the fuzzy set. In the case of 30 years old, the membership of 0.5 means that it is moderately but not strongly compatible with the notion of YOUNG. As the membership value moves toward zero the sample becomes less and less compatible with the semantics of the fuzzy set and, as the membership value moves toward one, the sample becomes more and more compatible with the fuzzy set's semantic property.

FUNDAMENTAL OF FUZZY SETS

Suppose that $X = \{x\}$ is a universe of discourse, i.e. the set of all possible elements to be considered with respect to a fuzzy concept. Then a fuzzy subset A in X can be characterized by the membership function $\mu_A: X \rightarrow [0,1]$. $\mu_A(x) \in [0,1]$ as the grade of membership of x in A, from zero for full nonbelongingness to one for full belongingness, through all intermediate values. So, a fuzzy subset A in X is described as a set of ordered pairs as follows:

$$A = \{(x, \mu_A(x))\}, |x \in X\} \tag{3}$$

When X is a finite crisp set set $\{x_1, x_2, \ldots x_n\}$, then A can be described as follows:

$$A = \{(x_1, \mu_A(x_1)), (x_2, \mu_A(x_2)), \ldots (x_n, \mu_A(x_n))\} \tag{4}$$

For instance, suppose someone wants to define the set of natural numbers "close to five". If $X = \{1, 2, \ldots, 9\}$, then the fuzzy set of "close to five" can be expressed as $A = \{(3,0.4), (4,0.8), (5,1), (6,0.8), (7,0.4)\}$, i.e., 3 and 7 belong to degree 0.4, 4 and 6 belong to degree 0.8 to the class of "close to five". Furthermore an alternative notation for $(x, \mu_A(x))$ is $\mu_A(x)/x$, where/denotes a tuple. Then the set A is the described as follows:

$$A = \mu_A(x_1)/x_1 + \ldots + \mu_A(x_n)/x_n = \sum_{i=1}^{n} \mu_A(x_i)/x_i \tag{5}$$

where + satisfies $a/x + b/x = \max(a,b)/x$, i.e., if the same element has two different membership degrees 0.8 and 0.6, then its membership degree becomes 0.8.

For instance, a fuzzy set "close to five" can be expressed as $A = 0.4/3 + 0.8/4 + 1/5 + 0.8/6 + 0.4/7$.

BASIC PROPERTIES OF FUZZY SETS

The basic operations of fuzzy sets are derived from classical set theory. Two fuzzy sets are equal (A = B), only if:

$$\mu_A(x) = \mu_B(x), \ \forall x \in X. \tag{6}$$

A fuzzy set A in X is a subset of a fuzzy set B in $X (A \subseteq B)$, only if:

$$\mu_A(x) \leqq \mu_B(x), \ \forall x \in X. \tag{7}$$

Some other concepts of cardinality, mainly defined as a fuzzy number, are also used.

The complement of a fuzzy set A in X is described $\neg A$ and defined as follows:

$$\mu_{\neg A}(x) = 1 - \mu_A(x), \ \forall x \in X \tag{8}$$

The intersection of two fuzzy sets A, B in X is described $A \cap B$ and defined as follows:

$$\mu_{A \cap B}(x) = \min(\mu_A(x), \mu_B(x)), \ \forall x \in X \tag{9}$$

The union of two fuzzy sets, A, B in X is described $A \cup B$ and defined as follows:

$$\mu_{A \cup B}(x) = \max(\mu_A(x), \mu_B(x)), \ \forall x \in X \tag{10}$$

In the fuzzy theory, the method of description of $\min(a,b) = a \wedge b$, $\max(a,b) = a \vee b$ are often used to show the relation of min or max between a and b. Therefore, the intersection of two fuzzy sets A, B in X and the union of two fuzzy sets, A, B in X are also described as follows:

$$\begin{aligned} \min(\mu_A(x), \mu_B(x)) &= \mu_A(x) \wedge \mu_B(x) \\ \max(\mu_A(x), \mu_B(x)) &= \mu_A(x) \vee \mu_B(x) \end{aligned} \tag{11}$$

FUZZY RELATIONS

Fuzzy relations — exemplified by "much larger than", "more or less equal", etc. — are clearly omnipresent in human discourse. Formally, if X = {x} and Y = {y} are two universes of discourse and μ_R: $X \times Y \rightarrow [0, 1]$, then a fuzzy relation R is defined as follows:

$$\sum_{U \times V} \mu_R(x,y)/(x,y) \tag{12}$$

For instance, suppose that X = {horse, donkey} and Y = {mule, cow}. The fuzzy relation "similar" may then be defined as follows:

$$\begin{aligned} R = \text{"similar"} &= 0.8/(\text{horse, mule}) + 0.4/(\text{horse, cow}) \\ &\quad + 0.9/(\text{donkey, mule}) + 0.5/(\text{donkey, cow}) \end{aligned} \tag{13}$$

This equation is to be read as, e.g., a horse and a mule are similar to degree 0.8, a horse and a cow to degree 0.4, etc.

We show a next simple example. When X = {1,2,3}, then "approximately equal" may be defined as follows:

$$\begin{aligned} R = \text{"approximately equal"} &= 1/(1,1) + 1/(2,2) + 1/(3,3) \\ &\quad + 0.8/(1,2) + 0.8/(2,3) + 0.8/(2.1) \\ &\quad + 0.8/(3,2) + 0.3/(1,3) + 0.3/(3,1). \end{aligned} \tag{14}$$

The member function μ_R of this relation can be described by equation (15):

$$\mu_R(x,y) = \begin{cases} 1 \text{ } when \text{ } x = y, \\ 0.8 \text{ } when \text{ } |x-y| = 1, \\ 0.3 \text{ } when \text{ } |x-y| = 2. \end{cases} \tag{15}$$

In matrix notation this can be represented as follows:

(16)

X \ Y	1	2	3
1	1	0.8	0.3
2	0.8	1	0.8
3	0.3	0.8	1

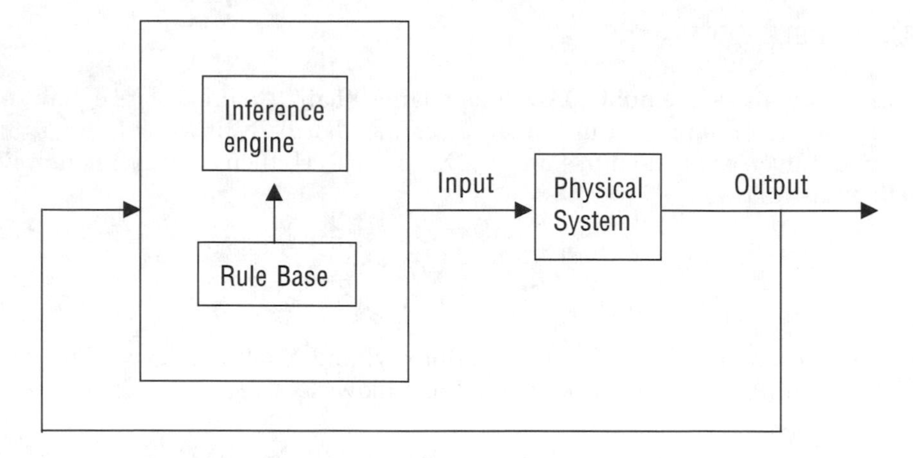

Figure 2.2 Architecture for fuzzy control.

Fuzzy relations are very important in fuzzy control because they can describe interactions between variables. This is particularly interesting in if-then rules.

FUZZY CONTROL

Mamdani and Assilian (1975) developed the first fuzzy rule-based control system implemented on a laboratory-scale steam engine. This control system acted as direct replacement for conventional control algorithms, based on proportional (P) or proportional plus integral (PD or PI, respectively) formalisms, commonly used in industrial control applications under the broad category of control.

The basic paradigm in fuzzy rule-based control is a frame-work based on rules of the form as follows:

If OA1 is — andOA2 is — and ... then CA1 is — and CA2 is — ..., (17)
If OA3 is — andOA4 is — and ... then CA3 is — and CA4 is — ...,

which map the observable attributes (OA1, OA2, ...) of the given physical system into its controllable attributes (CA1, CA2, ...). The terms left as blanks are linguistic terms, such as medium, slightly, or slowly. The controller structure in Figure 2.2 relates the architecture of a fuzzy rule-based control system to the conventional feedback architecture in control engineering.

The plant for which the controller was implemented comprises a steam engine and boiler combination. The simplified fuzzy controller for the steam engine is illustrated in Figure 2.3. The model of the plant used has two inputs: heat input to the boiler and throttle opening at the input of the engine cylinder, and two

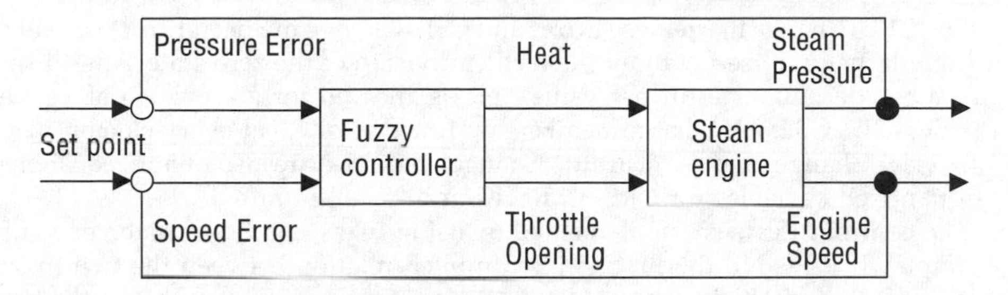

Figure 2.3 Fuzzy controller for steam engine.

outputs: the steam pressure in the boiler and speed of the engine. The definition of a fuzzy set permits one to assign values to fuzzy variables. In this application six (four input and two output) fuzzy variables are used: (1) PE = Pressure Error, defined as the difference between the present value of the variable and the set point, (2) SE = Speed Error, defined as in (1), (3) CPE = Change in pressure error, defined as the difference between present PE and last (corresponding to last sampling instant), (4) CSE = Change in speed error, defined as in (3), (5) HC = Heat Change (action variable), (6) TC= Throttle Change (action variable). These variables are quantized into a number of points corresponding to the elements of a universe of discourse, and values to the variables are assigned using seven basic fuzzy sets: (1) PB = Positive Big, (2) PM = Positive Medium, (3) PS = Positive Small, (4) = Nil, (5) NS = Negative Small, (6) NM = Negative Medium, (7) NB = Negative Big. Using these basic subsets and three operators defined earlier values such as "Not Positive Big or Medium" can be assigned to the variables.

The PE (Pressure Error) and SE (Speed Error) variables are quantized into 13 points, ranging from maximum negative error through zero error to maximum positive error. The zero error is further divided into negative zero error (NO – just below the set point) and positive zero error (PO – just above the set point). The subjective fuzzy sets defining these values are shown in Table 2.1.

Table 2.1 Quantized variables (*PE, SE*).

	–6	*–5*	*–4*	*–3*	*–2*	*–1*	*–0*	*+0*	*+1*	*+2*	*+3*	*+4*	*+5*	*+6*
PB	0	0	0	0	0	0	0	0	0	0	0.1	0.4	0.8	1.0
PM	0	0	0	0	0	0	0	0	0	0.2	0.7	1.0	0.7	0.2
PS	0	0	0	0	0	0	0	0.3	0.8	1.0	0.5	0.1	0	0
PO	0	0	0	0	0	0	0	1.0	0.6	0.1	0	0	0	0
NO	0	0	0	0	0.1	0.6	1.0	0	0	0	0	0	0	0
NS	0	0	0.1	0.5	1.0	0.8	0.3	0	0	0	0	0	0	0
NM	0.2	0.7	1.0	0.7	0.2	0	0	0	0	0	0	0	0	0
NB	1.0	0.8	0.4	0.1	0	0	0	0	0	0	0	0	0	0

*Light water reactor, without recycling of plutonium for the explanation of LWR in the original Table.

The CPE (Change in Pressure Error) and CSE (Change in Speed Error) variables are similarly quantized without the further division of the zero state. Apart from the above definitions a further value ANY is allowed for all four variables, i.e. PE, SE, CPE, CSE. ANY has a membership function of 1.0 at every element. The HC (Heat Change) and TC (Throttle Change) variables are also quantized. Tables of quantized variables for CPE, CSE, HC and TC are omitted.

The control rules were implemented by using fuzzy condition statements, for example "If PE is NB then HC is PB". Implied relation between the two fuzzy variables PE and HC is expressed in terms of the cartesian product of the two subsets NB and PB. To recapitulate, two algorithms were implemented in this application: one to compute the "heat change" (HC) control action. And other to compute the "throttle change" (TC) control action. Every rule in these algorithms is a relationship between the input variables PE, CPE, SE, CSE (in that order) and either HC or TC. The control actions are computed by present values for the input variables to the two algorithms. The input vectors are of course obtained by sampling the states of the steam engine at the sampling points. The output of either algorithm is obviously a fuzzy set which assigns grades (of membership) to the possible values of the control fuzzy variable.

For instance, a part of fuzzy rules according to HEATER ALGORITHM are shown as follows:

HEATER ALGORITHM

If PE=NB
and CPE = not (NB or NM)
and SE = ANY
and CSE = ANY
then HC = PB
Else

If PE = NB or NS
and CPE = NS
and SE = ANY
and CSE = ANY
then HC = PM

The above scheme containing the 24 rules was implemented on the PDP-8 computer and applied to the steam-engine plant. A fixed digital controller was also implemented on the computer and applied to the same plant for the purpose of comparison. The quality of control with fuzzy controller was found to be better each time than the best control obtained by the fixed controller as shown in Figure 2.4.

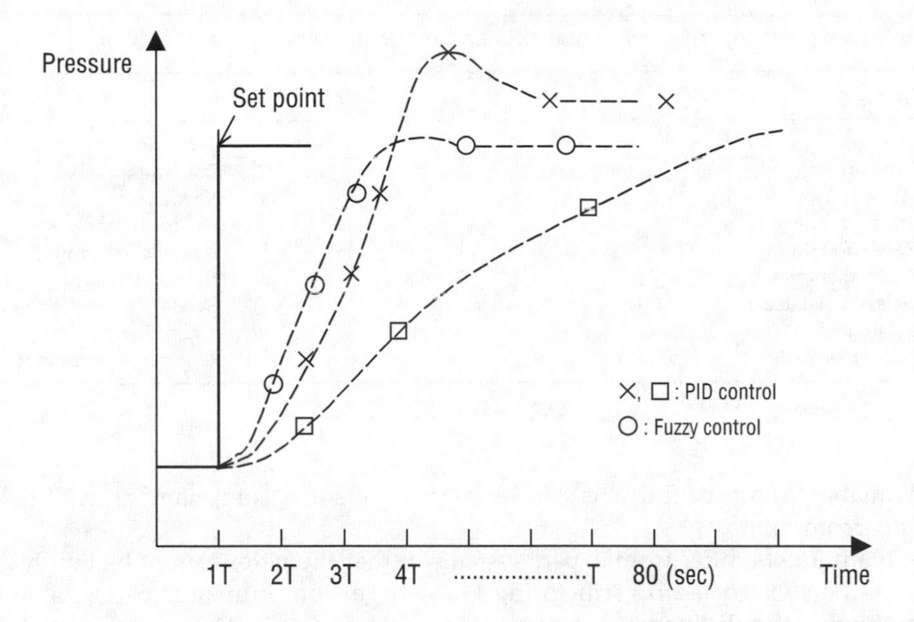

Figure 2.4 Comparison of PID control and fuzzy control for steam engine.

NEURAL NETS

Neural nets mimic learning processes and are named after the network of neurons formed in the human brain. Neural nets are made up of interconnections of devices, called neurons, and local external inputs. Considerable attention has been devoted in recent years to neural nets and neural network models since they show great promise for solving complex and nonlinear problems. Neural nets are inherently parallel processing machines and can solve problems more efficiently than serial digital computers, provided that problems are formulated in a form that enables parallel processing.

Neural nets have the ability to adapt and continue learning so as to improve their performance. They learn directly from examples, using a special learning algorithm. A user trains a neural net by presenting it with a series of examples of the inputs it will receive, paired with the outputs it will deliver. The neural net then learns the relationships between the input examples and the expected outcomes, and subsequently is able to predict outcomes from fresh input conditions. Neural nets are particularly suited to the solution of chemical process engineering problems such as fault detection and diagnosis, process design and simulation, process modeling and control, and quality control.

A wide variety of neural-net products are available, ranging from software tools for personal computers to integrated systems comprising of hardware chips and workstations. Hardware chips to build neural-net products are now becoming

Table 2.2 A comparison of neural nets and expert systems.

Neural nets	*Expert systems*
Example based	Rule based
Domain free	Domain specific
Finds rules	Needs rules
Little programming	Much programming
Easy to maintain	Difficult to maintain
Fault tolerant	Not fault tolerant
Needs a database	Needs a human expert
Fuzzy logic	'Crisp' logic
Adaptive system	Requires reprogramming

Source: AI Ware Inc.

available. Many neural nets can be run on personal computers as well as on supercomputers.

Neural nets differ from expert systems in that the latter cannot be taught but must be constructed by structuring knowledge from human experts. Table 2.2 highlights the distinction between neural nets and expert systems. Limitations inherent in current expert systems include the laborious nature of knowledge acquisition, the inability of the system to learn or improve its performance and the unpredictability of the system outside its immediate domain of expertise. A potential solution to those problems is the use of neural nets as will be described in the following sections.

SIMPLE NEURAL NET MODELS

Numerous neural net models have been proposed to simulate complex and nonlinear relations: here some simple models will be discussed. McCullough and Pitts (1943) presented a net of a threshold logic unit or neuron, so called because such a net is able to assume the binary values 1 and 0. This is shown in Figure 2.5.

Each neuron receives input signals, 1 or 0, modifies each of these signals by the weight of the inputs, sums the signals, and transmits the result, 1 or 0, as an output signal. Each neuron in such a net is described by the following equation.

$$\begin{aligned} z = f(u, h) &= 1, \quad u > 0 \\ &= 0, \quad u \leq 0 \\ u = w_1 x_1 + &\ldots + w_n x_n - h \end{aligned} \tag{18}$$

where z denotes an output, $x_i (i = 1, n)$ denotes an input, $w_i (i = 1, n)$ represents the weighting of the inputs, and h denotes a threshold value. The step function $f(u, h)$

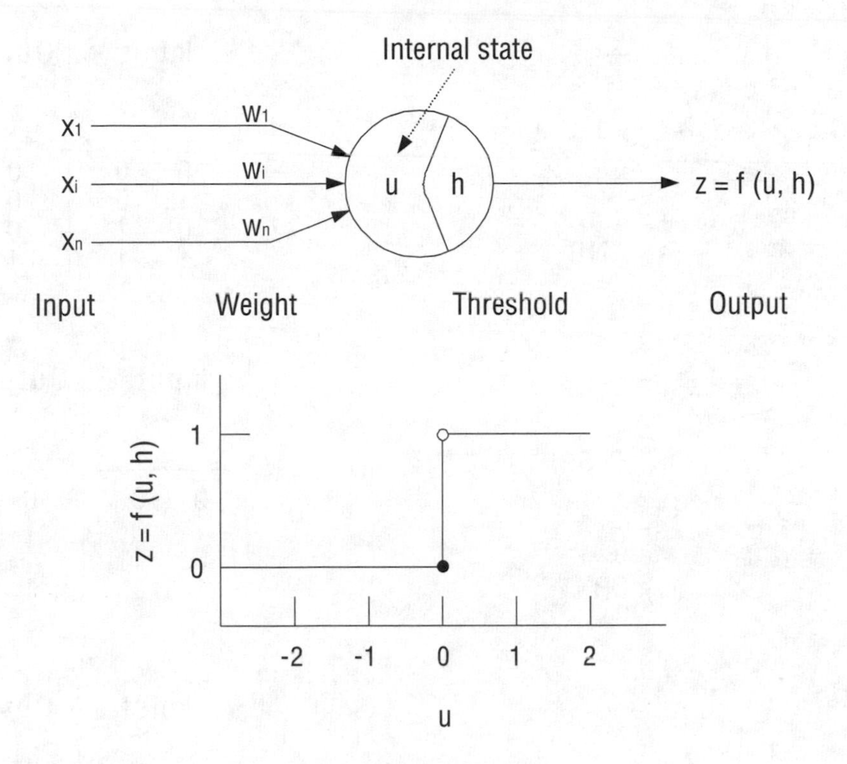

Figure 2.5 A threshold logic unit or neuron.

is defined to be equal to 1 when u is positive and equal to 0 when u is zero or negative. In demonstrating that the net is able to solve logic problems, McCullough and Pitts (1943) showed that difficult problems can be solved by interconnecting a net of simple processors in an appropriate way. Figure 2.6 shows examples of the simple neural net models for AND, OR and NOT calculations while the input-output relation for the exclusive OR(XOR) net is given in Figure 2.7. The net for XOR can be realized by introducing one hidden neuron as shown in Figure 2.7 and will be discussed in detail later.

NEURAL NET AND THE BACKPROPAGATION ALGORITHM

Figure 2.8 presents a typical neural net architecture. The circles denote neurons arranged in three layers input, hidden and output. Each hidden layer neuron is connected to each input and each output neuron. The connecting lines denote information flow channels between the neurons: these are called connections. Inputs to the net are stored in neurons represented by boxes. Inputs and outputs to the net have to be scaled into the range 0 to 1. The neurons in the input layer

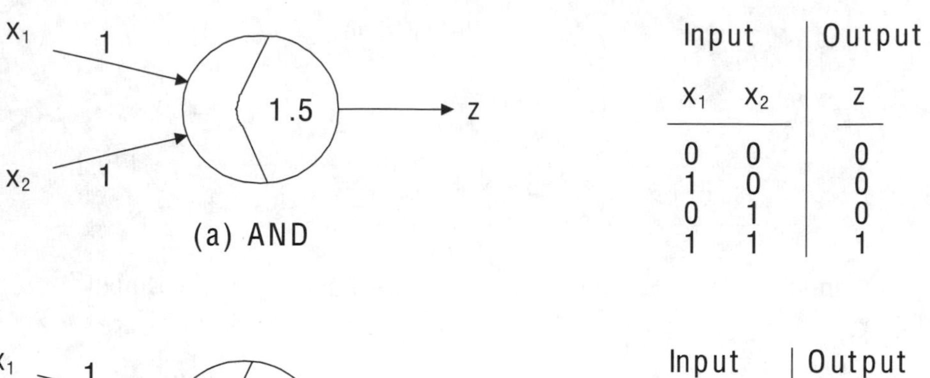

Input X_1	Input X_2	Output z
0	0	0
1	0	0
0	1	0
1	1	1

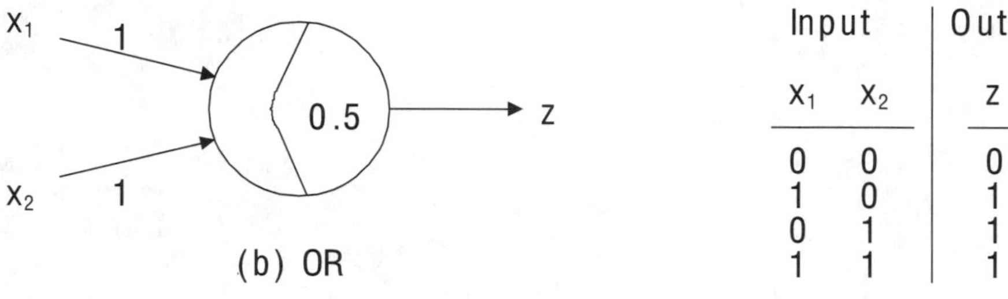

Input X_1	Input X_2	Output z
0	0	0
1	0	1
0	1	1
1	1	1

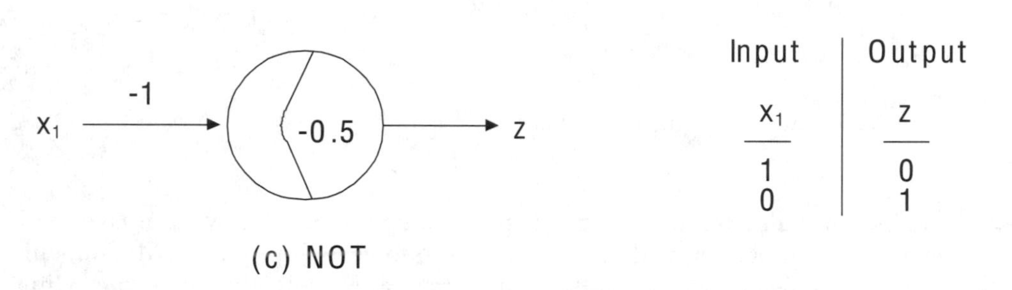

Input X_1	Output z
1	0
0	1

Figure 2.6 Simple neural net models.

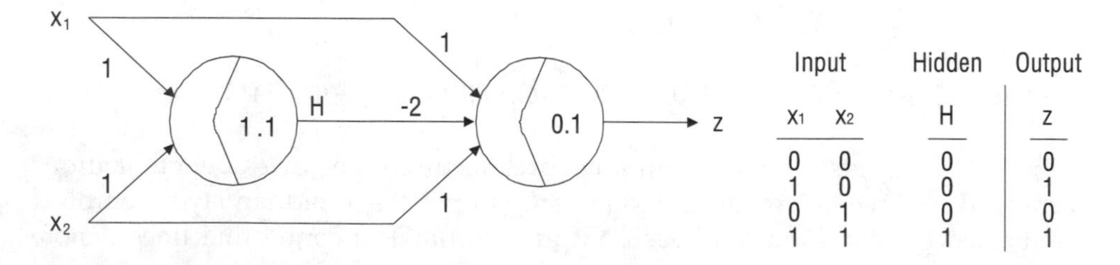

Input X_1	Input X_2	Hidden H	Output z
0	0	0	0
1	0	0	1
0	1	0	0
1	1	1	1

Figure 2.7 Multilayer net with hidden neuron for XOR.[2] (Reprinted with permission from MIT Press).

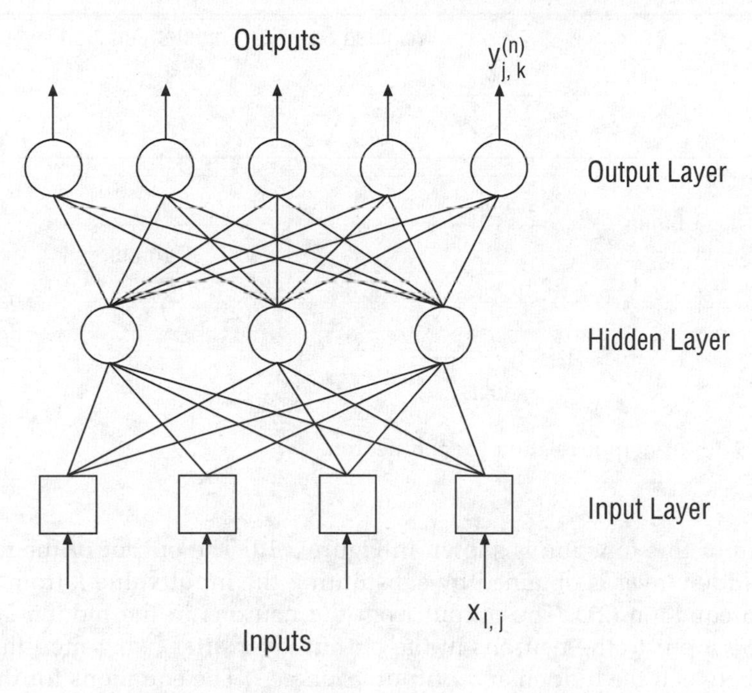

Figure 2.8 Three layer feedforward neural net.

simply store the input values while the hidden layer and output layer neurons each carry out two calculations.

Figure 2.9 shows one of the neurons in the network — the j-th neuron — used for these calculations. It is assumed that this is the hidden layer. The inputs to the neuron have N-dimensional vector x and a bias whose value is always 1. Each connection has a weight an $w_{i,j}$ associated with it. The first calculation takes a weighted sum I_j of the inputs as:

$$I_i = \sum_{i=1}^{N} W_{i,j} x_i + w_{N+1,j} \tag{19}$$

Then the output of the neuron O_j is calculated using a nondecreasing and differentiable transfer function. Usually a sigmoid function $f(z)$ is used as this transfer function:

$$O_j = f(I_j) \tag{20}$$

$$f(z) = \frac{1}{1+e^{-z}} \tag{21}$$

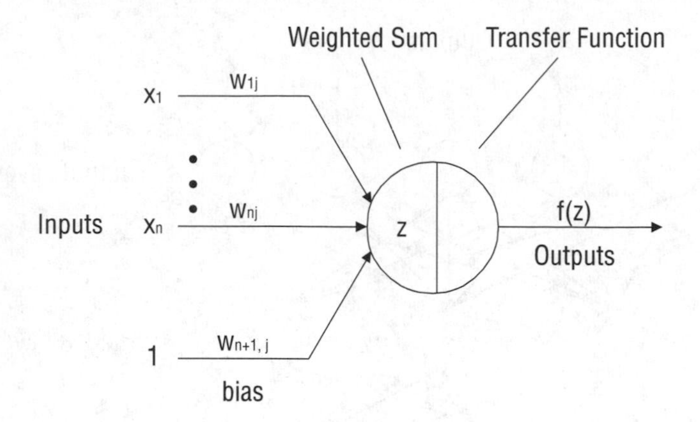

Figure 2.9 Input-output relation for *j*-th neuron.

The form of this function is shown in Figure 2.10. The output of the neuron O_j in the hidden layer is obtained by substituting the input value I_j from equation (19) into equation (21). The output from the neurons in the hidden layer then forms the input to the neurons in the output layer after being modified by the weight between the hidden and output layers $w_{j,k}$. The equations for the output layer are:

$$I_k = \sum_{j=1}^{M} w_{j,k} O_j + w_{M+1,k} \tag{22}$$

$$O_k = f(I_k) \tag{23}$$

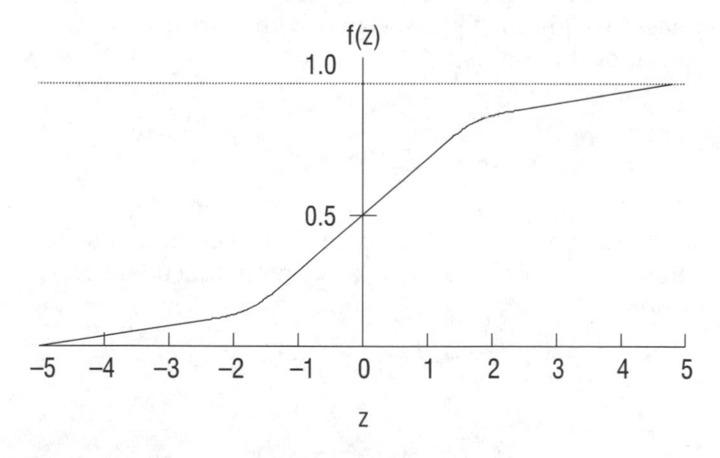

Figure 2.10 Sigmoid function (transfer function).

The neural net of Figure 2.8 learns by making changes in the weights of the connections. This is done by using what is known as a backpropagation algorithm and the generalized delta rule (GDR) for gradient calculation has been widely applied to various fields of engineering. The GDR is an iterative descent algorithm that minimize the mean square error E defined as:

$$E = \sum_{m=1}^{N} \sum_{k=1}^{P} (O_k^{(m)} - Y_k^{(m)})^2 \tag{24}$$

where N and P denote the number of training patterns presented to the input layer and the number of neurons in the output layer respectively, while $Y_k^{(m)}$ represents the desired target output of the k-th neuron and the m-th pattern. The GDR identifies an error signal associated with each neuron for each iteration involving pattern. The error signal of the k-th neuron of the output layer in the m-th learning $\delta_k^{(m)}$ is represented by the following equation:

$$\delta_k^{(m)} = (O_k^{(m)} - Y_k^{(m)}) f'(I_k) \tag{25}$$

where f' is the derivative of the transfer function, equation (21). Change in the weight $\Delta w_{j,k}^{(m)}$ is calculated by:

$$\Delta w_{j,k}^{(m)} = \eta \delta_k^{(m)} O_j^{(m)} + \alpha \Delta w_{j,k}^{(m-1)} \tag{26}$$

where η is the learning rate and α is the coefficient of the momentum term. Weights are updated using the expression:

$$w_{j,k}^{(m)} = w_{j,k}^{(m-1)} + \Delta w_{j,k}^{(m)} \tag{27}$$

The error signal for neurons in the output layer can be obtained by equation (25) but this cannot be done for neurons in the hidden layer(s). The backpropagation algorithm can improve this difficult problem by propagating the error signal backward (backpropagating) through the net. For the hidden layer(s), the error signal is presented by:

$$\delta_j^{(m)} = f'(I_j) \sum_{j=1}^{N} \delta_j^{(m)} w_{i,j}^{(m)} \tag{28}$$

The weight $w_{i,j}$ in the hidden layer are updated by using equations (26) and (27), except that the weights $w_{j,k}$ are replaced by the weights $w_{i,j}$. GDR calculates an error for each neuron in the output and hidden layer(s) using equations (25) and (27), and iteratively updates the weights of the hidden and output layers using equations (26) to (28), starting from the output layer and working backward to the input layer. Derivations for the expressions given here and described in detail by Rumelhart and McClelland (1986).

EXAMPLE — NEURAL NETWORK PREDICTION SYSTEM FOR CRITICAL TEMPERATURES

There have been various kinds of physical property prediction methods based on concepts of physical chemistry combined with statistics. Most of these consist of correlation equations with empirically determined parameters. Ohe (1992) described how the critical temperatures of substances can be determined directly by the association ability of a neural network trained by the back propagation method. There is a correlation between the critical temperature of hydrocarbons and the number of carbon atoms, the molecular weight, boiling point, and acentric factor. By giving the network these four parameters as input and the critical temperature as output for teaching, the network was 'trained' until the sum of error squared between actual output and ideal output became sufficiently small.

The neural network for this task contains 4 input nodes and 1 output node which represent the fundamental properties of the substance and its critical temperature, respectively. Table 2.3 lists the training data for the 35 substances involved. In all computations, the training rate parameter was set at 0.7 and the momentum term at 0.8.

The network error in prediction for training is computed as the sum of the output node errors for 34 of the 35 substances. The remaining substance was taken out of the training data to examine the network accuracy for prediction using the trained connection weights. Neural networks associate the non-trained data from the trained data. The best average performance was achieved by a network with 9 hidden units which attained 0.090 sum of trained error squared for 35 substances in 100,000 time steps. There was an insignificant increase in performance with the number of hidden units being between four and nine.

Comparison between the trained values calculated by the network and the observed values is included in Table 2.3. Hidden units numbered four and time steps 100,000. The average absolute error of training was 0.24% for the 35 hydrocarbons; the maximum error for training was –0.86% (obtained for ethane) and the minimum error was 0.012% (for butane). The ability of the network to train such numerical data is almost perfect.

During training the predicted substance was omitted from the example data so that in this case the prediction by the network was performed from data for the other 34 hydrocarbons. The average absolute error for the 35 hydrocarbons was 0.39%. On the other hand, the conventional Lydersen prediction method gave an average absolute error 0.59%. These errors show that the neural network based method is equivalent, and can be superior to conventional methods.

Table 2.3 Learning data for network to predict critical temperature.

No.	*Substance name*	*Carbon number*	*Molecular weight [–]*	*Boiling point [°C]*	*Acentric facter [–]*	*Critical temperature [K]*
1	Methane	1	16.0	–161.5	0.008	190.6
2	Ethane	2	30.1	–88.5	0.097	350.5
3	Propane	3	44.1	–42.0	0.152	369.8
4	Butane	4	58.1	–0.5	0.198	425.2
5	Pentane	5	72.2	–36.1	0.251	469.6
6	Hexane	6	86.2	68.7	0.300	507.7
7	Heptane	7	100.2	98.4	0.349	540.2
8	Octane	8	114.2	125.7	0.401	568.6
9	Nonane	9	128.3	150.9	0.445	594.6
10	Decane	10	142.3	174.1	0.488	617.6
11	Undecane	11	156.3	195.9	0.537	638.8
12	Dodecane	12	170.3	216.9	0.568	659.0
13	Tridecane	13	184.4	235.5	0.622	675.8
14	Tetradecane	14	198.4	253.6	0.643	694.6
15	Pentadecane	15	212.4	270.7	0.705	707.0
16	Hexadecane	16	226.4	286.8	0.775	717.0
17	Heptadecane	17	240.5	302.1	0.774	733.0
18	Octadecane	18	254.5	316.4	0.801	745.0
19	Nonadecane	19	268.5	330.0	0.827	756.0
20	Ethylene	2	28.1	–103.6	0.084	282.7
21	Propylene	3	42.1	–47.0	0.139	365.0
22	1-Butene	4	56.1	–6.3	0.208	419.6
23	1-Pentene	5	70.1	30.0	0.293	464.7
24	1-Hexene	6	84.2	63.5	0.296	504.0
25	1-Heptene	7	98.2	93.6	0.347	537.2
26	1-Octene	8	112.2	121.3	0.403	566.6
27	1-Nonene	9	126.2	147.0	0.383	601.0
28	1-Decene	10	140.3	170.6	0.491	617.6
29	1-Undecene	11	154.3	192.7	0.519	637.0
30	1-Dodecene	12	168.3	213.4	0.558	657.0
31	1-Tridecene	13	182.4	232.8	0.599	674.0
32	1-Tetradecene	14	196.4	251.2	0.645	689.0
33	1-Pentadecene	15	210.4	268.4	0.686	704.0
34	1-Hexadecene	16	224.4	285.0	0.725	717.0
35	1-Octadecene	18	252.5	316.4	0.801	745.0

As demonstrated, neural networks are able to acquire numerical knowledge of physical properties from examples of series of substances. The neural network's recall of trained data is nearly perfect with an average absolute error of 0.39%. The ability to learn from example and extract correlations from data makes neural networks ideal for the prediction of physical properties. It is believed that integration of neural networks with traditional correlation methods is highly effective in developing prediction systems using numerical data.

AREA OF FUZZY SYSTEMS AND ARTIFICIAL NEURAL NET APPLICATIONS

Nearly 30 years after the first paper by Zadeh (1965) introduced the concept of fuzzy set, fuzzy logic has finally emerged as alternative to classical – i.e., binary valued – logic in application ranging from industrial control to consumer products to aerospace and bioengineering. The direct conversion of operator experience into an automated solution is useful in most industrial automation and process control applications. The control in continuous processes — for example, in the chemical industry — is a special strength of fuzzy logic. One area of fuzzy logic control is to use the fuzzy logic system to determine the set points for low-level PID controllers. One example of such an application is a decanter control for a biochemical process in pharmaceutical industry in Austria. The methods of automated tuning for fuzzy logic controllers, meta-rule-based self tuning (Bare *et al.*, 1992) and reinforcement-based self learning (Lee, 1991) have been developed. They both demonstrated the capability to tune the parameters of fuzzy logic controllers to adapt to a wide range of changes in operating conditions. Fuzzy logic also proved able to solve control problems of chemical processes that have time-variant parameters. One example for this is the optimizing control of a hydrogenation plant. The hydrogenation reaction uses a catalyst that, due to an aging process, changes its properties over time (Yen *et al.*, 1995). Takagi and Sugeno (1983) proposed a fuzzy identification algorithm for modeling a human operator's control actions. They formulated the fuzzy control rules such that the consequence of a rule is a linear function of its preconditions. This provides a more systematic approach to the design of a fuzzy logic control. Fuzzy sets theory has been playing a remarkable role in control and knowledge-based systems as described above.

Reliability and safety analyses of systems is also one of the most important fields in which fuzzy sets theory may find applications (Onisawa *et al.*, 1995). Tanaka *et al.* (1983) proposed to employ the possibility of failure, viz. a fuzzy set defined in probability space. He presented approach based on fuzzy fault tree model, the maximum possibility of system failure is determined from the possibility of failure of each component within the system according to the extension principle. Kenarangui (1991) proposed event tree analysis under uncertainty. Fuzzy set logic is used to account for imprecision and uncertainty in data while employing event tree analysis. The fuzzy event tree logic allows the use of verbal statements for the probabilities and consequences; such as very high, moderate and low probability. Fault diagnostic system using information from fault tree analysis and uncertainty/impression of data has been developed by Yang *et al.* (1996). The system employed fault analysis and fuzzy failure analysis to construct a knowledge base.

Neural net technology is well suited to chemical engineering problems that require pattern recognition and data analysis including non-liner problems. The problems that have been considered can be classified into three categories, fault detection and diagnosis, process control and dynamic modeling.

An introduction to neural nets for chemical engineers is given by Bhagt (1990) who outlines neural net computation and simple examples while Samdani (1990) explains neural-net tool boxes and potential applications. Examples of fault detection and diagnosis of chemical plants have been described by Hoskins and Himmelblau (1988) for three continuous-stirred tank reactors in series while Watanabe *et al.* (1989) have discussed incipient faults in a chemical process involving heater, reactor and controller. The behavior of a fluidized catalytic cracking unit and a CSTR and reactor-distillation column have also been considered by Venkatsubramanian *et al.* (1989, 1990). Process control applications include: sensor failure of detection in a control system (Naidu *et al.*, 1990) diagnosing failure in a CSTR and controlling a highly nonlinear biosensor (Ungar *et al.*; 1990) and a chaotic time series Lorenz system and reactor temperature control of a CSTR (Ydstie, 1990). There have been a number of dynamic modeling applications such as the dynamic response of pH in a CSTR (Bhat *et al.*, 1990) and a steady-state CSTR, a dynamic pH CSTR and interpretation of biosensor data (Bhat *et al.*, 1990). A clear and concise introduction to neural nets is given in Lippmann (1987), neural nets are explained in detail by Rumelhart *et al.* (1986) and a number of texts are available for general reading.

A current and future research area is the use of hybrid fuzzy/neural nets systems and the unification of fuzzy and neural net modeling and control. Bringing the learning abilities of neural networks to automate and realize the design of fuzzy logic control systems has recently become a very active research area (Kosko, 1992). This integration brings the low-level learning and computation power of neural networks into fuzzy logic systems, and provides the high-level, human-like thinking and reasoning of fuzzy logic systems into neural networks. Rule identification and design is central to fuzzy systems. Learning systems, which identify rules from measured data, have been suggested and successfully implemented.

The use of the neural nets to aid engineering in developing fuzzy inference systems has also become the focus of much attention. Takagi and Hayashi (1991) point out that fuzzy reasoning presents particular problems: 1. the lack of a definite method for determining the membership function; 2. the lack of a learning function. They then go on to describe an approach for using neural nets to overcome these problems. The method is to investigate if-then rules by using neural networks to determine the membership functions of the antecedent and then determine the consequent component as the output for each rule. The approach used is to take raw data (say, in a control problem), apply a conventional clustering algorithm to group the data into clusters and to apply a neural net to this clustered data to determine the membership of a pattern within particular fuzzy sets. This approach is applied to two applications — estimation of the chemical oxygen demand density in Osaka Bay and the estimation of the roughness of a ceramic surface. Their method, in both cases, out performed more conventional methods. This combination of neural networks and fuzzy reasoning not only allows for automatic generation of membership function in certain applications but the use of fuzzy if-then rules and neural networks presents a potentially powerful paradigm.

CONCLUSION

This chapter presented an overview of the state of the art of fuzzy sets and neural nets and has indicated some of their applications in chemical engineering. Problems in chemical engineering are characterized by the need to be able to process incomplete, imprecise, vague or uncertain information. The problems faced are how to deal with uncertainty and unparticular linguistic terms that can not be defined exactly. The idea of a fuzzy set has been the key technologies for representing and dealing with vagueness in chemical engineering and for constructing adaptive systems. With their massive parallelism and learning capabilities neural nets appear to offer new and promising directions toward the better understanding and perhaps even the solution of the most difficult problems in chemical engineering. Fuzzy sets and neural nets are likely to be accepted and used widely if they solve problems that have previously been considered very difficult or even impossible to solve. The present challenge is to find the best way to fully utilize this new tool in the design, simulation and control of chemical processes.

REFERENCES

Bare, W., Mulholland, R. and Sofer, S. (1992) *IEEE Trans. on Automatic Control*, **35**, 2

Bhat, N. and McAvoy, T. J. (1990) *Comput. Chem. Engng*, **14**, 573

Bhat, N., Minderman, P. A., McAvoy, T. J. and Wang, N. S. (1990) *IEEE Contr. Syst. Mag.*, **8**, 4–24

Bhagt, P. (1990) *Chem. Eng. Prog.*, **86**(AUG), 55

Hoskins, J. C. and Himmelblau, D. M. (1988) *Comput. Chem. Engng*, **12**, 881

Kenarangui, R. (1991) *IEEE Trans on Reliability*, **40**, 120

Kosko, B. (1992) Neural Networks and Fuzzy Systems. Englewood Cliffs, NJ: Prentice Hall

Lee, C.C. (1991) *International Journal of Intelligent Systems*, **6**, 71–93

Lippmann, R. R. (1987) *IEEE Acoustics, Speech, Signal Proc. Mag.*, **4**, 4–22

Mamdani, E. H. and Assilian, S. (1975) *Int. J. Man- Machine Studies*, **7**, 1

McCullough, W. S. and Pitts, W. (1943) *Bull. Math. Biophys*, **5**, 115

Naidu, S. R., Zafiriou, E. and McAvoy, T. J. (1990) *IEEE Contr. Syst. Mag.*, **8**, 4–49

Ohe, S. (1992) *Sekiyu Gakkaishi (Journal of The Japan Petroleum Institute)*, **35**, 107

Onisawa, T. and Kacprzyk, J. (1995) Reliability and Safety Analyses under Fuzziness. Heidelberg: Physica-Verlag

Rumelhart, D. E. and McClelland, J. L. (eds) (1986) Parallel Distributed Processing, Vols 1 and 2. Cambridge, Mass: MIT press

Samdani, G. (1990) *Chem. Engng*, **97**(AUG), 39

Takagi, T. and Sugeno, M. (1983) *Proc. IFAC Symp. Fuzzy Inform.* Knowledge Representation Decision Analysis

Takagi, H. and Hayashi, I. (1991) *International Journal of Approximate Reasoning*, **5**, 191

Tanaka, H., Fan, L. T., Lai, F. S. and Toguchi, K. (1983) *IEEE Trans. on Reliability*, **R-32**, 453

Ungar, L. H., Powell, B. A. and Kamens, S. N. (1990) *Comput. Chem. Engng*, **14**, 561

Venkatsubramanian, V. and Chan, K. (1989) *AIChEJ*, **35**, 1933

Venkatsubramanian, V., Vaidyanathan, R. and Yamamoto, Y. (1990) *Comput. Chem. Engng*, **14**, 699

Watanabe, K., Matsuura, I., Abe, M., Kubota, M. and Himmelblau, D. M. (1989) *AIChEJ*, **35**, 1803

Yang, Z., Suzuki, K., Shimada, Y. and Sayama, H. (1996) *Journal of Japan Society for Fuzzy Theory and Systems*, **8**, 1073

Ydstie, B. E. (1990) *Comput. Chem. Engng*, **14**, 583

Yen, J., Langari, R. and Zadeh, L. A. (1995) *Industrial Applications of Fuzzy Logic and Intelligent Systems*, IEEE Press

Zadeh, L. A. (1965) *Fuzzy Sets. Information and Control*, **8**, 338

Zimmermann, H. J. (1991) Fuzzy set theory and its opplication, 2nd ed., Kluwer Academic Publishers

3. NEW TECHNIQUES TO PRODUCE FUNCTIONAL MATERIALS: CHEMICAL VAPOUR DEPOSITION

HIROSHI KOMIYAMA

Department of Chemical System Engineering, University of Tokyo, Tokyo, Japan

WHAT IS CVD?

In chemical vapour deposition (CVD) a solid is deposited from a chemically reactive vapour onto a solid surface or 'substrate'. The application of CVD is particularly important in the formation of thin films of semiconductors, insulators and metals. For example, a thin film of the compound semi-conductor gallium arsenide GaAs can be deposited as a layer on a substrate by the reaction between $Ga(C_2H_5)_3$ and AsH_3. This enables the construction of very large scale integrated (VLSI) circuits which contain up to one megabyte of memory on a silicon wafer chip about 1 cm^2 in area and comprise millions or transistors and capacitors. New processes to form diamond coatings at moderate pressures and temperatures are also based on the use of CVD. A more mundane example is the production of a 'gold' finish on fashion accessories and domestic furniture by depositing TiN on a metallic substrate.

WHY CVD?

The present 'age of materials' reflects the role of materials in governing the performance of devices which provide us with energy, artifacts and information. There are many illustrations of this. The necessity of energy saving leads us to develop more efficient car engines; one contribution to this is the evolution of the ceramic engine but this can only be achieved by developing cheaper and more reliable ceramics. The lifetime of robots and the accuracy of machine tools can be improved by new technologies to coat metals with diamond films. Application of optical fibres for data transmission depends on the production of quartz fibres of very high purity so as to reduce the number of relays between stations. In all these examples the need for improved materials is clear but the development of the appropriate materials and manufacturing techniques to produce them imposes severe constraints on future developments. The 'age of materials' is the response to such problems, recognising that the development of new materials provides the means to overcome many of the present limitations. CVD is one of the key techniques in the development of new manufacturing processes.

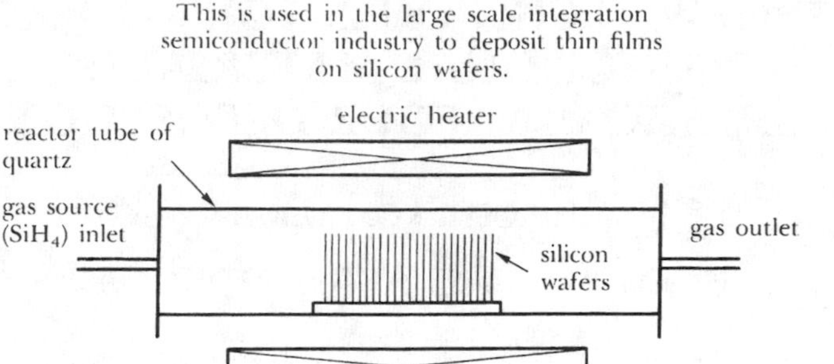

Figure 3.1 A typical CVD reactor.

FEATURES OF CVD

CVD has three great advantages. First it can produce a wide variety of substances because a large number of chemical reactions can be utilized, so giving many advantages over purely physical processes. Second, products of very high purity can be produced. The starting materials used in CVD are gases or vaporized liquids. By using techniques such as distillation, adsorption and crystallization that have been developed on an industrial scale by chemical engineers, gases and liquids can generally be purified to a much higher specification than solids and most very high purity materials are sold in the fluid state. Purity is indispensable for new materials and the raw materials required to produce the high specification products also have to be pure. It takes only a trace of nitrogen for example to contaminate diamond and give it a yellow colour. CVD using gaseous feeds thus has an immediate advantage.

The third advantage of CVD is its controllability. It is possible to deposit films with thicknesses controlled to a few atomic monolayers so that 'super lattices' can be built up from multilayer films, each composed of up to several tens of atomic layers. CVD also allows one to control the crystallographic form and morphology. Silane (SiH_4), for example, is the most common gas source used in silicon technology (see Figure 3.1). Depending on conditions monocrystalline, polycrystalline or amorphous structures can be produced. Thus at temperatures above 1000°C a monocrystalline silicon layer can be grown onto a single crystalline silicon wafer so that the crystal lattices in the layer are aligned with those in the wafer (the substrate); this is called epitaxial growth and is essential to form a p-n junction in highly integrated monolithic circuits. At around 700°C, SiH_4 is decomposed to form a polycrystalline form of silicon which can be used as a gate electrode and as an electrical bonding in integrated circuits. Below 700°C films will still grow, but too slowly to be industrially feasible. This requirement of adequate production rates is vital to ensure economic production and we will consider this later in more detail.

Figure 3.2 Photograph of a thin film deposition process by glow discharge plasma CVD.

To facilitate the decomposition of SiH_4 at low temperatures where thermal activation is inadequate for production purposes, various other methods of activation using energy sources such as plasmas, light and lasers have been tried. Some of these are now used industrially and plasma-CVD (Figure 3.2) is a key process for depositing amorphous hydrogenated silicon films to produce solar cells as used as power supplies for calculators, clocks and, very recently, electric cars.

Controllability therefore is a great advantage of CVD or the production of silicon films. It enables us to produce a precisely controlled atomic thickness and crystallographic orientation using SiH_4 as a common gas source and hence produce materials having a wide range of specific and different functions by careful control of process conditions.

PROBLEMS IN APPLYING CVD

The greatest limitation to the industrial exploitation of CVD is the low rate at which deposition takes place. The rate at which amorphous silicon films grow is about 0.5 nm s^{-1}. Since 0.1 nm is about the size of a hydrogen atom and the atom-to-atom distances in silicon are about 0.2 nm, this growth rate corresponds to 2 or 3 atomic layers per second. A silicon film thickness of about 10^3 nm is required in solar cells for them to absorb solar energy to convert to electricity;

using the above figures this would take about 2000 s (just over 30 min) and this is just about the maximum time that is practical for industrial production processes.

This rate of 0.5 nm s^{-1} is however totally impractical if we are trying to produce ceramic materials such as the blades of a gas turbine, parts of a car engine or tubes for high temperature applications. To deposit a layer 1 mm thick for example would take 2×10^6 s or about three weeks to produce. A possible structural ceramic material is Si_3N_4. CVD processes to produce thin films of this material from SiH_2Cl_2 and NH_3 have been commercialized in the wafer process for integrated circuits so it is not the synthesis of Si_3N_4 that is difficult, but rather the rate at which it can be deposited if we want to use it as a structural ceramic.

Why is it so difficult to increase the rate at which CVD takes place? The main reason is the formation of solid nuclei or powder in the gas phase. Chemical reactions are the essence of CVD and the rates of the reactions depend on temperature and the reactant concentrations. As would be expected, increasing temperature and concentration increases the film growth rate in most CVD processes. Beyond a certain limit, however, the growth rate starts to decrease because homogeneous nucleation is initiated in the gas phase, the resulting solid or powder formation consuming the gas phase reactants and so reducing the concentration on the growing surface of the film. Although homogeneous nucleation followed by powder formation is the basis of an important technology for the production of fine powders from the gas phase, it is an awful nuisance in film processing because it not only degrades the quality of the film but it also limits the maximum rate of film growth.

How then can we solve the apparent growth rate limitation that seems to be inherent in CVD? By applying basic and creative chemical engineering we can see how the problem has been overcome and also the way we developed a solution.

MECHANISM OF POWDER FORMATION IN CVD

Let us imagine that CVD could be used to produce both films and powders. Films are produced when the gaseous components diffuse, adsorb on the substrate surface and there react to form the product which nucleates and grows. On the other hand powder is formed when the reactions, nucleation and growth take place in the gas phase. The balance between film and powder production thus depends on whether nucleation occurs on a substrate or in the gas. We have studied the mechanism of powder formation in the gas phase and as an example let us look at the formation of TiO_2 which has uses as diverse as a white pigment and cosmetic powders. The production process is illustrated schematically in Figure 3.3.

The gaseous reactant is $Ti(C_3H_7O)_4$ in which four isopropyl ligands are attached to a titanium atom. This is a stable, volatile and moderately reactive compound, so it is a good raw material for CVD. TiO_2 forms according to the reaction:

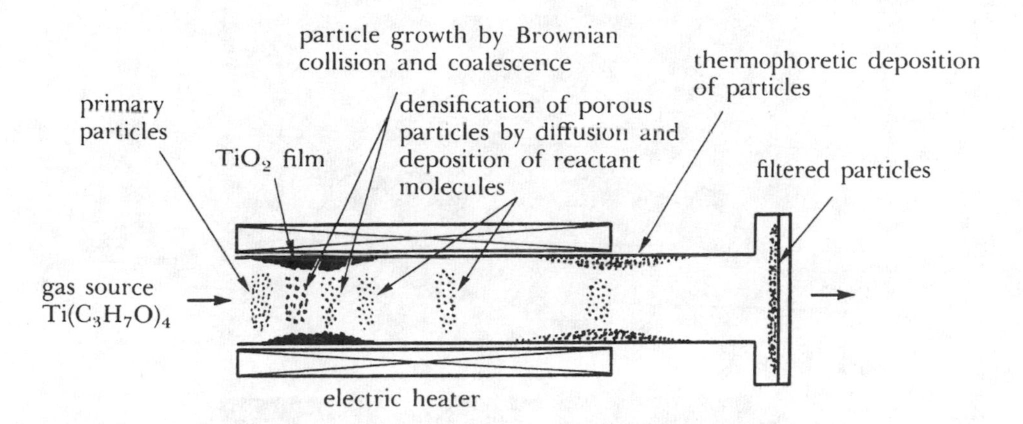

Figure 3.3 Conceptual view of stimultaneous formation of particles and films by a CVD process using a hot wall reactor system.

$$Ti(C_3H_7O)_4 - TiO_2 + 4C_3H_6 + 2H_2O$$

Some of the $Ti(C_3H_7O)_4$ which is fed to the tubular reactor is adsorbed on a growing TiO_2 film on the inner wall and partially decomposed to evolve C_3H_6 and H_2O. The resulting TiO_2 is incorporated into the film. The unreacted $Ti(C_3H_7O)_4$ desorbs back into the gas where it reacts further to form TiO_2 which nucleates and forms powders. The mechanism of powder formation is complex. Partially decomposed gas molecules condense, polymerize, form clusters and subsequently grow by colliding with each other as a result of Brownian motion. Figure 3.4 shows some of the TiO_2 particles produced in this way. Film growth and powder formation thus take place in parallel and the relative importance of the two processes can be changed by altering the processing conditions.

The ratio of desorption from the TiO_2 film to decomposition directly on the film determines the particle-to-film ratio of TiO_2. In this particular case the activation energy for desorption is greater than that for the decomposition reaction and so deposition on the film shows a maximum at a certain temperature, above which powder formation dominates. The concentration of $Ti(C_3H_7O)_4$ has a similar influence to temperature.

The growth rate limitation of CVD is thus inherent in, and is an inevitable consequence of the fundamental mechanism of CVD.

FROM FROST TO SNOW

Frost formation is a sort of CVD process; gas phase water molecules diffuse to a surface and are deposited and frozen there. Snow however comprises ice crys-

1.

2.

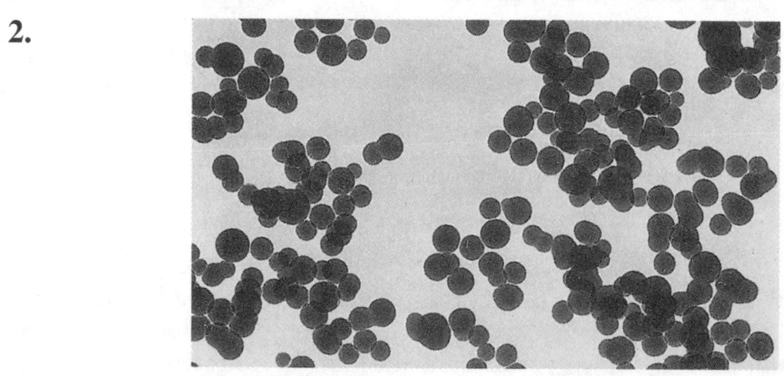

3.

Figure 3.4 Examples of particles prepared by CVD.
1. TiO_2 ultrafine particles (average diameter 0.02 μm)
2. TiO_2 particles (average diameter 0.13 μm)
3. particles of unique morphology (average diameter 1.5 μm)

tals or powder and deposits much more rapidly than frost. If snow, or powder formation, is an inevitable consequence of the high temperatures and concentrations necessary to speed up the CVD process, is it possible to collect the clusters and powders together on the substrate and so greatly increase the rate at which deposition takes place? This might then give us a way of using CVD processes for bulk ceramics where production requires a rate at least two orders of magnitude greater than conventional CVD processes. Although this might sound a rather crude idea, it was the basis of the developments that we made in my research group.

HOW TO PRODUCE SNOW POWDERS ... THERMOPHORESIS

So how can we devise a process to collect powders on the substrate? Small particles only settle very slowly in a fluid because their large surface-to-volume ratio causes great resistance to movement. Dust particles that are visible in a beam of light don't settle but dance along in the wind. We can see dust and other particles that are larger than the wave length of light — between about 0.4 and 0.7 μm. The particles that need to be collected on the substrate in a 'snowing CVD' are smaller than 0.1 μm and may be as small as 1 nm so they will never settle as a result of a gravitational force.

We thought that thermophoresis could perhaps be used to enhance the settling rate. Thermophoresis results from the Brownian motion of particles. Small particles in a fluid move randomly because they are constantly in collision with molecules of the fluid. In an isothermal field the time-averaged momentum that is imparted to a particular particle by the many molecular collisions that it undergoes is equal in all directions — it is balanced isotropically. Consequently,

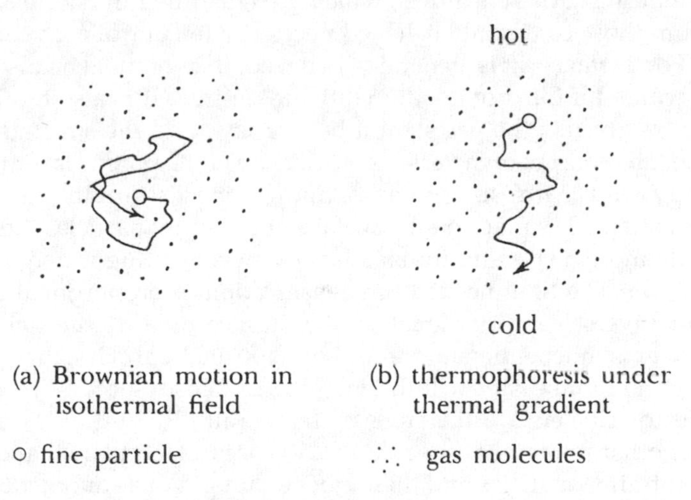

Figure 3.5 Brownian motion and thermophoresis of fine particles.

the most probable position of a particle after some time is its original position (see Figure 3.5). In a thermal gradient the molecules that arrive from the hotter side of the field exert more energy on a particle with which they collide than those from the colder side because the thermal velocity of a molecule is larger at higher temperatures. As a result a particle in a thermal gradient moves from the hotter to the colder region. This phenomenon is called thermophoresis and it is effective for particles smaller than about 1 μm. In the CVD reactor shown in Figure 3.3 a radial thermal gradient exists near the reactor exit because of the cooling of the gas stream and, as a result, thermophoresis occurs and particles are deposited on the wall in that part of the reactor.

PARTICLE-PRECIPITATION CVD (PPCVD)

Using the concept of thermophoresis enabled us to develop a rapid growth process which we call particle-precipitation CVD. It embodies three important features:

1. clusters and particles are formed in the gas phase which is maintained at a high concentration and temperature;
2. the substrate on which the film grows is kept cooler than the gas phase so as to enhance thermophoretic deposition of the clusters and particles; and
3. the interstices between the deposited particles are filled by surface growth resulting from the surface reaction, so increasing the density of the film. The surface must not be cooled too much so as to ensure that the surface reaction proceeds at a reasonable rate.

The ideas underlying these features, which are illustrated in Figure 3.6, are very different from those commonly believed necessary to control conventional CVD processes. For instance it is generally believed that homogeneous nucleation should be avoided if films of good quality are desired. It is also thought that to achieve the best results the gas should be at a low, and the substrate at a high temperature. The concept of increasing the density of the porous layer was arrived at by analogy with the frosting process. Snow deposits rapidly but with a low density. If frosting takes place on the surface of the snow particles then the snow layer is made more dense and forms a layer of ice. Frosting is analogous to the surface reaction. The time needed for densification is proportional to the void volume and inversely proportional to the surface area. If the density of the deposited layer is independent of the size of deposited particles, then the smaller the particles the more easily will the increase in density occur. Precipitation of fine particles or clusters are therefore preferable in PPCVD.

There is another important aspect to PPCVD; it is a technique that can be used to produce ultrafine particles and these can be thought of as another type of new material with particular applications in the production of new ceramic materials.

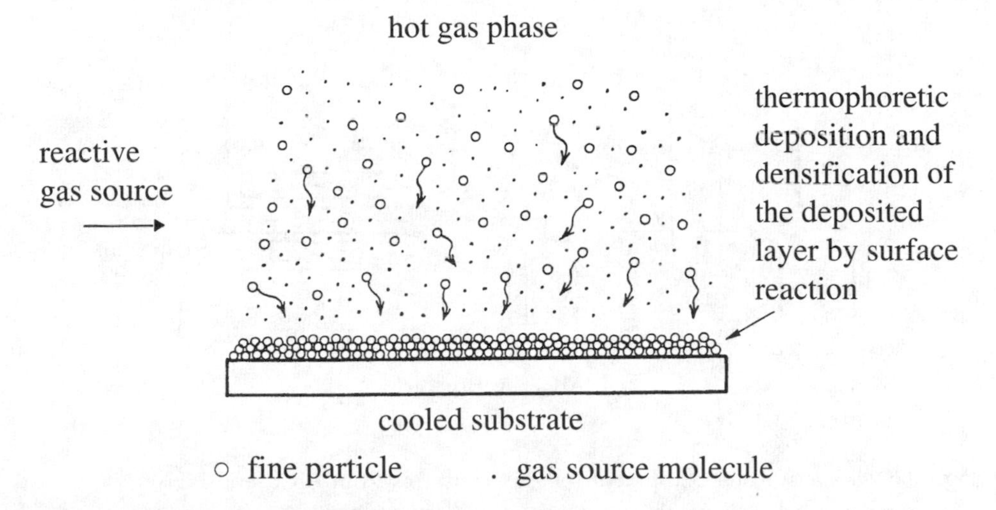

Figure 3.6 Conceptual view of particle precipitation CVD.

Traditional production of ceramics involves a series of processing steps including powder formation and sintering which have been used to produce pottery for thousands of years. In the age of new ceramics such processes give rise to severe problems because high temperature-resistant materials are difficult to sinter. High temperature durability is linked to strong chemical bonding between atoms and the diffusion rate in the solid is small. Since sintering proceeds by the diffusion of atoms to cause surface disappearance and volume shrinkage, solids of low diffusivity are hard to sinter.

This problem can be solved by using ultrafine particles which have short diffusion path lengths. What is more, ultrafine particles have lower melting points so that the melting point of a 3 nm gold particle can be as low as about 500°C compared to the melting point of pure gold which is 1053°C. Because of these advantages, ultrafine particles have great potential, but they also have disadvantages, particularly those arising from the huge specific surface area which they possess. As a result it is difficult to prepare the necessary presintered particle layers with the required high packing densities and to prevent surface contamination, particularly with oxide layers. PPCVD overcomes these disadvantages; although the deposited layer may be of low packing density this is increased by the subsequent chemical reaction while the one-stage process in the absence of air greatly reduces the chance of contamination.

SYNTHESIS OF AlN, TiO_2 AND ZrO_2

With these potential advantages of PPCVD, development work was initiated to

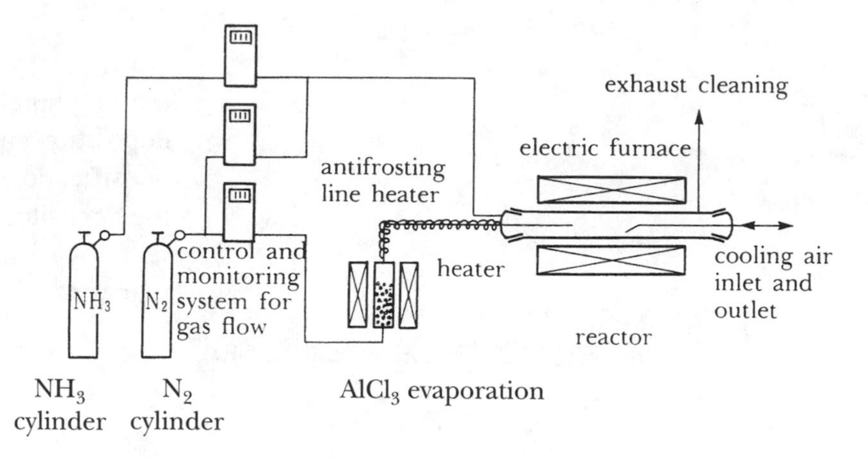

Figure 3.7 Experimental equipment for AlN synthesis from $AlCl_3$ and NH_3.

prepare some of the exciting new ceramic materials by this technique. Figures 3.7 and 3.8 show the equipment used to produce plates of AlN. The stoichiometry of the reaction used in the process is

$$AlCl_3 + NH_3 - AlN + 3HCl$$

The reactor, which was made of quartz and so able to resist the reaction temperature of 800°C, was about 50 mm in diameter and was heated electrically from the outside. The two reactant gases, $AlCl_3$ and NH_3 were fed separately into the reactor while the substrate was placed on a hollow holder so that it could be cooled by passing air through the holder. Many experiments had to be performed

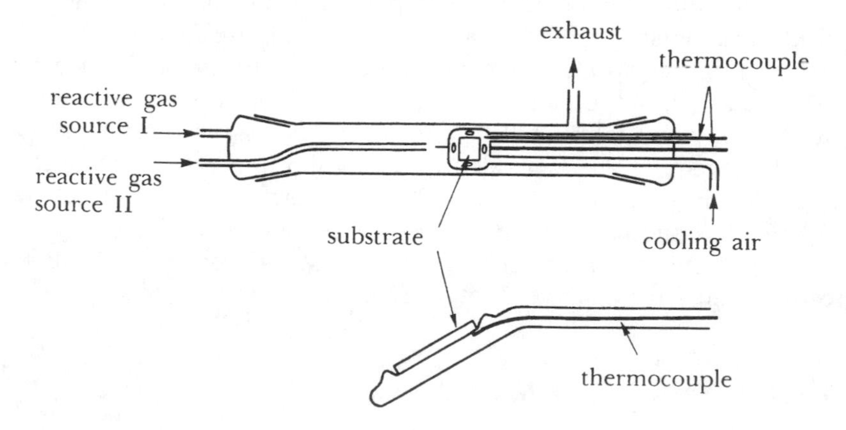

Figure 3.8 Reactor for particle precipitation CVD.

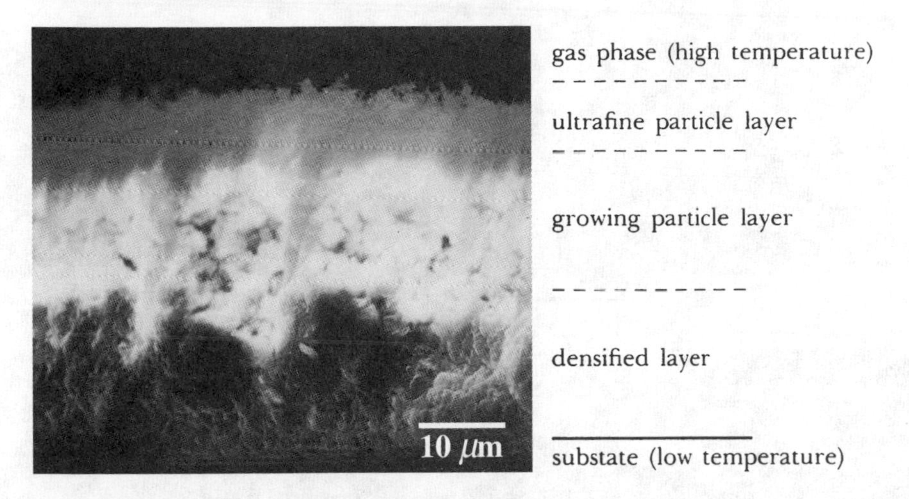

Figure 3.9 Scanning electron micrograph clearly showing the postulated mechanism in particle precipitation CVD (the example is of AlN formation).

in order to explore the effects of variables such as the reactant concentrations, gas and substrate temperatures and the geometry of the reactor. Many combinations of these variables are possible and a combination of mathematical analysis, careful and lengthy experimentation and confidence in the concepts were all required in order to make any progress. We were particularly concerned to see whether the film would grow on the cooled substrate or deposition would take place on the reactor walls where the temperature was higher and the reaction rates greater. The previous studies, however, had convinced us that an increase in temperature would enhance powder formation rather than film growth, that the physical principles of thermophoresis were correct and that at a properly selected temperature particles deposited on the substrate would grow by surface reactions.

Figure 3.9 is a scanning electron micrograph (SEM) of a cross section through one of the first successful films that we grew by this technique and so is a very special photograph for my research group. A dense layer can be seen adjacent to the substrate with a particle layer near the gas phase. It appears that the layer density increases with increasing layer depth. PPCVD could be made to work!

We have now succeeded in producing dense ceramic films of AlN, TiO_2 and ZrO_2; some examples are shown in Figure 3.10. The crystallographic structure can be controlled to produce amorphous, crystalline or oriented crystalline materials. Table 3.1 compares the film growth rates achievable by PPCVD with those obtained using conventional CVD. Growth rates by PPCVD can reach 100 nm s^{-1} (that is 360 μm h^{-1} or 0.36 mm h^{-1}). This is 2 or 3 orders of magnitude greater than conventional techniques and holds out the possibility of producing ceramic plates of 1 mm thickness by PPCVD.

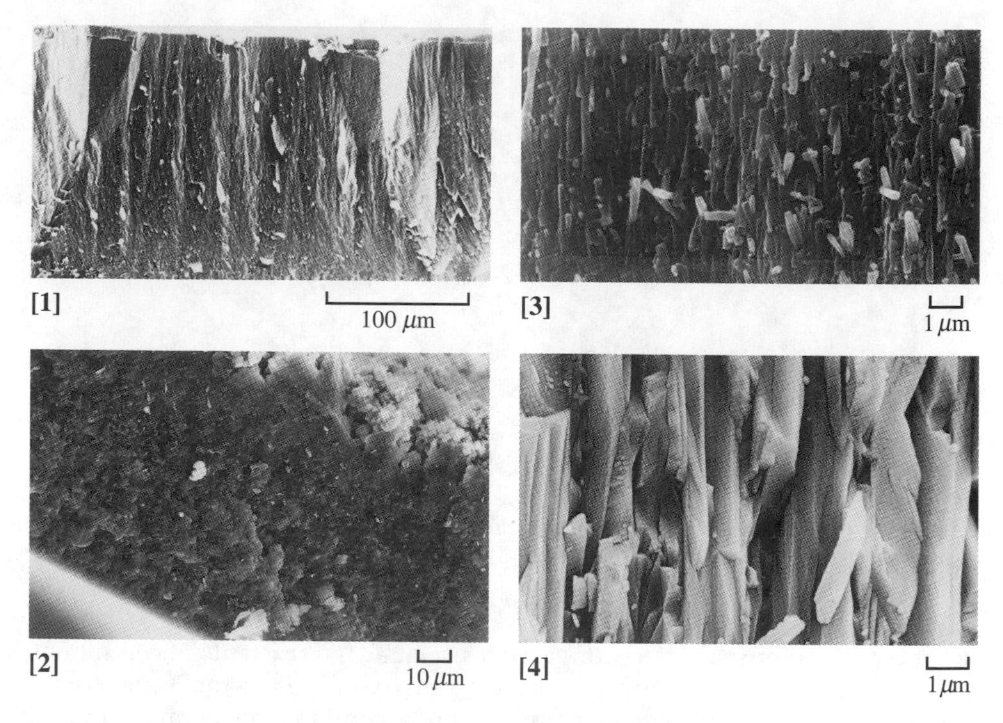

Figure 3.10 Ceramic plates prepared by particle precipitation CVD.
[1]: Fractured surface of AlN plate
[2]: Microstructure of AlN plate
[3]: Crystallographically oriented microstructure of AlN plate (a part of [1])
[4]: Crystallographically oriented microstructure of TiO_2 plate

Table 3.1 Comparison of the growth rates achieved by PPCVD with those of conventional CVD methods.

Workers	*Temperature, K*	*Growth rate, μm/h*
Noreika *et al.* (1)	1173–1623	4.0–12
Bauer *et al.* (2)	973–1573	0.6–50
Yim *et al.* (3)	1273–1373	12–15
Chu *et al.* (4)	1073–1473	0.5–1.0
Pauleau *et al.* (5)	673–1173	–0.3
PPCVD		200–400

A 0.5 mm thick plate of AlN prepared by PPCVD is colourless and transparent whereas AlN plates available for sale and produced by other methods are opaque or translucent with a green, yellow or black colour. Pure AlN is essentially colourless and transparent to light in the visible range. Any colour originates from impurities while the translucence results from the grain and porosity in the microstructure. PPCVD is thus able to produce a material at a rapid rate that is both highly pure and dense.

CVD AND CHEMICAL ENGINEERING

Chemical engineering concepts have been central to the development of CVD techniques. One or more chemical reactions are at the heart of a CVD process and a knowledge of the chemistry is important. The chemistry, however, is only a small fraction of what is needed. We need to understand nucleation on the substrate and in the gas phase, solidification, transport of molecules, the formation of clusters and reaction intermediates, the roles of temperature and pressure and their spatial distribution in the reactor, and so on and so on. Furthermore, in plasma and laser processes, high velocity electrons, plasma chemistry and photochemistry also play a part. A systematic understanding of all these factors is at the heart of chemical engineering in general and of chemical reaction engineering in particular and it is the overall and comprehensive view that enables the whole system to be optimized. Chemical engineering provides a methodology for treating complicated systems that comprise chemical and physical processes, so clarifying the key issues in the overall process.

CVD is an example of what can be achieved by applying the systematic methodology of chemical engineering. It acts as a focus for the other disciplines that are needed to understand and develop the process, so producing a product that can contribute to economic activity. Although the essential principles required to make this contribution are comparatively few in number, it is the ability to interpret a complicated process using these few principles that distinguishes the role of chemical engineering.

REFERENCES

1. Noreika, A. J. and Ing, D. W. (1968) *J. Appl. Phys*, **39**, 5578
2. Bauer, J., Biste, L. and Bolze, D. (1977) *Phys Status Solidi a*, **39**, 173
3. W. M. Yim, Stofko, E. J. Zanzucchi, P. J. Pankove, J. I. Ettenberg, M. and Gilbert, S. L. (1973) *J. Appl. Phys*, **44**, 292
4. Chu, T. L. and Kelm Jr., R. W. (1975) *J. Electrochem. Soc.*, **122**, 995
5. Pauleau, Y., Bouteville, A., Hantzpergue, J. J., Remy, J. C. and Cachard, A. (1982) *J. Electrochem. Soc.*, **129**, 1045

4. DEVELOPMENT OF OPTOELECTRONIC MATERIALS

MASAKUNI MATSUOKA

Department of Chemical Engineering, Tokyo University of Agriculture and Technology, 24-16, Nakacho-2, Koganei, Tokyo, Japan

OPTICS AND OPTOELECTRONICS: INTRODUCTION

Nothing travels faster than light in a vacuum. It travels almost a million times faster than sound passing through the air. The difference in arrival time between the flash of lightning and the thunder that follows is familiar to us all.

Light has long been used in communications. It could well be said that light was the only medium that enabled long distance communication before the invention of electric communication. Signal fires and arm signals are the oldest types of optical communication methods while information transmission by the use of mirrors may also be very old. In the 18th century towers having three arms mounted on the top were employed for communication using a combination of arm positions to represent different words (Figure 4.1). They were built every few miles by Napoleon and covered the whole of Western Europe and North Africa as a telecommunication network which, at its zenith, reached 5000 km in total length (Ohgaki, 1982).

A common feature of there technologies was the use of natural light such as the sun or a flame as their light source and of air as the transmission medium. Use of the technologies was therefore restricted by the weather while the distance over which communication was possible was also limited to be within sight distance, even after the invention of telescopes. Another characteristic of these methods was that the information tended to diffuse in many directions so that complicated and high density information could not be transmitted in such natural ways.

At the present time most information is transmitted and stored as electrical signals, for example as radio waves. These are similar to light waves, both being electromagnetic waves. Light has higher frequencies than do radio waves and a wider range of frequencies can therefore be made available so as to send more information than can be achieved using radio waves. Furthermore due to its higher frequencies, and hence shorter wavelengths, light beams can be focused more sharply than can radio waves.

Communication technologies have a history of employing shorter and shorter wavelengths; in Marconi's time the wavelength was about 10 km while now it is as short as 0.1 μm. The advantages of using light in place of radio waves include higher transmission densities, better directional control, faster transmission, smaller

Figure 4.1 Communication tower used in Napoleaon's time.

effects from noise caused by the fluctuation of magnetic or electric fields and, at least in principle, no requirement for electric energy to be used to the transmission process.

The field of optics has developed very rapidly as a result of using lasers as light sources in place of natural sunlight, as well as by employing optical fibres rather than the air as the transmission medium. Using new materials light has become controllable and can be transmitted faster and further with great reliability. Further improvements can be expected to result from the development of new materials and by improvements in manufacturing processes.

Light transmission is equivalent to sending information. Consequently the basic problem is how to convert light into information. Both on-off control and frequency control may be possible but in order to provide light with a high information density, accurate control of light over extremely short time scales is needed. This is usually done electrically and so the interface between light and electricity seems to be an essential component of such transmission systems. An electrical signal (a flow of electrons) is converted into light, transmitted and the light or optical signal is then re-converted into electricity. This sequence is outlined in Figure 4.2.

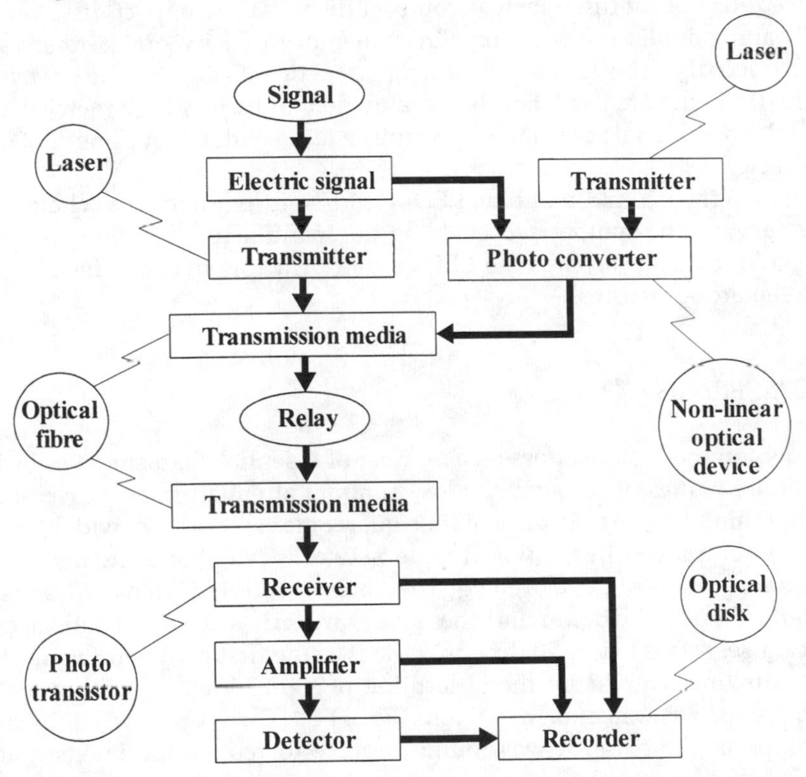

Figure 4.2 Outline of telecommunication by optoelectronics technology.

A further crucial difference between the technologies of the signal fire age and those of today is found in the recording of the transmitted or received signals and in particular the high information density that can now be achieved.

The technology used to transmit, treat and record large volumes of information with the aid of light is referred to as optoelectronics. In what follows the developments of optoelectronic materials will be described, particularly optical fibres and growth of nonlinear optical crystals.

LIGHT EMITTING AND LIGHT ACCEPTING MATERIALS

Conversion from electricity to light can be done by the use of either semiconductor lasers or light emitting diodes. Compound semiconductors such as GaAs and AlGaAs are typical materials that are used for this purpose. Crystals of these two materials have properties that enable them to emit light with a wavelength of 0.84 μm and 0.78 μm respectively. The resulting laser beams can be focused to micrometer or submicrometer dimensions so that they can be used for writing onto optical discs. In particular substantial numbers of AlGaAs lasers have been produced for optical discs such as compact discs (CD), compact discs-rewritable (CD-R) and minidiscs (MD). For higher memory density optical discs such as digital video discs (DVD), semiconductor lasers of AlGaInP having a wavelength of 0.635~0.650 μm are used. For the development of further high memory density optical discs, materials of GaN which emits a laser with a wavelength of 0.40 μm are expected.

Thanks to the invention of blue LEDs (light emitting diodes) a whole range of colour can now be synthesized enabling the manufacture of white lamps and/or full-colour displays. The blue LEDs are in principle made of InGaN/AlGaN double-hetero structures.

OPTICAL FIBRES

Transmission media must possess a number of essential characteristics including appropriate values of refractive index, minimal attenuation and ease of manufacturing thin fibres. At the present time quartz glasses are most widely used and have been employed in the world wide networks telephone circuits.

In the early stage of developing quartz fibres, the loss of intensity was as much as several thousand dB/km and these early materials were not satisfactory for practical use. A loss below 20 dB/km was set as the first goal to be achieved. The loss was mainly caused by the absorption of light by either trace amounts of impurities such as transition metal ions (Fe^{2+}, Fe^{3+}, Cu^{2+}, Cu^{3+} etc.) or by hydroxide groups in the glass. It was found possible to reduce the intensity loss by removing these impurities from the feed stock used in the manufacturing process and by processing in a closed environment so as to avoid contamination by

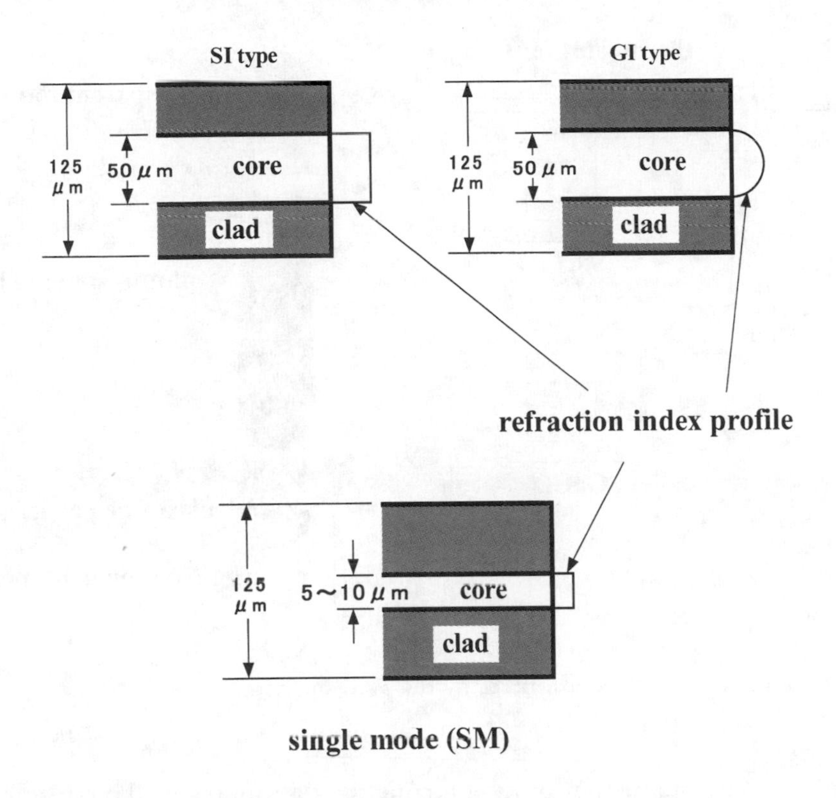

Figure 4.3 Different type of quartz optical fibres.

impurities during fabrication. In this way the Corning Company, for example, reduced the loss to 20 dB/km by using the chemical vapour deposition (CVD) method. Further developments reduced the loss still further down to 0.2 dB/km and so contributed to the practical use of optical fibres. This enabled long distance transmission with relays more than 100 km apart.

Quartz glass optical fibres consist of a core through which the light passes and a cladding layer, the outer diameter of which is typically 125 μm. Light beams are totally internally reflected at the interface between the core and the cladding and so pass along the fibre. To increase the refractive index of the fibre, small amounts of Ge or B are added to quartz glass. As illustrated in Figure 4.3 optical fibres can be classified into single mode (SM) and multimode (MM) types depending on the core diameter, the former having a diameter as small as 5 to 10 μm while in the latter it is about 50 μm. Multimode fibres include the step-index (SI) type in which the refractive index is uniform across the radius of the fibre and the gradual index (GI) type in which it is distributed in a parabolic manner.

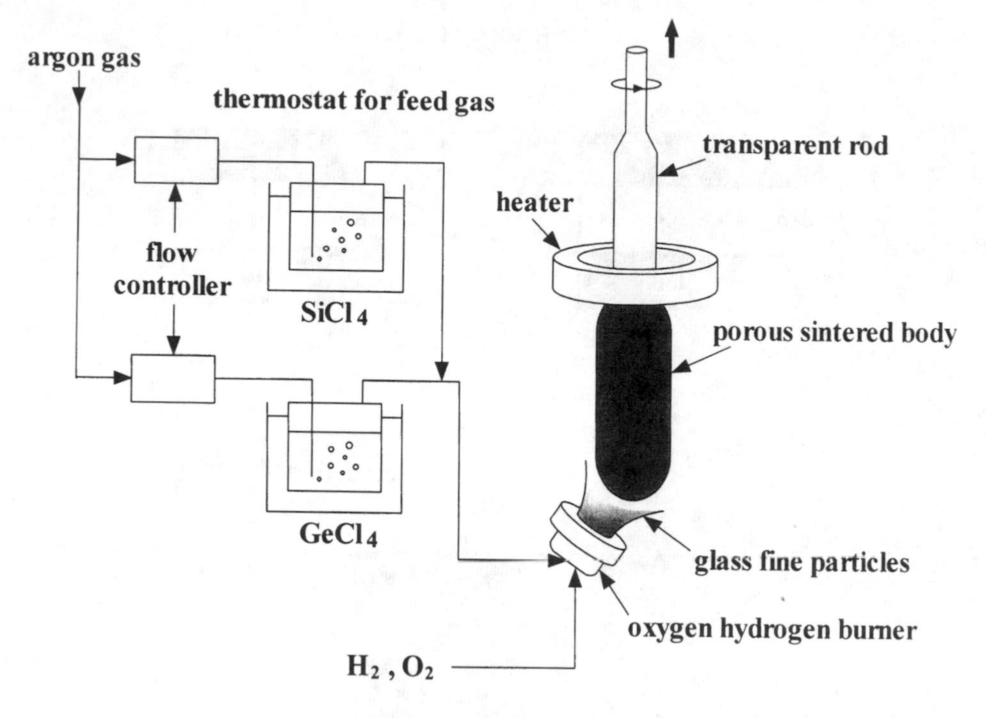

Figure 4.4 Production of optical fibres by the VAD method.

A continuous production process for quartz glass fibres of the GI type was developed in Japan and is called the VAD process. The process is outlined in Figure 4.4. A gas stream made up of $SiCl_4$ as the main component and $GeCl_4$ which is added to increase the refractive index, is mixed with Ar gas and fed to an oxygen-hydrogen burner. Here a hydrolysis reaction takes place in the flame to form glass particles consisting of SiO_2 and GeO_2:

$$SiCl_4 + 2H_2O = SiO_2 + 4HCl$$
$$GeCl_4 + 2H_2O = GeO_2 + 4HCl$$

Both the $SiCl_4$ and $GeCl_4$ used in the reaction must be purified from 99.9 to 99.999 percent by distillation before use, the normal boiling points of the chlorides being 57 and 84°C respectively. Increasing the GeO_2 concentration increases the refractive index of the glass as shown in Figure 4.5. Changes in the vapour composition and the reaction temperature of each nozzle enable glass rods to be produced that have an appropriate radial refractive index profile. These porous rods are then put in a furnace at 1600~1700°C so that they become transparent and by making them into fine threads optical fibres are finally produced. A typical

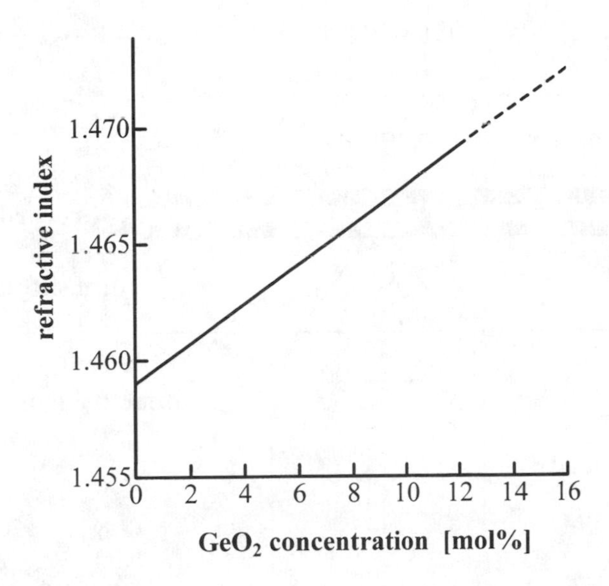

Figure 4.5 Dependence of the refractive index of quartz on GeO_2 concentration.

composition for the core material of a single mode fibre is 90% SiO_2 and 10% GeO_2 while the cladding layers are made of pure SiO_2.

Polymethylmethacrylate resin (PMMA), known as organic glass, has also been used to produce plastic optical fibres. Although the cost is about half that of quartz, the attenuation is larger and consequently its use has to be restricted to special fields of short distance transmission, say within 1000 m; future uses are most likely in cars, aircraft and domestic machines. Many types of plastic optical fibres have been studied and developed; they usually consist of PMMA as a core material and fluorine-containing polymers as a cladding. Typically the core diameter is between 100 and 1000 μm and the light loss may be as large as 100 to 1000 dB/km. The thermal and corrosion resisting properties are also generally inferior to those of quartz glass although plastic materials having identical thermal properties have been developed. The advantages of the plastic fibres such as relatively easy fabrication and manipulation, lower costs and lighter and flexible properties may lead to their wider use in the field of near-distance communication or transmission.

For practical use, a coating is needed to protect both quartz and plastic fibres and to increase tensile strength. A wide variety of materials including plastics, ceramics and metals such as aluminum have been evaluated. As a result of the development of coating technology over the fibre surface with plastic resin or carbon layers, a tensile strength as high as 5 GPa has been established, which has improved the long term stability and reliability of the fibres.

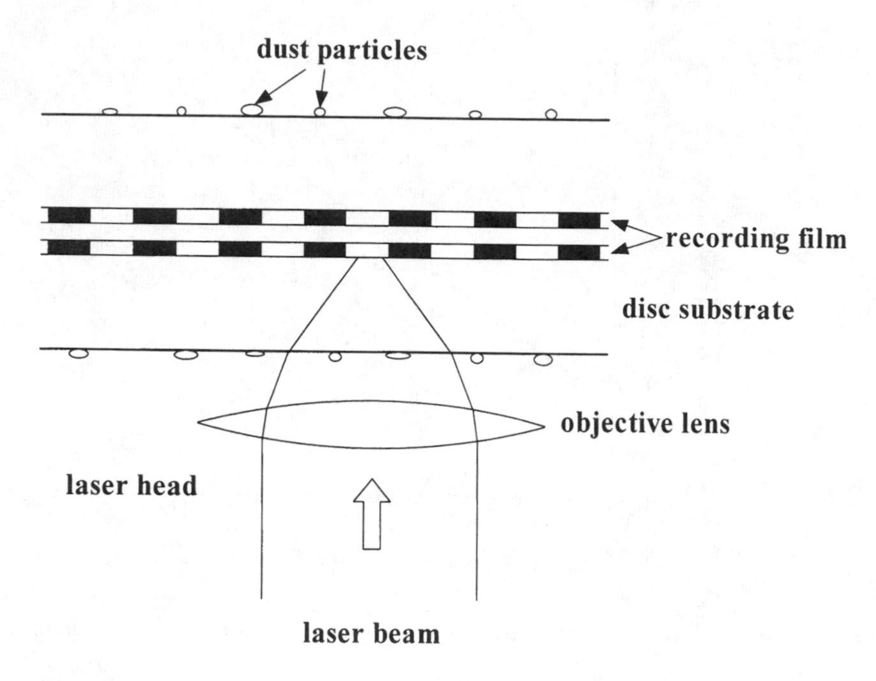

Figure 4.6 Principle of the optical disc.

OPTICAL MEMORY CELLS

Optical memory materials record and store information by reacting under light irradiation and then can release it when required. Photographic films are the best-known example of such materials. Development of optical discs is still progressing vary rapidly and practical uses of such discs are now common-place in compact discs (CD) or videodiscs (VD). Optical discs possess the great advantages of high memory densities and contactless reading between the disc and the head. A laser beam focuses on the memory layer located below the disk surface. The spot diameter of the beam on the disc surface can be as large as 1mm and so there is no need for very careful attention to the removal of dust, foreign particles and scratches on the surface (Figure 4.6).

There are in principle two types of commonly rewritable optical recording media: magneto-optical and phase-change recording media. Magneto-optical discs are made of amorphous RFeCo films where R=Tb, Cd or Dy. These materials have a relatively low Curie point, so that a sudden rise and fall of local temperature in a thin magnetic layer above the Curie point caused by irradiation of laser beams, results in inversion of local magnetization. Very small areas, perhaps 0.7 μm in diameter can be magnetized as desired, and subsequently detected on reading.

Phase-change media use the amorphous-crystal phase change produced by local heating by laser beams. Materials such as GeTe, Sb_2Te_3 and compositions along the pseudo-binary GeTe, Sb_2Te_3 tie line are used for this purpose (Suzuki, 1995). Since the detection of the phase depends on the difference of reflectivity between crystal and amorphous phases, the difference must be high enough to provide high contract. This is the main reason that the above materials are adopted for this purpose. In addition the material should be stable even after multiple cycling of overwriting, up to as many as perhaps 10^6 cycles.

LIGHT CONVERTING MATERIALS

Laser light is coherent and strong so that it behaves in a different way to normal light when it passes through transparent media. Some materials generate light having a wavelength of one-half or one-third that of the incident light wavelength. Other materials increase the incident wavelength by a factor of two or three. Materials showing these characteristics are called nonlinear optical materials. Use of these specific characteristics enables conversion of light frequencies without the need to adopt any other energy sources or conversion devices. Application of nonlinear optical materials to communication or information transmission are therefore being investigated particularly with the expectation of higher densities and improvement in reliability. Traditional nonlinear optical materials are semiconductors or inorganic materials such as GaAs, $NH_4H_2PO_4$ (ADP) and $LiNbO_3$ but organic materials often have superior characteristics.

The relation between the intensity of the electric field (E) and the intensity of the polarization (P) is linear when the electric field is weak but as the intensity increases a second term in the expression relating these two parameters and which is proportional to E^2 can no longer be neglected. The variation of the field intensity of a strong laser is a function of frequency having the form of $\sin(\omega t)$ so that during transmission of the laser light through such materials polarization depending on $\sin^2(\omega t)$ occurs, indicating that light having frequencies of 2ω is generated. This polarization in turn causes generation of an electromagnetic wave having a wavelength of 2ω. For even higher intensities of the laser light, higher order terms proportional to E^3 or higher would appear, and light having 3ω or higher frequencies result.

Compared with standard inorganic crystals, such second harmonic generating (SHG) properties of organic substances are highly efficient and in addition they have a rapid response due to the contribution of electrons in the molecule. The features common to SHG active organic substances are:

1. they are molecular crystals in which the intermolecular forces are weak and of the van der Waals type;
2. the molecular skeletons are made of pi electron conjugated systems, and
3. the electron supplying groups (donors) and electron accepting groups (acceptors) are attached to the pi electron systems.

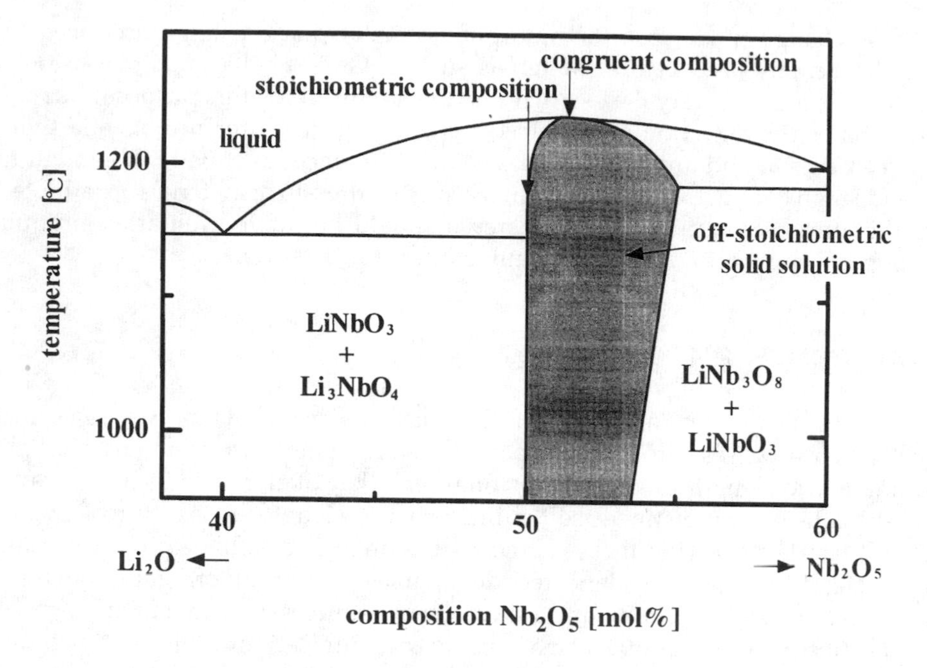

Figure 4.7 Solid-Liquid phase diagramme for LiO-NbO_3 system showing the formation of off-stoichiometric solid solutions.

Many organics have been reported to be SHG active substances; they include methyl-4-amino-aniline (MNA), methyl-(2,4-dinitrophenyl)-aminopropanate (MAP) and (poly)diacetylenes. Extensive surveys are being made using computer-aided molecular design techniques to discover new materials.

Single crystals of nonlinear optical materials have been grown either from the vapour, solution or melt phases. Inorganic materials are generally grown from the melt, although solution growth is also used for materials having good solvents. On the other hand, all of the three types of growth can be applied to organic crystals since organics generally have higher sublimation pressures and lower melting points than inorganic materials.

As a representative inorganic material with nonlinear optical properties as well as photorefractive effects which cause variation of refractive index induced by light, $LiNbO_3$ (LN) crystals have been grown mainly from binary melts comprising Li_2O+Nb_2O_5. Although wafers of LN as large as 7.5 cm are available at a reasonable price, there are problems arising from the phase equilibrium. As seen from Figure 4.7, LN has a congruent melting point which is located at a slightly different composition from the stoichiometric composition of LN. The congruent composition corresponds to the composition Li_2O: Nb_2O_5 = 48.5: 51.5. The resulting excess of Nb ions in the lattice causes vacancies in the anion sites in order to maintain the total electrical neutrality of the crystal. This causes difficulty in growing single crystals having the uniform stoichiometric composition.

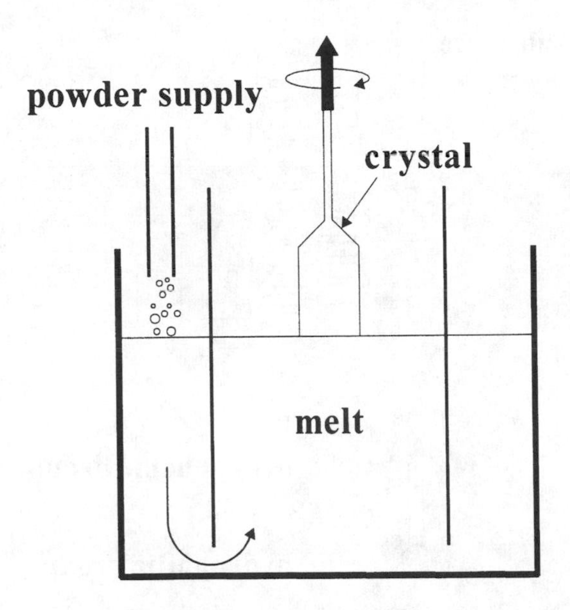

Figure 4.8 Double crucible Czochralski apparatus with continuous powder supply.

The presence of the vacancy leads to inferior photorefractive effects, so a technique to grow crystals with the stoichiometric composition has been developed (Kitamura *et al.*, 1992), (Kan *et al.*, 1992). Figure 4.8 illustrate a double crucible apparatus in which a LN crystal is pulled. It is basically a CZ method, but incorporates continuous feeding of material having the stoichiometric composition into the outer annulus in order to compensate for the consumption of material by crystal growth in the inner crucible. Both the inner and outer crucibles are connected at the bottom through holes and the stoichiometric melt is continuously fed from the outer crucible. The composition in the inner crucible is about 60 mol% Li_2O and from this grows the single crystal of LN at its stoichiometric composition.

Another example of crystal growth of a non-linear optical crystal is cesium lithium borate (CLBO: $CsLiB_6O_{10}$) (Mori *et al.*, 1995). Single crystals as large as $14 \times 11 \times 11$ cm^3 grown in three weeks by a flux crystallization technique has been reported. A single crystal of CLBO with dimensions $29 \times 20 \times 22$ mm^3 has been obtained in four days from the melt by the top-seeded Kyropoulos method.

For growth from solution selection of the solvent to be used is a key factor in growing crystals having a favourable shape. Thorough purification of both solvent and crystallizing component before growth is also essential. The shape of organic materials often changes with the solvent. Thus prism-like m-CNB crystals result from acetone solution whereas needle-like or plate-like crystals are grown from other organic solvents. These changes are due to different interactions between the solvent molecule and the crystal interfaces.

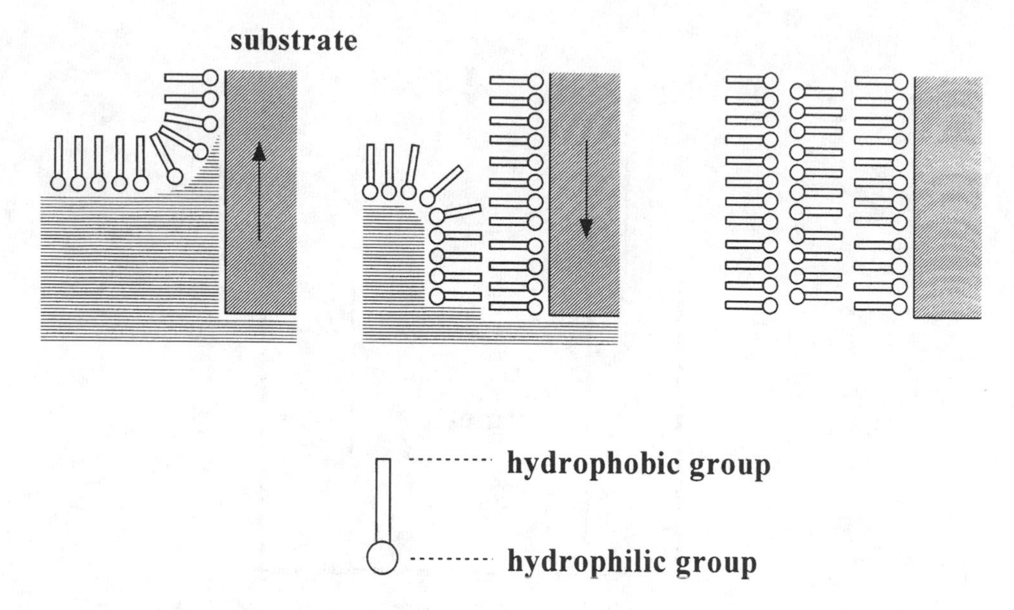

Figure 4.9 Fabrication of LB films.

In order to be incorporated into a device, crystals need to be of small cross sectional area to allow the laser to pass through, and to be of a length sufficient for conversion. This implies that larger single crystals are not specifically required. Nevertheless large crystals have advantages in practical applications and so techniques to grow large and high quality single crystals are constantly being refined.

LANGMUIR-BLODGETT FILMS

One of the techniques that has been examined to prepare a thin film of optical converter involves Langmuir-Blodgett (LB) films. Organic substances that are insoluble in water and contain both hydrophilic and hydrophobic groups are dissolved in organic solvents and placed on the surface of water in a trough. The solvents are then allowed to evaporate. By applying an appropriate surface pressure an organic monolayer with the hydrophilic groups directed downward into the water phase is formed. A glass plate, the surface of which is hydrophilic, is then dipped into the trough so that molecules accumulate on the plate with their hydrophilic groups attached to the surface. On dipping the plate again into the trough another layer is formed but this time with the hydrophobic groups directed toward the plate surface because they attach to the hydrophobic groups already on the surface. This operation, which is illustrated in Figures 4.9 and 4.10,

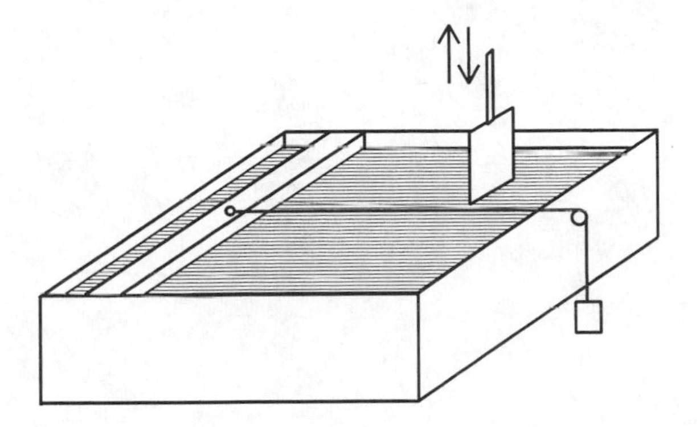

Figure 4.10 Apparatus for LB film fabrication.

may be repeated several times. The thin films thus prepared are referred to as Langmuir-Blodgett films after the names of the inventors of the technique. Production of the films always results in the formation of successive molecular layers lying in opposite directions, so that the film will have no overall nonlinear activity. However if nonlinear active diacetylene having carboxyl groups at both ends of the molecule is combined with optically inactive fatty acid so that they are laid in order, the resulting structure eliminates symmetry and so nonlinear activity can be expected. Succeeding polymerization will stabilize the molecules and, in this way, waveguides with ultrathin membranes may be produced.

Optoelectronics have developed as a result of the production of new materials or new principles. From the chemical engineering viewpoint there are many problems for practical application, particularly in the development and improvement of new materials. The development and optimization of new processes for the production of new materials in particular still needs great efforts.

REFERENCES

Kan, S., Sakamoto, M., Okano, Y., Hosokawa, K. and Nakai, S. (1992) $LiNbO_3$ single crystal growth by the continuous charging Czochralski method with Li/Nb ratio control. *J. Crystal Growth*, **119**, 215–220

Kitamura, K., Yamamoto, J. K., Iyi, N., Kimura, S. and Hayashi, T. (1992) Stoichiometric $LiNbO_3$ single crystal growth by Double crucible Czochralski method using automatic powder supply system. *J. Crystal Growth*, **116**, 327–332

Mori, Y., Kuroda, I., Nakajima, S., Sasaki, T. and Nakai, S. (1995) New nonlinear optical crystal: Cesium lithium borate. *Appl. Phys. Lett.*, **67**, 1818–1820

Ohgaki, T. (1982) Hikari Erekutoronikusu (Optoelectronics), pp. 2, Korona-sha

Suzuki, T. (1995) High density optical storage materials. *Oyo Buturi (Appl. Phys.)*, **64**, 208–219

5. CHEMICAL ENGINEERING IN POLYMER PROCESSING

KURT KOELLING

Department of Chemical Engineering, The Ohio State University, 140 W. 19th Ave., Columbus, Ohio 43210, USA

POLYMERIC MATERIALS AND PROCESSING

Polymeric materials possess extremely useful properties which make them ideal candidates for a broad range of engineering applications. Polymers have been synthesized over the past 100 years to be used in specific applications where certain chemical, thermal, mechanical, optical, or electrical properties are desired. Polymers can be subdivided into two main classes of materials including thermoplastics and thermosets. Thermoplastic polymers can be processed into final products by the following steps of heating, melting, forming in the liquid state, and then solidifying or freezing the polymer into a final shape. Thermoplastic materials may be solidified by cooling and remelted by heating. This heating/cooling cycle can be repeated many times without significant loss of properties. Several common examples of thermoplastic materials include polystyrene, polyethylene, polypropylene, and polycarbonate. Thermoset polymers are initially liquid resins which are chemically reacted or set into final shapes. Thermoset materials may not be melted or reprocessed. Upon heating, chemically reacted thermosets will soften but will not flow or lose their shape. The most common thermoset materials include epoxies, polyurethanes, and cross-linked polyesters.

Polymer processing has been defined as the "engineering specialty concerned with operations carried out on polymeric materials or systems to increase their utility" (Bernhardt and McKelvey, 1958). Polymer processing involves the conversion of polymeric materials into final products. Chemical engineers have traditionally played a key role in the development and optimization of polymer processing operations. This chapter overviews several of the key polymer processing technologies and products which have been developed in recent years. These include 1) continuous extrusion processing operations and 2) batch or cyclic molding processes.

EXTRUSION TECHNOLOGY

Nearly 60% of the world's manufacture of plastics is processed using extrusion technologies which are used to produce polymeric fibers, films, sheets, tubing, piping, insulated wire and a wide variety of profile extruded shapes. These products have one thing in common, which is that they are all extruded as a

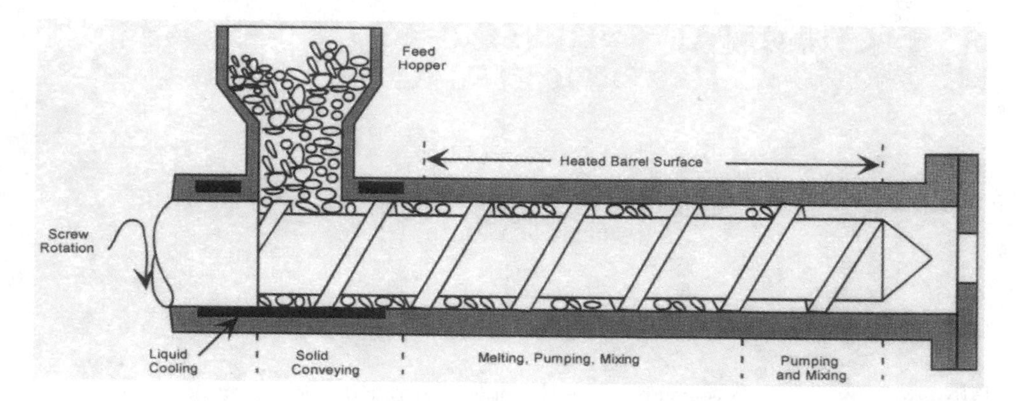

Figure 5.1 Schematic of a single screw plasticating extruder, after Tadmor and Gogos (1979).

continuous product with constant cross-section. The function of the extruder is to process large quantities of solid plastic pellets into a homogeneous liquid polymer melt and then pump the melt through a die at a constant flow rate. As the polymer exits the die, the molten polymer is cooled and formed into a specific cross-sectional shape. Figure 5.1 shows a schematic of a typical single screw extruder. Small plastic pellets are fed into the hopper of the extruder where they are conveyed through the extruder by the rotating screw. The plastic pellets are compacted and melted in the compression section of the extruder. The metering section of the screw pumps the homogeneous polymer melt at high pressures through the forming die at the exit of the extruder. The specific design of the die and type of forming equipment after the die determine the final product's cross-sectional shape.

Fiber Spinning

Man-made fibers are produced by a process of extrusion through a spinneret die that is known as spinning. The spinneret die is a plate containing many small holes through which molten or dissolved polymer is extruded under pressure.

The most common industrial spinning processes are melt spinning, dry spinning and wet spinning. In melt spinning, a polymer melt is extruded through a spinneret and then cooled through its glass transition or melting temperature to form a fiber. Melt spinning is the most economical spinning process, but can only be used when the polymer is stable at temperatures above its melting point. In dry spinning, a polymer solution is extruded through a spinneret into a chamber where the solvent is evaporated and the fiber solidifies. Wet spinning is the extrusion of a polymer solution through a spinneret into a liquid nonsolvent where the polymer precipitates to form a fiber.

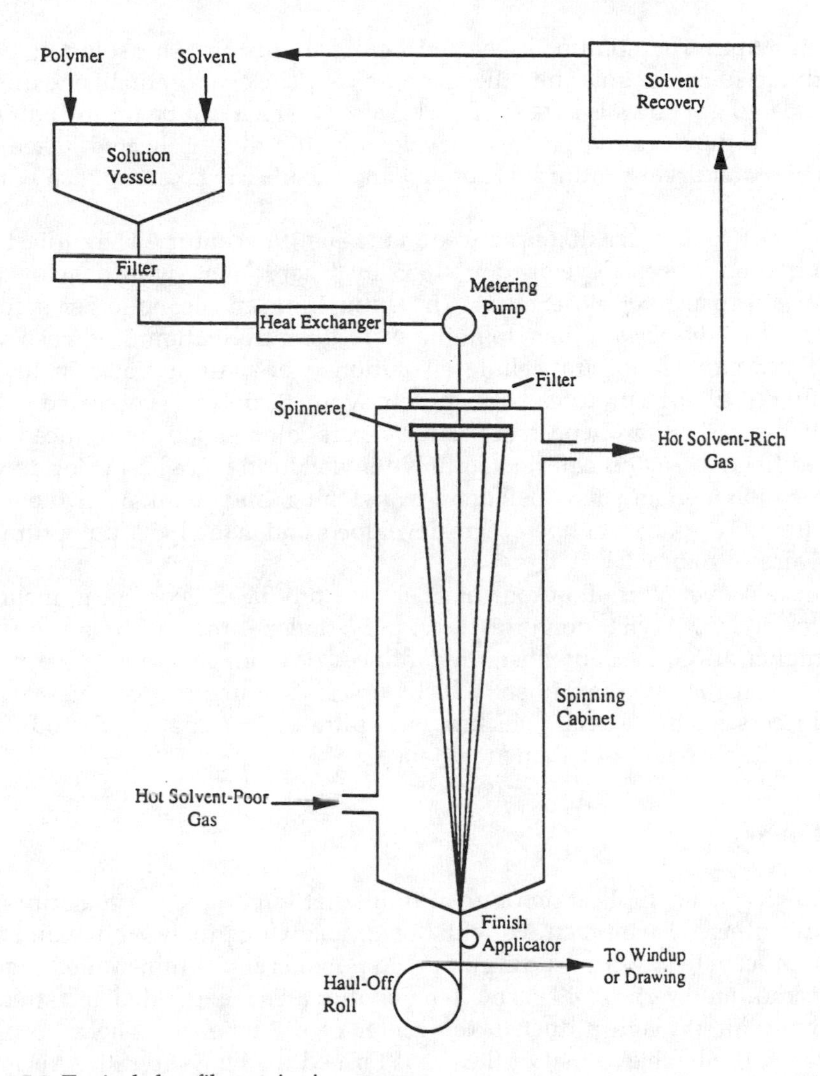

Figure 5.2 Typical dry fiber spinning process.

The spinneret dies used in melt spinning, where pressures can reach 7500 psi, are about 10 mm thick. Individual hole diameters are in the range of 0.175–0.75 mm at the exit. The spinnerets for solutions, where pressures are lower, range in thickness from 0.25–1.25 mm. Individual hole diameters are in the range of 0.025–0.2 mm. The number of holes in a spinneret may range from a few to several thousand. The holes must be arranged to provide even cooling to all parts of the spinneret face.

Figure 5.2 shows a schematic of a fiber dry spinning process. The equipment for solution delivery consists of a solution vessel, heat exchanger, metering pump, and spinnerets. The solution is heated before extrusion. The solution passes through the spinneret holes into the spinning cell. The evaporation zone through which heated gas flows is a vertical enclosed cell which can be up to 10 m high. In this zone the fiber morphology develops as the fiber is heated, dried, and stretched. After drying, a finish is applied and the fiber is taken up on a windup reel.

Man-made fibers were first produced in the late 19th century. The earliest fiber spinning process was developed in 1885 by Chardonnet, who developed an artificial silk using cellulose nitrate dry-spun from an alcohol-ether solution. Further early advances in fiber spinning were the introduction of viscose rayon in 1892 and cuprammonium cellulose solutions wet-spun into acid in 1897.

The first fibers to be produced from purely synthetic polymers occurred in 1912. In the 1930s the first synthetic fiber products were commercially introduced. They included fibers based on chlorinated poly(vinyl chloride), a vinyl chloride-vinyl acetate copolymer, and two polyamides, nylon-6,6 and nylon-6. In the 1940s, acrylic fibers and polyester fibers were developed, and in the 1950s polypropylene fibers were introduced.

At present over 40 million tons of fibers are produced every year, including polyester, nylon, acrylic, polypropylene, polyethylene, pvc, and viscose fibers. Other higher priced specialty fibers are produced on a smaller scale. These include the elastane (Lycra spandex) fibers, which possess elastomeric properties, and the aramid fibers, such as Kevlar, which possess ultra-high strength and modulus as well as high chemical and flame resistance.

Film Blowing

Film blowing is an application of great industrial importance. A majority of all polymeric films are manufactured with the film blowing process shown in Figure 5.3. The process involves extruding a molten polymer into a thin-walled tube die which is mounted vertically. This molten polymer tube is expanded and stretched by blowing air through a duct in the center of the tube die. The ratio of the diameter of the bubble to that of the die is termed the blow-up ratio. Air is also blown over the outside of the polymeric film bubble to cool and solidify the polymer film. This continuous polymer film bubble is then taken up with a pair of pull rolls which stretch the film in the flow direction. The blow-up ratio and the rate of take-up controls the film thickness and the amount of molecular orientation developed in the film. Molecular orientation strongly influences the mechanical properties, including tensile modulus, tensile strength, and tear and puncture resistance. The most common polymers used to produce films are polyethylene, polypropylene, and polyethylene terephthalate. End markets for films include plastic bags, food packaging, medical packaging, industrial liners, and electronics insulation.

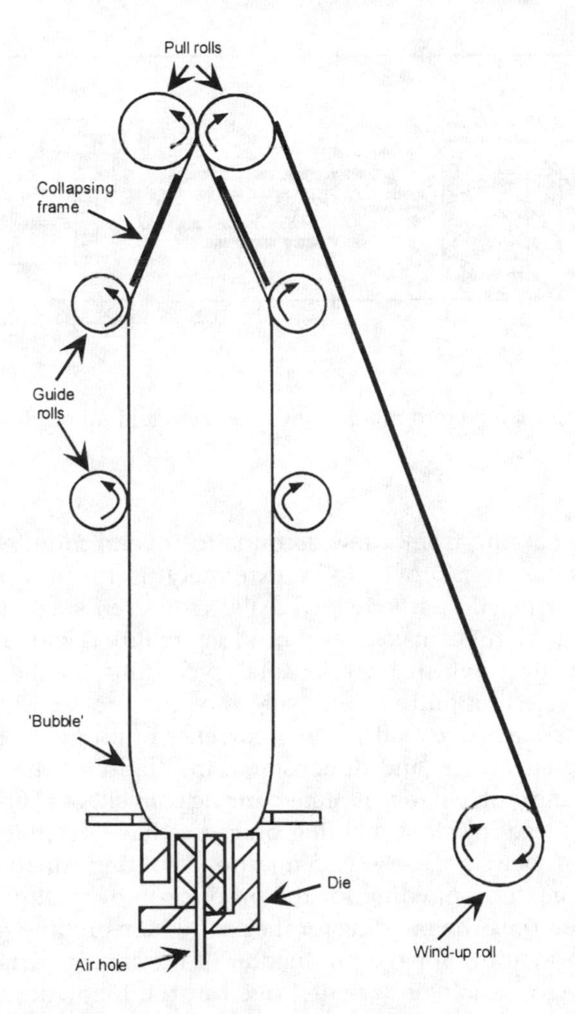

Figure 5.3 Film blowing die, film "bubble", and take-off equipment for the film blowing process, after McCrum *et al.* (1997).

ADVANCED INJECTION MOLDING TECHNOLOGIES

The conventional injection molding process involves the injection of a hot molten polymer under high pressure into a cooled mold as shown in Figure 5.4. In the mold the polymer cools and solidifies into its final shape. The process is capable of producing complex shaped products with short cycle times, high dimensional accuracy, and low cost. Typical injection molded parts range in cross sectional size from 1 mm^2 to approximately 1 m^2 with thicknesses in the order of 1 to 5 mm

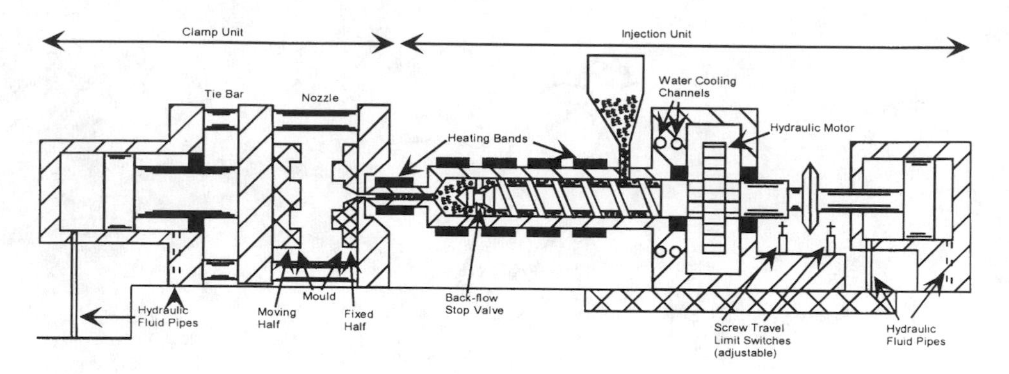

Figure 5.4 Schematic of a reciprocating screw injection molding machine, after McCrum *et al.* (1997).

with cycle times ranging from a few seconds to several minutes. The injection molding process has been proven to be extremely useful in rapidly producing large quantities of plastic parts at low costs. However, several aspects of the process still lead to significant costs and product quality issues. These include 1) the long cooling times which increase total cycle time, 2) the high pressures required during injection and packing stages of the process, 3) complex mold designs which are expensive, and 4) the occurrence of product defects including warpage, surface blemishes, and dimensional instability of the plastic part.

To overcome these problems encountered during conventional injection molding, several new advanced injection molding processes have been developed. These include processes such as gas-assisted injection molding for the production of hollow parts, co-injection molding for the production of multi-material molded parts, injection-compression molding for the production of thin-walled parts, and reaction-injection molding for the production of thermoset parts. Each of these advanced molding technologies are described in the following sections.

Gas-assisted Injection Molding

Gas-assisted Injection Molding (GIM) is a truly revolutionary plastic molding process. This innovative technology has been claimed to be one of the most important developments in the injection molding field since the invention of the reciprocating screw machine in the 1950s. This process consists of a partial injection of polymer melt into the mold cavity followed by an injection of high pressure nitrogen gas to create hollow sections in the plastic part as shown in Figure 5.5. This process is capable of producing large, stress-free parts with high quality surface finishes. In addition, the GIM process can substantially reduce operating expenses through reduction in mold design costs, reduction of clamp tonnage, reduction of material usage and reduction in cycle time for thick sections.

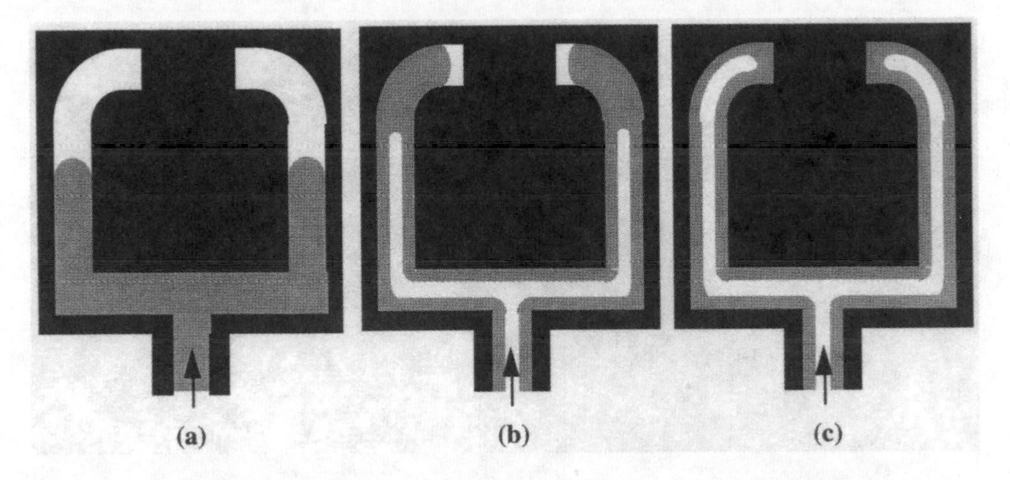

Figure 5.5 Schematic of the gas-assisted injection molding process. A hollow part is produced through a series of steps: a) Injection of polymer melt to fill approximately 60 to 90% of the mold cavity; b) Injection of high pressure gas through the same location of polymer melt injection; c) Mold filling is completed to produce a hollow part.

Consumer products are continuing to be produced in larger sizes using these advanced injection molding technologies. The production of large television cabinets, computer monitors, automotive bumpers and print copiers would not be possible without the advent of gas-assisted injection molding. GIM is being used to produce 85% of the world supply of one-piece plastic television cabinets ranging in size from 27 inch to 45 inch.

Co-injection Molding

Plastic processors are beginning to use multi-material injection molding to produce parts with recycled plastics and to develop parts with special features. The co-injection molding process (or sandwich process) produces parts with a "core" material which is completely covered by a "skin" material. Figure 5.6 illustrates how the process begins with the injection of the skin polymer into the mold cavity, which is immediately followed by the injection of the core polymer. With the proper choice of materials, mold design and processing conditions, a molded part can be produced with none of the core material being visible on the surface of the part. This process is gaining attention in the United States as the public increases pressure on industry to recycle plastic materials. Through this process, the properties of both materials can be optimized to improve product quality and reduce material costs.

The automotive industry is beginning to use the co-injection molding process to produce parts with a recycled plastic as the "core" material, sandwiched with

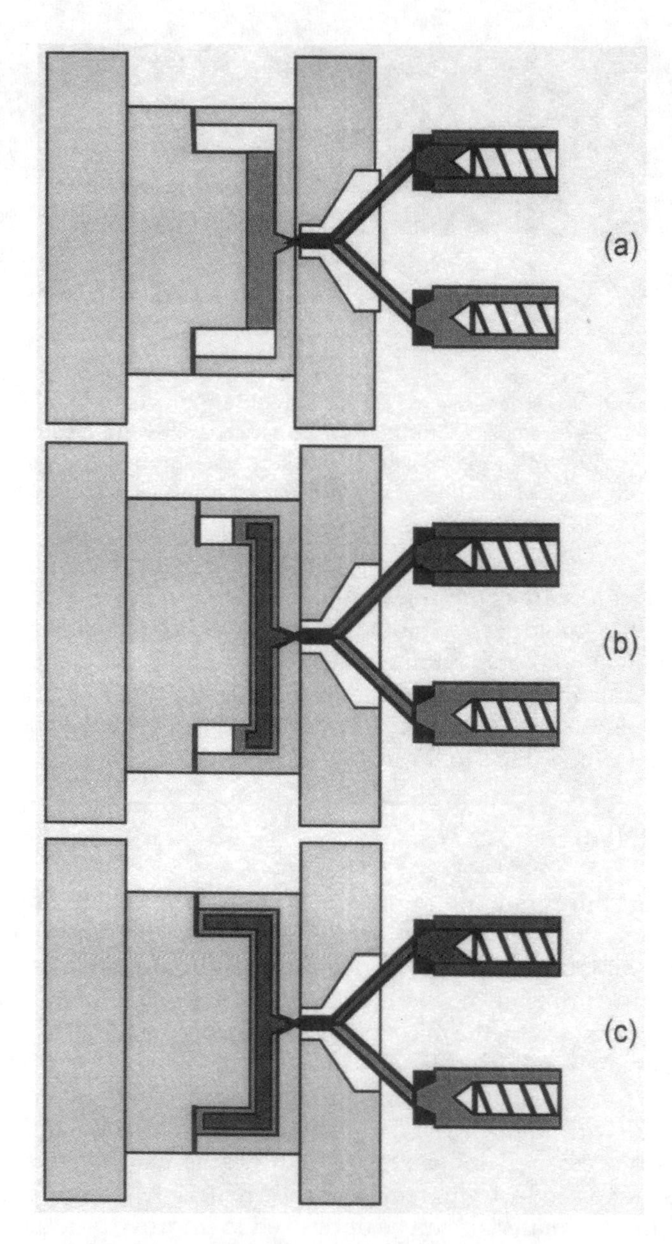

Figure 5.6 Co-injection molding process, after Battenfeld (1996). A core/skin plastic part is produced through a series of steps: a) injection of skin polymer into the mold; b) injection of core polymer into the mold; c) complete filling of mold cavity.

a decorative, expensive "skin" material to reduce material cost and improve surface finish. Applications include the use of a foamed core polymer to reduce weight and improve surface. A foamed or rubbery polymer has also been incorporated into the skin layer to create co-injection molded parts which are strong, due to the core polymer, but have a cushion feel, due to the foamed or rubbery skin layer. Several rules of thumb must be followed when choosing combinations of skin and core materials for the co-injection molding process. Both materials should have similar shrinkage behavior during cooling and should have good adhesion between the polymers so that delamination does not occur. Co-injection molding is being used for the production of most automobile headlight and taillight assemblies as well as for many door handles and keyboard panels.

Injection Compression Molding

Injection compression molding is a technique that is very well suited for the production of thin wall plastic parts. The goal of injection compression molding is the ability to produce lightweight thin parts without high levels of molded-in stresses. As the name implies, injection compression molding involves the steps of first injecting polymer into a mold cavity which is partially open, followed by compression of the polymer between the closing mold cavities. This process is capable of producing large molded parts which have extremely thin walls.

The art of "thinwalling" has developed into somewhat of a market standard within the portable electronics industry. Cellular phones lead the way with wall thickness recently reported down to 0.20 mm. Notebook computers are pushing for 1.0 mm wall thickness, down from 1.5 mm to 2.0 mm. However, material and process advancements directed at these markets are quickly spreading to other industries. The compact disk (CD) industry and digital video disk (DVD) industry are using injection compression molding to produce polycarbonate optical discs with low molded-in stresses. Without injection compression molding, these industries would not be able to produce standard DVDs or larger diameter optical disks. The potential for part cost saving, both by lowering the cycle times and reducing material consumption, has become attractive to other markets. Other advantages associated with thin wall molding include lower processing temperatures and lower part shrinkage.

Reaction Injection Molding

Reaction injection molding (RIM) is a process used to manufacture large parts with thermoset resins. Figure 5.7 shows a schematic of the RIM process where two low viscosity liquid reactants (A and B) are injected through a mixhead and then into the mold cavity. Micro-mixing of the liquid reactants occurs in the mixhead through high pressure impingement of the two liquid jets. Components A and B must react rapidly so that the polymer crosslinks and sets into the final

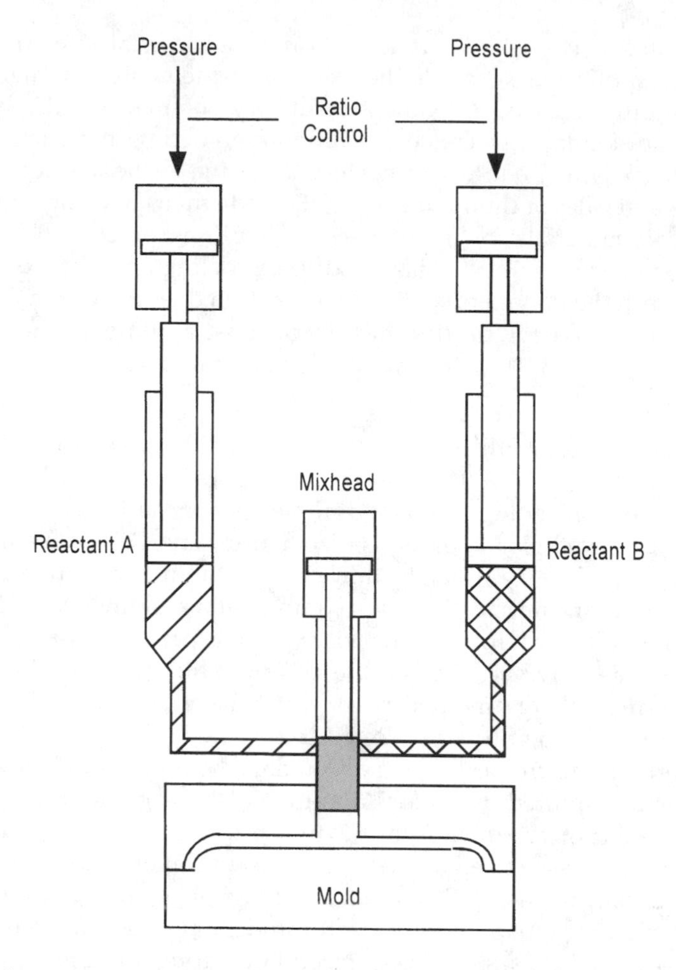

Figure 5.7 Schematic of the reaction injection molding machine. When the mixhead valve opens, two liquid reactants (A and B) flow at high pressure into the mixhead chamber. The two components mix by impinging flow and begin to rapidly polymerize as they flow into the mold cavity, after Macosko (1989).

shape. After the part has cured enough to become solid, the part can be ejected from the mold cavity and the next molding cycle can begin. A variation of this process, termed reinforced reaction injection molding (RRIM), involves the use of short reinforcing fibers or other fillers which are incorporated into the resin. The fiber reinforcements dramatically improve mechanical properties of the final product. Materials which are most commonly used in the RIM process include polyurethanes, Nylon 6, and epoxies. RIM and RRIM have been succesfully used to produce large automotive body panels, including bumpers, fenders, door panels, and front and rear facias.

CONCLUDING REMARKS

Chemical engineers play an important role in the synthesis and processing of polymeric materials. The optimization of polymer processing operations requires an understanding of many chemical engineering fundmentals. These include heat transfer for the melting and solidification of polymers, fluid mechanics for the pumping and flow of polymers, mass transfer for the devolatilization and drying of polymers, and reaction kinetics for the synthesis of thermoplastic polymers and curing of thermoset polymers. The transportation and communication industries continue to demand lighter polymeric materials with improved thermal, mechanical, optical, and electrical properties. Chemical engineers are well suited to make major contributions in developing new polymeric materials and processes to meet these challenges.

REFERENCES

Battenfeld of America, Inc. (1996) "Multi-material Technology." W. Warwick, RI, September

Bernhardt, E. C. and McKelvey, J. M. (1958) *Modern Plastics*, **35**, July, 154

Macosko, C. W. (1989) Fundamentals of Reaction Injection Molding. New York: Hanser Publishers

McCrum, N. G., Buckley, C. P. and Bucknall, C. B. (1997) Principles of Polymer Engineering, 2nd edn. Oxford, New York: Oxford University Press

Tadmor, Z. and Gogos, C. G. (1979) Principles of Polymer Processing. New York: John Wiley and Sons

6. CATALYSIS: ENERGY, ENVIRONMENT, AND ECONOMICS

UMIT S. OZKAN and RICK B. WATSON

Department of Chemical Engineering, The Ohio State University, 140 W. 19th Ave., Columbus, OH 43210, USA

Catalysis encompasses a very large part of our everyday lives. From home to chemical industry, from environmental protection to agriculture, catalysis plays a crucial role for energy, the environment, and the economy. While the first application of catalysts by humans was in the form of enzyme catalysis through the use of yeast, the first use of inorganic materials as catalysts dates back to 1835. Today, catalysis is a very important interdisciplinary science, with nearly 90% of all chemical production involving catalysis. A brief introduction to catalysis is presented, including catalytic materials, catalyst preparation and development, and catalyst characterization.

The mainstream of catalytic technology lies in refining and use of fossil fuels, chemical production, and environmental protection. These mainstays in catalysis are discussed and some common examples of catalysis in everyday life are presented. As we enter a new decade, catalysis will face challenges to keep up with the growing needs of the society. Areas that need urgent attention for the future and the scientific challenges are discussed. These include methods such as advanced catalyst design, nano-scale preparation, and advanced analytical characterization methods. High impact areas also include selective oxidation, alkane activation, enhanced use for the environment, alternative feedstocks, and a continued push for 100% selective processes.

WHAT IS CATALYSIS?

A catalyst, in the simplest of terms, has the ability to initiate or accelerate a chemical reaction without itself being consumed in the process. In 1835, J. J. Berzelius first used the Greek word "catalyst", that means "decomposition", in relation to some phenomena he observed in chemical transformations (Bond, 1987). In modern times, the definition and use of catalysts have become much clearer. The term has evolved to include such materials as metals, metal-oxides, nitrides, sulfides, carbides, organometallic compounds, and also liquid-based materials such as mineral acids, surfactants, and biological enzymes. Catalysis is the study and practice of catalytic chemical reactions, reaction kinetics, reactor design, catalyst synthesis, and catalyst characterization. It is an integral part of the chemical manufacturing industry, with around 90% of all chemical products involving catalysis at some point of production. However, the impact of catalysis is not limited to the chemical industry, as it finds application in energy use, environmental protection, and food and drug industries.

Catalytic reactions can be considered in two groups: 1) Homogeneous and 2) Heterogeneous. Reactions in which the catalyst shares the same phase as the reacting species are referred to as homogeneous catalytic reactions. The best-understood classes of homogeneous catalysis lie in acid-base catalysis. Here, proton transfers catalyze reactions and the solvent usually participates. Examples of acid catalysis include the gas phase dehydrogenation of alcohols in the presence of HBr. Enzyme catalysis and catalysis in solution by organometallic complexes also constitute important examples of homogeneous catalysis.

The second class of catalysis encompasses reactions in which the catalyst does not share the same phase as the reactants. These reactions may include gas/solid, liquid/solid, and gas/liquid phases. A well-known example of heterogeneous catalysis is the production of ammonia from hydrogen and nitrogen gases. If hydrogen and nitrogen are brought together at high temperature and pressure, we don't see any ammonia formation. When they are brought into contact over a heterogeneous catalyst such as iron metal, the reaction proceeds readily, leading to ammonia production at appreciable rates. Although both homogeneous and heterogeneous catalysts are used in industry, the latter is much more common and will be the main focus of our discussion.

HOW DOES A CATALYST WORK?

In heterogeneous catalytic reactions, one or more of the reactants attach to the surface and the chemical reaction involves at least one reactant participating as a surface species. In a heterogeneous reaction, the nature of the catalyst surface is of utmost importance. When we talk about the catalyst surface, what we are referring to is not the exterior surface of a catalyst particle, but the interior surface of the pores, which make up close to 50% of the total volume of a particle, as catalysts are almost always highly porous materials. Figure 6.1 represents the different pore sizes found in catalyst pellets. Pores with diameters greater than 50 nm are called *macropores*, those with diameters between 50 and 2 nm are called *mesopores*, and the ones with diameters smaller than 2 nm are referred to as *micropores*. The interior surface area of the pores is several orders of magnitude larger than the exterior area of a particle. Typical industrial catalysts may have surface areas ranging from 50 to 1000 m^2 per gram of catalyst. The surface atoms (or group of atoms) where reactant molecules are attached are called *active sites*. Commonly, only the first few layers of atoms present on the surface of the catalyst will behave as the *active sites*. The attachment of molecules on the catalyst surface is called *adsorption*. What facilitates adsorption is the higher free energy of the surfaces compared to the bulk of the solid. Since atoms on the surface have lower coordination than the bulk atoms (*i.e.*, they don't have neighboring atoms on one side), they may have free bonds on the surface, which are available for the

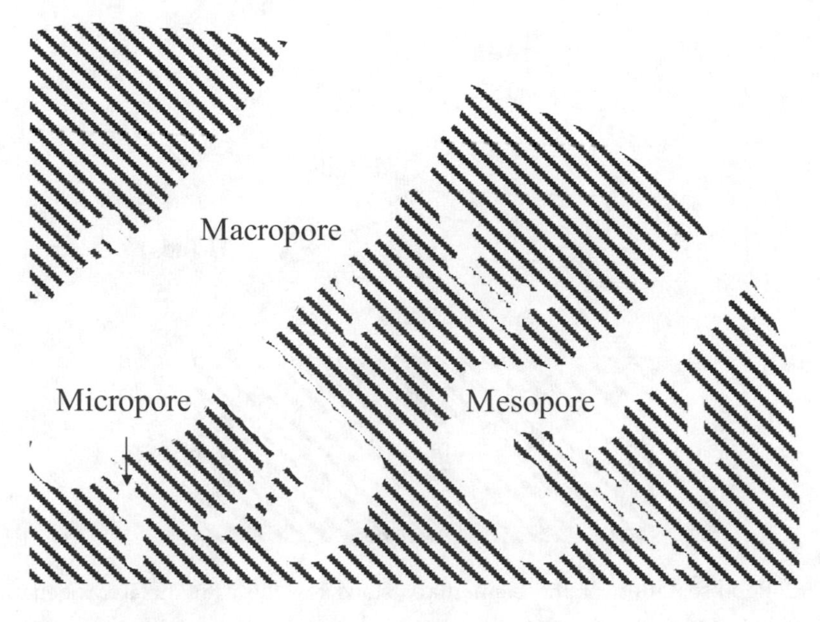

Figure 6.1 Different pore sizes found in catalyst pellets.

incoming molecules, allowing them to attach to the surface. The detachment of an adsorbed species from the surface is called *desorption*.

The steps involved in a heterogeneously catalyzed reaction are 1) diffusion of reactant molecule(s) through the pores to the catalytic sites, 2) adsorption of reactant(s), 3) reaction and/or bond re-arrangement on the surface, 4) desorption of reaction product, and 5) diffusion of product out of the pores. The adsorption, surface reaction, and desorption steps are represented in Figure 6.2, using as an example catalytic hydrogenation of ethylene on a supported nickel catalyst surface, where hydrogen adsorbs dissociatively. Any one of the steps mentioned above may be the slowest, and hence, the overall rate determining step in a catalytic reaction. Catalysis takes place when the catalyst creates an alternative path of lower activation energy by which the reaction can proceed. The simplified energetics of a catalytic reaction are shown in Figure 6.3. It is important to keep in mind that the presence of a catalyst does not change the state of a chemical equilibrium and will "speed up" both the forward and reverse reactions along the same path. In other words, if the product yield of a reaction is controlled by thermodynamic equilibrium, the use of catalyst will have no effect on the yield; it will only accelerate the approach to equilibrium by lowering the energy level of the activated intermediate.

Catalytic materials can range from mineral acids and solids to biological enzymes. Heterogeneous catalysts are most often metals, or metal oxides, nitrides, or sulfides. Often, the catalytic materials are "deposited" on a refractory support

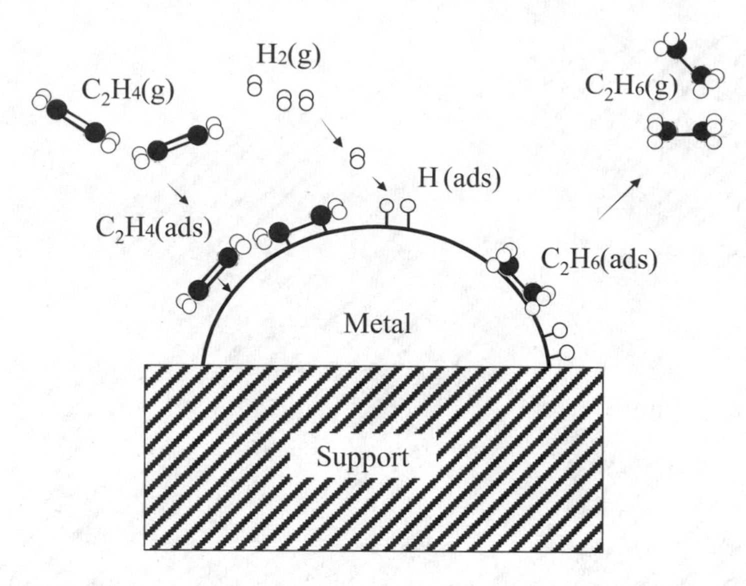

Figure 6.2 Representation of the elementary steps for the catalytic hydrogenation of ethylene.

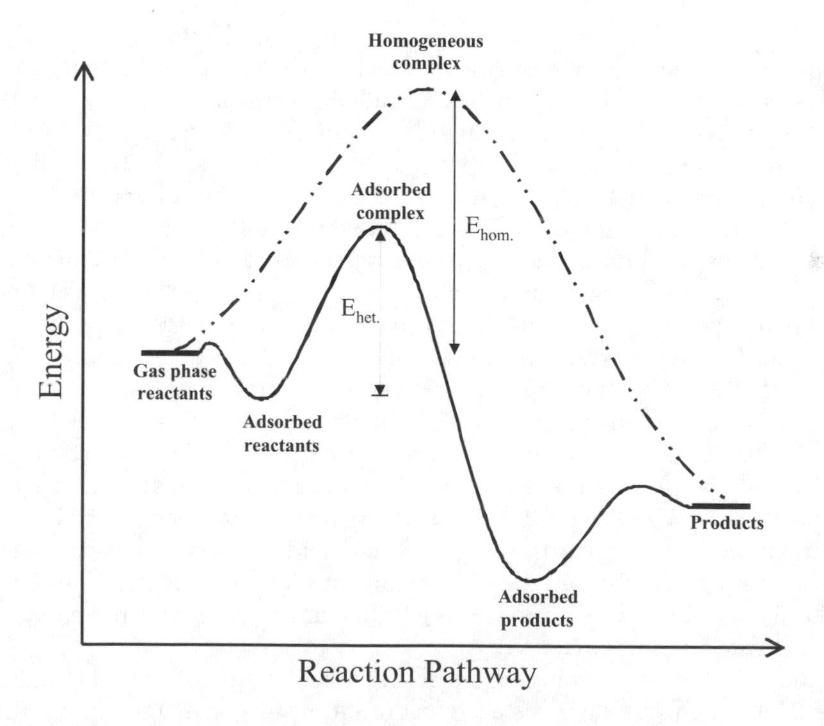

Figure 6.3 Simplified energy diagram of a reaction pathway.

material. Catalyst supports usually allow a fine dispersion of the active material by providing a larger surface area. For example, expensive metals, such as platinum, are supported to maximize the accessibility of the active sites to the reactants. Furthermore, catalyst supports also provide the thermal or mechanical stability that the catalytic materials may lack at high reaction temperatures. However, it is not realistic to think of catalyst supports as inert "carriers" for the active phase as they very well may contribute to the overall activity of the catalyst. The surface structure of the support may directly affect the nature of the active sites by providing a specific crystallographic orientation or a preferred coordination environment to the catalytic material. Especially in some supported metal catalysts, the strong interaction between the support and the metal is crucial for the catalytic action. The support materials most commonly used are alumina, silica, activated carbon, titania and, to a lesser extent, zirconia, chromia, and magnesia.

In addition to the active phase and the catalyst support, another component often present in a catalyst is the *promoter*. A promoter can be defined as a substance which, when added to a catalyst as a minor component, changes the performance of the catalyst significantly. While *textural promoters* are used to improve a physical characteristic of the catalyst such as dispersion or resistance to sintering, *chemical promoters* are used to enhance the activity or the selectivity of the catalyst. Examples of chemical promoters include alkali and alkaline earth metals. Although such promoters usually don't have a significant catalytic activity by themselves, they may electronically interact with the catalyst, changing its adsorption/reduction behavior and hence facilitating the adsorption and/or dissociation of a reactant molecule on the catalytic site. Promoter effects in catalysis are quite complex and often not well understood.

CATALYST DEVELOPMENT

Catalyst Preparation

Catalyst preparation is often considered an art form because of the complexity of the relationship between catalytic properties and preparation parameters. A good catalyst needs to be active and selective for the desired reaction. It should have a long active life span and be chemically and mechanically stable. It should have high surface area and proper pore structure. In addition to properties essential for the reaction, it should also have proper heat and mass transfer and flow characteristics for easy use in catalytic reactors. The cost and environmental compatibility are also major considerations in catalyst development. Ideally, it should not be difficult or costly to regenerate the catalyst.

The most common method of preparing catalysts is precipitation or co-precipitation of the active phase from solutions of the metal salts. It is possible to prepare the support and the active phase together or to deposit the active phase on the

Table 6.1 Catalyst preparation techniques.

Precipitation, co-precipitation	Preparing pure or mixed components from solution
Washcoating	Rinsing support material with solution containing active component
Wet impregnation, incipient wetness	Allowing active components to "diffuse" into the support from concentrated or dilute solution
Chemical vapor deposition (CVD)	Depositing active component from gaseous state
Sol-gel methods	Controlled inorganic polymerization on a nano-scale
Solid-state reactions	Phase transformations, synthesis from solid precursors
Reactive treatments	Ion exhange, oxidation, reduction, or sulphidation of catalytic precursors
Use of organometallic precursors	Use of organometallic precursors as catalyst templates

support at a later stage. Other steps may include separation of the precipitate, drying, heat treatment, and activation. More advanced synthesis techniques, such as sol-gel preparation, solid-state reaction, or use of organometallic precursors, may provide better control of the final properties of the catalyst. Different techniques used for preparing industrial catalysts as well as "model" catalysts are presented in Table 6.1. Catalysts are usually "shaped" in the form of cylindrical pellets, pellets with holes, spheres, or even with irregular geometries. The support form known as *monolith* or *honeycomb*, which is used in most automobile catalysts, offers a different geometry for minimizing pressure drop through the engine by using parallel, non-connecting channels. The factors that dictate the catalyst shape and size are not only flow properties and mass and heat transfer characteristics, but also the type of the catalytic reactor to be used. For further information regarding common industrial supports and preparation methods, the reader is referred to Stiles (1987).

From Laboratory to Industrial Process

Although in the past, most industrial catalysts have been developed empirically, catalyst development is rapidly moving towards molecular design of catalysts "custom tailored" for specific reactions using fundamental principles. Catalysis science, even though it has been practiced for nearly a century (perhaps now more than ever), presents many challenging opportunities for major breakthroughs. While catalyst development is largely an interdisciplinary science requiring skills from many practices, chemical engineers are involved in every phase of the

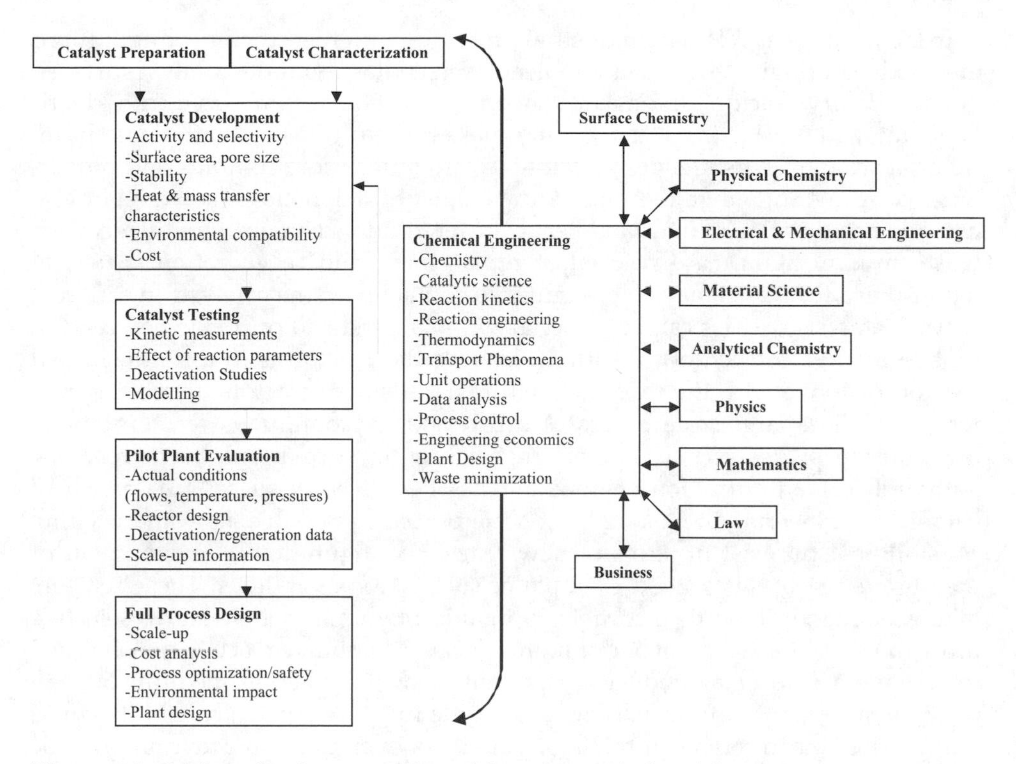

Figure 6.4 Catalyst development: From laboratory to industrial process.

process, often interfacing among different disciplines. Figure 6.4 shows the different stages involved in taking a catalyst from the laboratory to the industrial process and the participation of different fields.

At the base of the catalytic science is the *activity* and *selectivity* of a catalyst for a particular reaction. A very important part of the catalyst development is the testing of catalytic materials in laboratory–scale reactors. Reaction experiments are performed to test the catalytic performance of the materials, to elucidate *reaction schemes*, to obtain *kinetic parameters*, and to establish *relationships* between the catalytic performance and the physical and chemical characteristics of the catalysts. The reaction studies may range from experiments performed over very carefully prepared model surfaces under high-vacuum, to studies that try to mimic the conditions found in industrial-scale catalytic reactors. These studies, combined with detailed characterization of the catalysts, provide the fundamental understanding about the nature of the active sites and distribution of catalytic functions among these sites. Laboratory reaction studies also provide information about stability and the deactivation process of a catalyst exposed to reaction conditions.

In laboratories as well as in industrial processes, different reactor configurations are used for bringing the reactant molecules in contact with the catalyst surfaces. For laboratory reactors, the major concern is to be able to observe true kinetic behavior without having it masked by mass or heat transfer limitations. In industrial scale reactors, the design variables are much more complex and numerous, requiring not only a thorough knowledge of the kinetics involved, but also careful consideration of mass and heat transfer, fluid mechanics, energy recovery, waste minimization, ease of catalyst replacement and regeneration issues, to name a few. Reactants may be brought into contact with the catalysts in different ways. Catalyst particles can be stationary as in a fixed bed or trickle bed reactor, mobile relative to each other without leaving the reactor as in a fluidized bed reactor, or move with the reactant mixture as in slurry reactors or moving-bed reactors. While large-scale processes utilize mostly continuous reactors, many low-volume processes may operate in batch as in stirred-tank batch reactors. Although catalytic reaction engineering has been a very successful branch of chemical engineering in making large scale processes possible, new and exciting possibilities still exist in terms of new reactor configurations. Development of membrane reactors is a good example of such new possibilities. These reactors take advantage of semi-permeable inorganic membranes that allow selective diffusion and/or reaction of one or more species. Membrane reactors may be used to remove a feed or a product component from the reaction medium or may combine a separation function with a catalytic function. Such a capability could make it possible to better control the selectivity or even overcome thermodynamic limitations (Ross and Xue, 1995).

CATALYST CHARACTERIZATION

Knowledge of the chemical, structural and surface characteristics of a catalyst is essential for understanding the catalytic phenomena taking place on its surface. The rapid evolution of catalysis from an empirical practice to an interdisciplinary science in the last two decades is, for the most part, due to development of highly sensitive and sophisticated characterization techniques. Some of these techniques are aimed at measuring the physical properties. Others seek an understanding of the chemical characteristics.

Physical Characterization

When the reaction at the surface of the catalyst is the controlling step in a reaction, the rate of reaction may be directly proportional to surface area. Thus, the measurement of surface area and pore volume is the first step in characterizing catalyst behavior. Surface area is largely determined using gas adsorption techniques. The well-known BET method, developed by Brunauer, Emmett, and Teller in 1938, is a technique that is still used today. In the BET method, an inert gas, usually nitrogen or krypton, is physically adsorbed onto a clean catalytic surface to yield an adsorption *isotherm* graph. An isotherm shows the amount of gas adsorbed versus the pressure of adsorbing gas at a constant temperature. From the isotherm and the area occupied by the adsorbing molecule, one can calculate the surface area of the catalyst. Information about pore size, shape and volume can also be extracted from adsorption-desorption isotherms. Another method for pore volume and pore size distribution is mercury porosimetry. The volume of mercury "pushed" into pores is plotted against the pressure and from this, the size distribution of the pores can be calculated. Other physical properties of interest include mechanical strength (resistance to crushing, abrasion and attrition), flow characteristics of the catalysts (agglomeration) and particle size distribution.

Chemical Characterization

Chemical characterization refers to techniques used to examine the compositional, structural, morphological, and surface properties of catalysts. Quantitative analysis of the elements present in the catalyst is one of the first steps in chemical characterization. Techniques such Atomic Absorption Spectrometry (AAS) and X-Fluorescence (XRF) can be used for this purpose. While AAS uses the absorption of incident radiation by the vaporized atoms of the elements being analyzed, XRF is based on the emission of characteristic secondary X-rays (fluorescence) when the sample is irradiated with an X-ray beam.

Determination of the crystal structure (*i.e.*, the geometry of the way the atoms are positioned in the solid) and identification and quantification of phases can be achieved using X-ray diffraction. Structural and morphological information can also be obtained by use of electron microscopy techniques such as Scanning (SEM) and Transmission Electron Microscopy (TEM). Both techniques use well-focused electron beams to generate an image. In the case of TEM, the electrons that are transmitted through the sample are used to provide the contrast needed for the image. In SEM, electrons "back-scattered" from the surface or the ones emitted by the sample due to interaction with the incident beam are used to generate the image using a photo-multiplier tube. SEM can be very useful in characterization of size, shape and surface morphology of crystalline and amorphous particles. Figure 6.5 shows a stereo pair of scanning electron micrographs of molybdenum trioxide crystals used for three-dimensional imaging (Hernandez and Ozkan, 1990). TEM, on the other hand, can be used to gain information about the atomic arrangement and defects in the lattice. Figure 6.6 shows a TEM image of a sample, which is made up of mixed oxides of silicone and titanium prepared by a sol-gel technique (Watson and Ozkan, unpublished work). One can see the lattice spacing of ~3.8Å, which corresponds to the anatase phase of titania, indicating partial segregation of this phase from the Si-Ti network. Electron

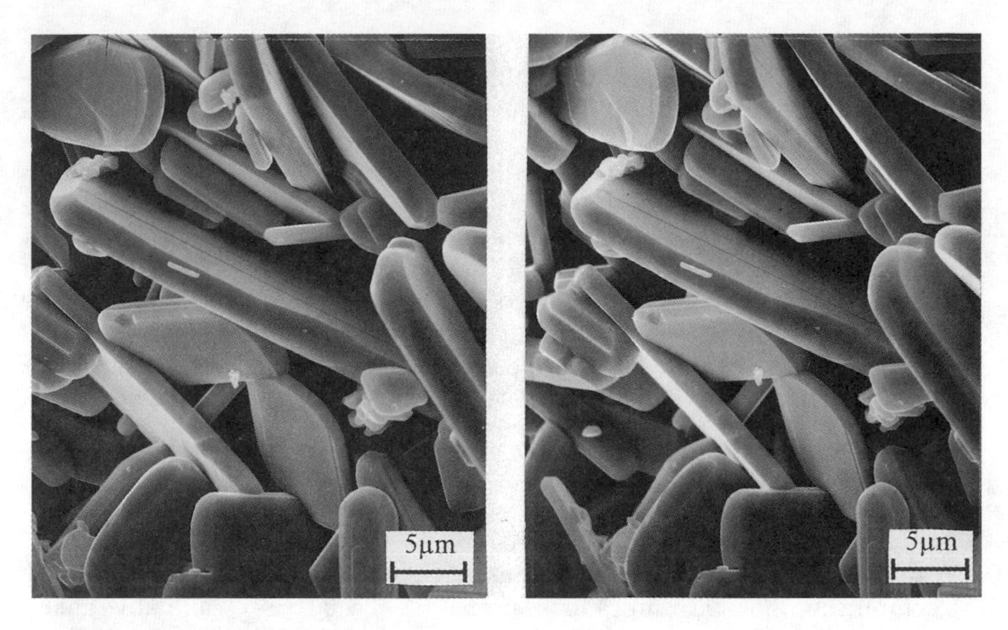

Figure 6.5 Stereo SEM images of MoO_3 crystals.

microscopy techniques are often coupled with other techniques such as Energy Dispersive X-Ray Analysis (EDXA), Low Energy Electron Diffraction (LEED), and Electron Energy Loss Spectroscopy (EELS) for compositional mapping of the region being imaged, to examine the structure of non-amorphous surface layers or defects, and for determining chemical and electronic structure of the surface.

The real strength of catalyst characterization lies in the availability of highly surface-sensitive instrumental techniques. Many of these techniques are based upon the interaction of the surfaces by an incident radiation or a beam of electrons or ions. Such interactions may include absorption, emission, transmission, scattering and sputtering. Information obtained through such techniques includes oxidation state and coordination environment of surface species (*e.g.*, X-ray photoelectron spectroscopy, Auger Electron Spectroscopy, Secondary Ion Mass Spectroscopy) and molecular structure, bond length and bond strength of surface species (*e.g.*, Fourier-Transform Infrared Spectroscopy, Laser Raman Spectroscopy). Another way of classifying the characterization techniques is based on the mode of excitation of the molecules with the incident radiation. Examples are techniques based on vibrational mode (Laser Raman Spectroscopy, Infrared Spectroscopy) and spin mode (Nuclear Magnetic Resonance and Electron Spin Resonance Spectroscopies). Another group of characterization techniques examines catalyst properties and behavior as a function of temperature, such as Thermo-Gravimetric Analysis (TGA) (weight changes of the catalyst as a function of temperature), Temperature Programmed Desorption (TPD) and Temperature Programmed

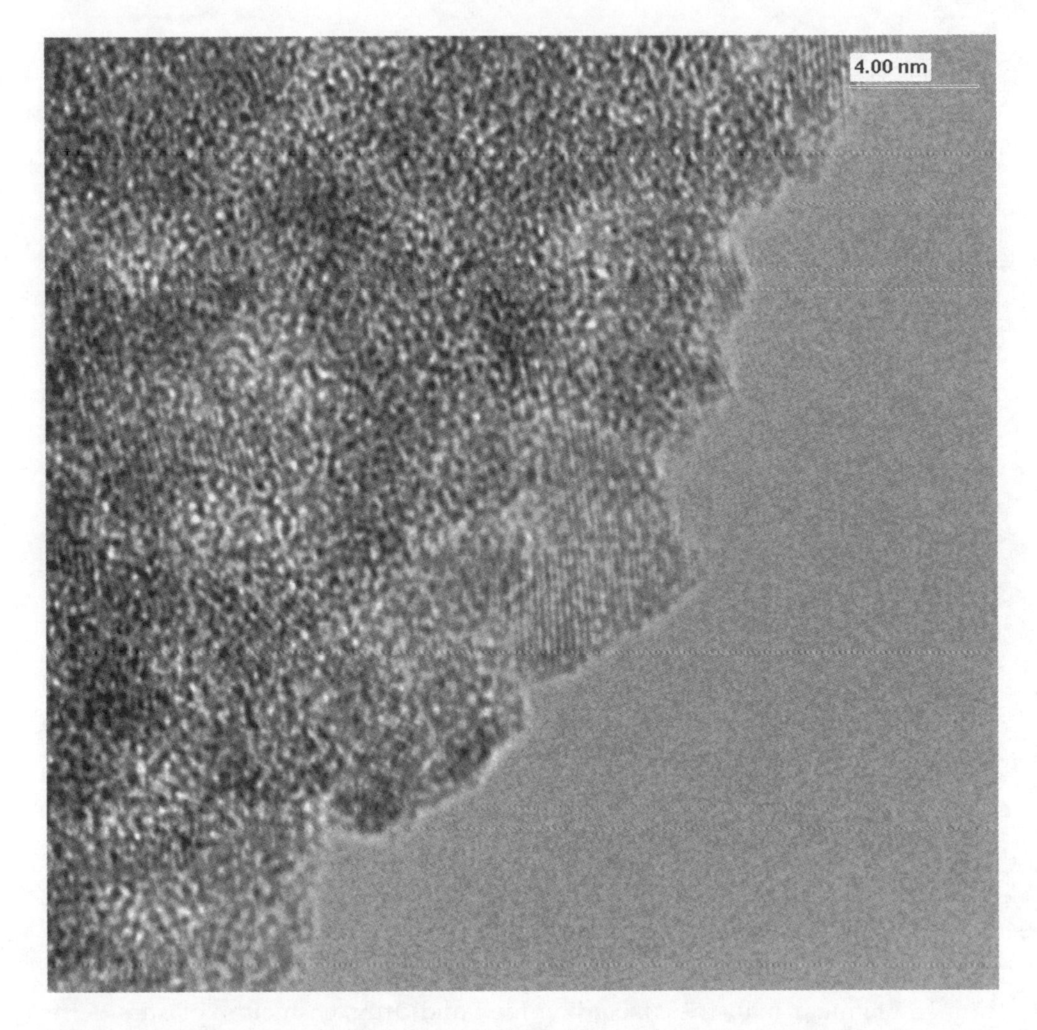

Figure 6.6 TEM image of a SiO_2-TiO_2 mixed oxide catalyst support.

Reduction (TPR) (adsorption/desorption and reduction rates of the catalyst as a function of temperature). Also included among the thermal analysis techniques is Differential Scanning Calorimetry (DSC) that can measure heat of adsorption, desorption, and reaction.

Among all the characterization techniques available to the catalysis researcher, the *in-situ* techniques provide the most powerful tools. These techniques, which allow characterization of catalysts and/or adsorbed species under reaction conditions, include Fourier Transform Infra-red Spectroscopy, Laser Raman Spectroscopy, Nuclear Magnetic Spectroscopy, and Electron Spin Resonance Spectrometry, to name a few. Recent advances have made controlled-atmosphere electron microscopy (ESEM) characterization also possible.

While the techniques mentioned above remain essential tools in the arsenal of catalysis researchers and practitioners, new techniques are opening up exciting new avenues. Some examples of these recent developments include Scanning Tunneling Microscopy (STM) and Atomic Force Microscopy (AFS) techniques that allow topographic mapping and spectral analysis of the surface at *atomic resolution*. While both techniques present many challenges in terms of data interpretation and analysis, they offer the highest three dimensional resolution of surfaces. Another new and exciting technique is the Extended X-Ray Absorption Fine Structure (EXAFS). This technique uses high energy X-rays obtained from a synchrotron facility to get structural information about the surfaces at an *atomic scale* and is amenable to *in-situ* characterization. While the limited availability of large accelerators (synchrotron) is a drawback for the wide application of this technique, it remains as a unique *in-situ* technique that can provide structural and coordination information with atomic resolution.

One should also mention that several other techniques exist that are aimed at understanding the properties and the functionality of the catalyst surfaces, such as surface acidity measurements, time-on-stream activity studies, and transient and steady-state isotopic labeling experiments, to name a few. Isotopic labeling can also be combined with vibrational spectroscopy techniques for identification of adsorbed species. A fundamental understanding of the chemistry that takes place on the catalyst surfaces will only be possible through use of several of these techniques coupled with detailed kinetic studies.

CATALYSIS APPLICATIONS

Everyday Examples

Seldom realized, catalysis touches every part of our daily life. Catalysis is used in many familiar industries, including food and drug, cosmetics, plastics, fertilizers, and automotive, to name a few (Bowker, 1998). Hydrogenating vegetable oil over a nickel catalyst produces margarine. Enzyme catalysis allows corn millers to use their by-product to produce a fructose sweetener present in many drinks and foods. A large percentage of the world population depends on livestock as their main source for food. The livestock depend heavily on grains for feed. In turn, the grains, and every farm product for that matter, depend on fertilizer. When we consider the reaction processes involved in fertilizer production, we see that catalysis plays a major role in every step. First natural gas (mostly methane) is converted to carbon monoxide and hydrogen in a steam reforming reaction over nickel catalysts. Water gas shift reaction converts CO and H_2O to hydrogen and CO_2 over iron catalysts. After being separated from CO_2, hydrogen is used to hydrogenate nitrogen over promoted iron catalysts to produce ammonia. Part of the ammonia is oxidized to NO_2 over platinum/rhodium catalysts and further reacted with water to form nitric acid. The final step is the reaction of ammonia with nitric acid to produce ammonia fertilizer (NH_4NO_3).

Besides the food industry, catalysis is used to improve the quality of our lives in many familiar ways. Formaldehyde is produced from methanol over an iron-based catalyst. Formaldehyde is a pre-cursor for many types of glue that are used in furniture, carpet, and many construction materials production. Today's world heavily depends on plastics and polymers. Plastics and synthetic fibers all involve catalysis in their production. These materials, in turn, are used in packaging, automotive industry, construction, and the production of clothes. Titania catalyst added to wall materials in our homes can use the energy from the sun (photocatalysis) to oxidize organic pollutants such as formaldehyde. Gold is beginning to be used as a low-temperature oxidation catalyst for removing carbon monoxide and odors in homes. It is quite clear that catalysis plays a major role in improving the quality of our everyday lives.

Energy and Catalysis

While examples of how catalysis touches our daily lives can be found very easily, it finds the mainstream in the use of fossil fuels for energy. Particularly, catalysis is at the heart of petroleum refining. Following distillation, most of the crude oil fractions (from naptha to resid) go through catalytic hydrotreating processes to remove organically bound sulfur, nitrogen, and oxygen. Especially hydrodesulfurization and hydrodenitrogenation reactions are the first steps in environmental catalysis since they remove sulfur and nitrogen before they get oxidized to sulfur dioxide and nitrogen oxides. Hydrocracking, reforming, isomerization, and alkylation reactions, which are important components of petroleum processing, are all catalytic. Steam reforming and water gas shift reactions, which are used for production of synthesis gas and hydrogen, proceed only with the use of catalysts. Hydrotreating will be discussed in the Environment and Catalysis section. Catalytic cracking is one of largest scale catalytic processes. It is the conversion of large petroleum molecules to lower molecular weight products (Gates, 1992). The operating conditions can vary greatly depending on the desired product. The catalysts are generally acidic zeolite catalysts. Zeolites are crystalline materials known for their regular pore structure (3–10Å in diameter), which makes them well suited to "crack" the C-C bonds of large petroleum fractions on their cationic acid sites. Isomerization is a process to improve the octane rating of fuels for burning in internal combustion engines (and also a major source for aromatic compounds for chemical production). In reforming, the bonding in the light fraction of petroleum distillates is re-arranged over a catalyst to form branched-chain isomers and aromatics. Most active reforming catalysts consist of platinum dispersed over alumina. Here, the dispersion of platinum is key to performance.

Many of these reactions are also used for processing coal derivatives. In short, whether we are using natural gas to heat our homes, gasoline to run our cars or electricity to maintain our way of living, all forms of energy production require catalysis. It is important to note that, in addition to all forms of energy sources resulting from the use of fossil fuels, almost all the chemicals produced from these fuels come about by catalytic processes (hydrocarbons, alcohols, aldehydes, plastics, synthetic fibers, polymers, and resins, to name a few).

Environment and Catalysis

Globally there are major pollution issues that result from the consumption of raw materials. Emissions from fossil fuel combustion are the main source of air pollution, with sulfur dioxide and nitrogen oxides being the two most important pollutants since they are the main causes of acid rain. Also of great concern is the emission of carbon dioxide, which is reported to be responsible for 60% of the greenhouse effect (US EPA). Atmospheric pollutants from mobile and stationary sources include unburned hydrocarbons, nitrogen oxides, sulfur dioxide, carbon oxides, polyaromatic hydrocarbons, and chlorinated hydrocarbons. As the energy demand of the world population increases, it is inevitable that finding effective solutions to environmental problems will have a much greater urgency in the years to come.

The most important impact of catalysis in the last three decades has been in the area of environmental protection. Perhaps the most familiar catalytic technology for pollution control is the automotive catalytic converter. The catalyst involved is a complex one, consisting of Pt, Pd, and Rh metals supported over alumina pellets or washcoated over a monolith structure. Many other components are added, such as ceria and other oxides, to help store oxygen on the catalyst for engine cycles in which oxygen concentration in the exhaust is low. The term "three-way catalyst" refers to the three reactions that are carried out over catalytic converters, namely oxidation of carbon monoxide and unburned hydrocarbons, and reduction of nitrogen oxides. The catalysis involved in three-way catalysts is very demanding, since these catalysts need to be able to perform the three functions through very adverse conditions such as large temperature and concentration fluctuations, heavy particulate concentrations, constant road vibration, and many on/off cycles. Nonetheless, the 3-way automotive catalyst has been very successful in reducing automobile emission of hydrocarbons, carbon monoxide, and nitrogen oxides by more than 95% (Jacoby, 1999). However, as will be discussed in the Future Directions section, environmental concerns will continue to push for better technologies.

Stationary combustion sources pose a major threat to the environment through emission of SO_2, NOx, and volatile organic compounds (VOCs). As mentioned in the previous section, perhaps the most effective way to reduce pollution is the pre-combustion removal of the pollutants or potential pollutants. Catalytic hydrotreating (*e.g.*, hydrodesulfurization, hydrodenitrogenation) is a good example of such a strategy. In these processes, hydrogen is used to break the C-S or

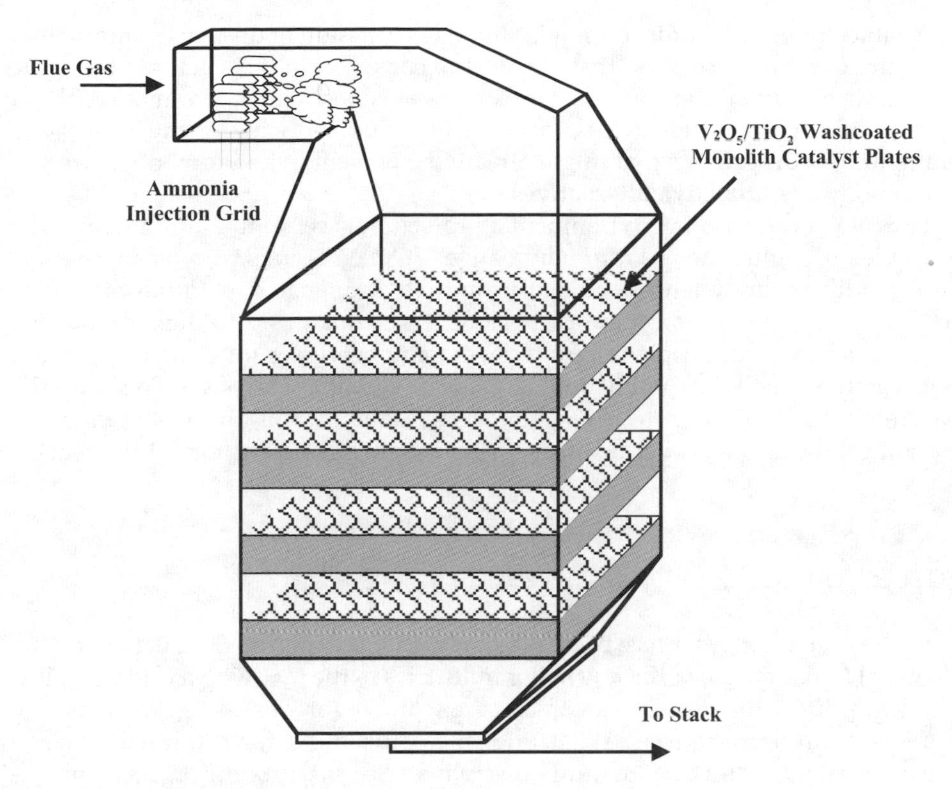

Figure 6.7 A typical NO_x-SCR unit.

C-N bonds in the sulfur or nitrogen containing heteroatom compounds, hence removing sulfur and nitrogen before they get oxidized to SO_2 and NOx. Hydrotreating catalysts usually consist of alumina-supported molybdenum or tungsten sulfides with cobalt or nickel promoters.

As desirable as it is, preventing pollution before pollutants are formed may not always be possible. For example, most stationary combustion fuel may still contain sulfur and nitrogen. In the case of NOx emissions, even if the fuel is completely free of nitrogen, NOx may still form at high combustion temperatures by "fixation" of atmospheric nitrogen. Stationary sources account for more than 60% of the total NOx emissions (Cooper and Alley, 1994). Selective catalytic reduction (SCR) of NOx emissions from stationary combustion sources is a major application of catalysis that came about in the last 15–20 years. The present technology uses ammonia as a reducing agent, converting NO and NH_3 to N_2 and H_2O over vanadia-titania catalysts. Figure 6.7 shows a simplified diagram of an SCR unit consisting of an ammonia injection grid and a stack of monolith blocks supporting the vanadia-titania catalyst. SCR catalysts have to perform with high resistance to SO_2 poisoning and high hydrothermal stability. Major shortcomings of the present technology include ammonia slippage (emission of un-reacted ammonia) and ammonia salt formation that can lead to corrosion in the stack gas lines, the cost, transportation and storage problems associated with ammonia. With all these issues, current research is focusing on other reducing agents to replace ammonia. Hydrocarbons that might already be present in an industrial stack gas (*e.g.*, C_1-C_3) are attractive alternatives.

The role of catalysis in environmental protection is certainly not limited to the examples presented here. From eliminating VOC emissions to replacing refrigerants with environmentally friendly ones (hydrogenated chloroflorocarbons (HCFCs) instead of chlorofluorocarbons (CFCs), which lead to depletion of the ozone layer in the atmosphere), from developing processes for clean combustion (*e.g.*, catalytic methane combustion) to using enzymatic catalysis to speed up the natural oxidation or pyrolysis of organic pollutants in soil and water, environmental catalysis is working towards a healthier environment for future generations.

FUTURE DIRECTIONS OF CATALYSIS

It is clear that catalysis has been playing a pivotal role in the success of the chemical industry, one of the few major industries in the U.S. with a trade surplus. Many products that we have come to rely upon for our everyday lives are produced through catalysis. What does the future hold for catalysis? Can we consider catalysis to be a "mature" discipline with little room for major breakthroughs? The answer is an emphatic "no". Catalysis will be the key to meeting the societal needs of the future through alternative fuels, alternative feedstocks for chemicals, "enantioselective synthesis" of pharmaceuticals and increasing application of catalysis to environmental protection. In reality, all these areas are very closely related. The incentive behind the search for alternative fuels is partly environmental concerns and partly the desire to decrease the dependence of our energy needs on fossil fuels. Now that carbon oxides are also known to pose a threat to the environment, there is a great deal of interest in fuels with a low carbon to hydrogen ratio. Methane offers the best possible C/H ratio, explaining the large volume of research focused on catalytic combustion of methane recently. Of course, the ultimate clean fuel is hydrogen. Hydrogen may very well be the main energy source of the future, if catalysis can come up with an answer to the difficult and challenging question of producing hydrogen efficiently. One focus of interest is the catalytic methanol dissociation to hydrogen and CO. The relative ease of storing and transporting methanol makes this a desirable option even for fueling automobiles. For example, hydrogen can be used in fuel cells (an electrocatalytic device for converting chemical energy to electrical energy) to generate electricity on board a vehicle. The major car manufacturers are looking into fuel cells as the best way to combat global warming and the world's dependence on fossil fuels. At the heart of the fuel cell are the porous catalytic electrodes.

Catalytic materials for the fuel cell electrodes include platinum, nickel, zirconium, and many other materials having the desired electron transfer capabilities. This is a large area of surface-catalysis that will certainly gain much more attention in the future. It is quite possible that in the near future catalytic or photocatalytic decomposition of water to hydrogen and oxygen may become the main technology for producing hydrogen. Through such a process, we can truly talk about "zero-emission" vehicles.

New catalytic technologies will be needed in the utilization of non-traditional energy sources, such as energy from biomass. The utilization of biomass involves forming the methane for electricity production. In this process, the waste is anaerobically (without oxygen) digested to produce methane by an enzymatic catalyst. The landfill gas is largely a mixture of carbon dioxide and methane. With experimental landfill facilities already underway in some countries, it is conceivable that this renewable energy might be able to compete with traditional sources in the future. Catalysis in the utilization of solar and nuclear energies is another exciting possibility. In this process, nuclear or solar energy is converted to "chemical energy" using a catalytic reaction that is highly reversible and highly temperature sensitive. At high temperatures, the solar or nuclear energy is used to "push" the endothermic reaction. At lower temperatures, the equilibrium is shifted to release energy through the reverse (exothermic) reaction (Parmon, 1997).

The future is also likely to bring increased use of catalysis in small-scale production of specialty chemicals. New catalytic processes are likely to be small-volume batch processes aimed at producing "value-added" products. An application area of utmost importance is production of pharmaceuticals by "enantioselective catalysis," which refers to selectively catalyzing the synthesis of the "correct" chiral form of the same molecule over the "wrong" chiral form. Chirality refers to the "handedness" of two isomers, which are non-identical mirror images of each other (similar to the left and the right hand). Often, one enantiomer may have the desired pharmaceutical properties while the other may be not only inactive, but even toxic. A well-known example is the molecule called thalidomide, which has a sedative effect if it has the "correct" enantiomer configuration, but can cause severe birth defects if it is the "wrong" enantiomer. The two enantiomers of thalidomide are shown in Figure 6.8 (NRC, 1992). Although most of the pharmaceutical use of catalysis has been through homogeneous catalysis, the separation difficulties of the product from the catalyst makes heterogeneous catalysis a very desirable alternative.

A growing application in which catalytic principles are put to use is the catalytic sensor, which is a device for monitoring the concentration of a specific gas in a mixture. A catalytic sensor can be considered to be an electrochemical cell, which consists of a catalytic material with two electrodes attached to the two opposite surfaces. One surface is in contact with the sample gas. The other surface is in contact with the reference air. The catalytic interaction of the sample gas with the catalyst surface (*e.g.*, adsorption, surface reaction) changes the electronic proper-

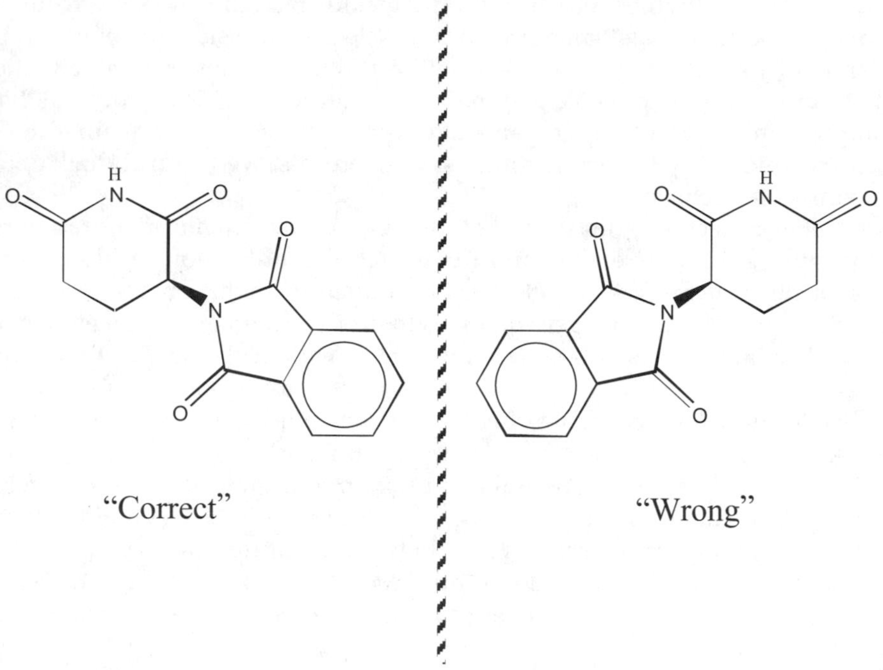

"Correct" "Wrong"

Figure 6.8 Mirror-image enantiomers of thalidomide.

ties of the surface, creating a potential between the two electrodes. This potential varies with the relative abundance of the gas being monitored. Since the introduction of the oxygen sensor in the 1970s to the automobile industry, the use of catalytic sensors has grown and will continue to develop many new applications for catalysis and material science in monitoring emissions from mobile and stationary sources (Farrauto, 2000). Under development are many sensors to detect or to quantify the concentration of hydrocarbons, NOx and CO emissions, and many other pollutant molecules.

In the area of environmental protection, the challenges and opportunities are even greater. They range from developing new emission control catalysts to meet the stricter environmental regulations to replacing liquid acids used in alkylation reactions (*e.g.*, H_2SO_4, HF) with solid acids. Potential applications for environmental catalysis include developing catalytic converters to be used in diesel-fueled vehicles, designing catalysts that can decompose NO to N_2 and O_2 without needing a reducing agent, using catalysts at home to eliminate carbon monoxide at ambient temperatures, using enzyme catalysis for environmental remediation, and finding catalytic solutions to SO_2 emission control. Further emphasis is likely to be placed on catalytic abatement of CO_2 emissions. And of course, the ultimate goal would be to develop processes that emphasize by-product and waste minimization. Any effort to increase the catalytic selectivity is also an effort in environmental protection and optimum utilization of resources.

Several reports, reviews and books written by university and industry researchers in recent years have all reiterated the importance of catalysis for the future (NRC, 1992, Armor, 1996, CCR, 1997, Farrauto and Bartholomew, 1997, and Courty and Chauvel, 1996). Considering the many critical societal needs for which catalysis will be called upon, it is clear that catalysis will only become more prevalent in the future.

What are the new tools and techniques being developed to help catalysis researchers and practitioners in meeting the demands of the future? One of the areas that will need the most attention is that of catalyst optimization. Through automation and computer methods, faster means of catalyst preparation and screening are being developed. The use of *combinatorial chemistry* (a time-minimized, computer-aided combination of experiments), has been essential for the pharmaceutical industry in the design of drug molecules. This technique is likely to become widely used in the field of catalysis as well.

Characterization of catalysts and adsorbed species through *in-situ* techniques (*e.g.*, Diffuse Reflectance Fourier Transform Infrared Spectroscopy (DRIFTS), EXAFS) will continue to be very important in understanding all catalytic phenomena, including nature of the active sites, chemistry of the surface interactions, stability, deactivation, solid state and surface diffusion of species. Fast analytical techniques that can detect short-lived intermediates (*e.g.*, time-resolved spectroscopy) are very important in elucidating reaction mechanisms. In addition to *in-situ* characterization, techniques that examine surfaces at atomic resolution (*e.g.*, AFM, STM, EXAFS) are likely to find wider application in the coming years.

One of the most important recent developments comprises the advances made to use molecular simulations to interpret structural properties of microporous materials, surface structures, adsorption behavior of probe molecules, and metal-support interactions at both the atomic and electronic levels. These simulations are based on either geometrical or chemical/physical models in which the size, shape, or chemical environment of the molecule or atom of interest form the basis of modeling. These theoretical studies, when combined with experimental data, can contribute significantly to the atomic level characterization of the local active sites and the mechanisms of molecular diffusion and reaction steps *at* the active site (Catlow *et al.*, 1997). With the rapid developments in computers and software analysis, this too will become a more common practice in catalysis. The new insight gained by theoretical modeling and calculations needs to be combined with experimental studies to lead the way to "molecular architecture" of catalysts for the desired reaction.

Furthermore, with advanced preparation methods, nano-scale engineering of catalyst surfaces with desired structure and properties will be possible. Controlling surface properties at the atomic level is becoming more attainable with increased use of unique "organometallic precursor templates" and chemical vapor deposition techniques.

Another field that is fertile for advancement is the reaction engineering aspect of catalysis. Innovative reactor concepts (membrane reactors, very small contact times, cyclic operations) and advances in *in-situ* separation can be the key factors in improving selectivity and overcoming thermodynamic limitations.

The application of these evolving experimental and theoretical techniques will meet the challenges present in many catalytic fields, ranging from selective oxidation to alkane activation, from by-product and waste minimization, to olefin polymerization, from use of alternative feedstocks to catalytic sensors.

SUMMARY

Catalysis is one of the most fascinating areas of chemical engineering, which truly combines chemistry knowledge with engineering expertise. Catalysis is fundamental to the three most important concerns of modern society, namely *energy*, *environment* and *economics*. With the help of new preparation methods, advanced characterization techniques and molecular simulation tools, we are closer than ever to molecular design of catalysts from first principles. Equipped with these new tools, catalysis is ready to meet the challenges of the future.

FURTHER READING

Armor, J. N. (1996) Global overview of catalysis: The United States of America. *Applied Catalysis A*, **139**, 217–228

Bond, G. C. (1987) Heterogeneous Catalysis: Principles and Applications, 2nd edn. New York, USA: Oxford University Press

Bowker, M. (1998) The Basis and Application of Heterogeneous Catalysis. New York, USA: Oxford University Press

Brunauer, S., Emmett, P. H. and Teller, E. (1938) Adsorption of gases in multimolecular layers. *Journal of the American Chemical Society*, **60**, 309–319

Catlow, C. R. A., Ackermann, L., Bell, R. G., Gay, D. H., Holt S., Lewis, D. W., Nygren, M. A., Sastre, G., Sayle, D. C. and Sinclair, P. E. (1997) Modelling of structure, sorption, synthesis and reactivity in catalytic systems. *Journal of Molecular Catalysis A*, **115**, 431–448

Cooper, C. D. and Alley, F. C. (1994) Air Pollution Control, A Design Approach, 2nd edn. Illinois, USA: Waveland Press Inc

Council for Chemical Research, U.S. D.O.E., and ACS (1997) Catalyst Technology Roadmap Report: Build Upon Technology Vision 2020: The U.S. Chemical Industry. ACS Workshop, Washington D.C., USA

Courty, P. R. and Chauvel, A. (1996) Catalysis, the turntable for a clean future. *Catalysis Today*, **29**, 3–15

Farrauto, R. J. and Heck, R. M. (2000) Environmental catalysis into the 21st century. *Catalysis Today*, **55**, 179–187

Gates, G. C. (1992) Catalytic Chemistry. New York, USA: John Wiley & Sons, Inc

Hernandez, R. A. and Ozkan, U. S. (1990) Structural specificity of molybdenum trioxide in C_4 hydrocarbon oxidation. *Industrial & Engineering Chemistry Research*, **29**, No. 7, 1454–1459

Jacoby, M. (1999) Getting auto exhausts to pristine. *C&EN (Chemical & Engineering News)*, **77**, No. 4 (Jan. 25), 36–44

National Research Council (1992) Catalysis Looks to the Future. Washington D.C., USA: National Academy Press

Parmon, V. N. (1997) Catalytic technologies for energy production and recovery in the future. *Catalysis Today*, **35**, 153–162

Ralph, T. R. and Hards, G. A. (1998) Fuel cells: Clean energy production for the new millennium. *Chemistry & Industry*, **1998**, No. 9 (May 4), 334–335

Ross, J. R. H. and Xue, E. (1995) Catalysis with membranes or catalytic membranes? *Catalysis Today*, **25**, 291–301

Satterfield, C. N. (1996) Heterogeneous Catalysis in Industrial Practice, 2nd edn. Florida, USA, Krieger and Malabar

Stiles, A. B. (1987) Catalyst Supports and Supported Catalysts: Theoretical and Applied Concepts. Boston, USA: Butterworth Publishers

United States Environmental Protection Agency (*www.epa.gov*)

7. SUPERCRITICAL FLUID TECHNOLOGY

TADAFUMI ADSCHIRI

Tohoku University, Department of Chemical Engineering, Aoba-ku, Sendai, 980-8579 Japan

INTRODUCTION

In the development of industrial chemical processes, solvent selection is of great importance, but generally speaking, there is a mindset that solvents must be used as liquids. Of particular interest in recent years are the attempts that have been made in replacing liquid organic solvents with supercritical fluids, since the physical properties of these solvents can be broadly controlled with temperature and pressure.

Water is a major natural resource of the earth and thus can be considered as the most environmentally benign solvent. At high temperatures and pressures, around critical point of water (374°C, 22.1 MPa), its properties become like those of a polar organic solvent. Supercritical water exhibits not only similar functions to organic solvents but also other specific features, as explained later in this chapter.

Supercritical dioxide, whose critical temperature is 31.2°C, can be used as a solvent for processes that operate at lower temperatures. Supercritical carbon dioxide has some advantages, compared with liquid organic solvents: nontoxic, odorless, inexpensive, no residual solvent problem, suitable for processing thermally labile compounds, and high selectivity in separation. For these reasons, industrial development has greatly advanced in supercritical carbon dioxide extraction processes chiefly in the areas of food and pharmaceuticals.

In this chapter, unique functions of the supercritical fluid solvent are explained. Then, various applications using these key features will be introduced. Finally, I will discuss the potential of supercritical fluid technology for creating industries of the future.

WHAT ARE SUPERCRITICAL FLUIDS?

Whether a fluid is in a gaseous phase or liquid phase depends on the balance of kinetic energy of its molecules and the energy of the intermolecular forces. When the intermolecular energy governs to a greater extent over the molecular kinetic energy, a fluid will be a liquid with a certain measure of order, but when the kinetic energy is dominant, it will be a gas with a random structure. Kinetic energy grows as temperature rises, and the intermolecular energy increases as the distance between molecules becomes shorter (i.e., as the density increases). Thus,

Table 7.1 Comparison of fluid properties.

Property	*Gas*	*Supercritical fluid*	*Liquid*
Density [kg/m^3]	0.6 ~ 1	200 ~ 900	1000
Viscosity [Ps·s]	10^{-5}	10^{-5} ~ 10^{-4}	10^{-3}
Diffusivity [m^2/s]	10^{-5}	10^{-7} ~ 10^{-8}	$< 10^{-9}$
Thermal conductivity [W/mK]	10^{-3}	10^{-3} ~ 10^{-1}	10^{-1}

in general, a liquid will become a gas when the temperature is raised, and a gas will liquefy when it is compressed. However, above a certain temperature, molecular kinetic energy is greater no matter what the distance between molecules, and the substance will become a non-condensable fluid. This temperature, which is called critical temperature, is unique to each substance, and a fluid at or above its critical temperature is a supercritical fluid.

Fluid properties including viscosity, diffusivity, thermal conductivity, dielectric constant, ionic product, are determined by the molecular interactions, and thus directly related to the distance of molecules in a fluid, namely the fluid density. Because the density of a supercritical fluid is between that of a gas and a liquid, the above mentioned properties are also usually between those of the gas and liquid. Some examples of the water properties are shown in Table 7.1. The PρT relation and the temperature and pressure dependence on dielectric constant are presented in Figures 7.1 and 7.2, respectively. These figures show that near the critical temperature, the physical properties change greatly. Characteristics of the solvent change greatly in accordance with the change of physical properties, as will be explained below.

KEY FEATURES OF SUPERCRITICAL FLUIDS

Phase Behavior and Phase Equilibrium

The controllability of the phase behavior and the phase equilibrium are key features of supercritical fluid separation, extraction and reaction processes. When a supercritical fluid is a solvent in a multi-component systems, interactions between molecules of different kinds change greatly near the critical point, and as a result, the phase behavior also changes significantly.

High pressure phase behavior of binary systems can be categorized into six types (Scott, 1972), but here I will explain only one example which is relevant to supercritical water reactions. The phase behavior of water-gas systems and water-hydrocarbon systems near the critical points are shown in Figures 7.3 (Franck, 1981) and 7.4 (Schneider, 1972), respectively. In the higher-temperature regions (right hand side) of the curves (critical points of the mixtures or critical loci),

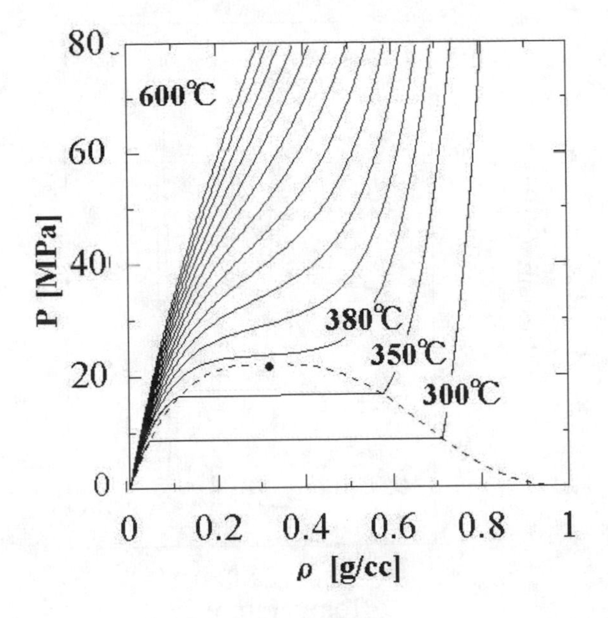

Figure 7.1 Temperature – density – pressure relation.

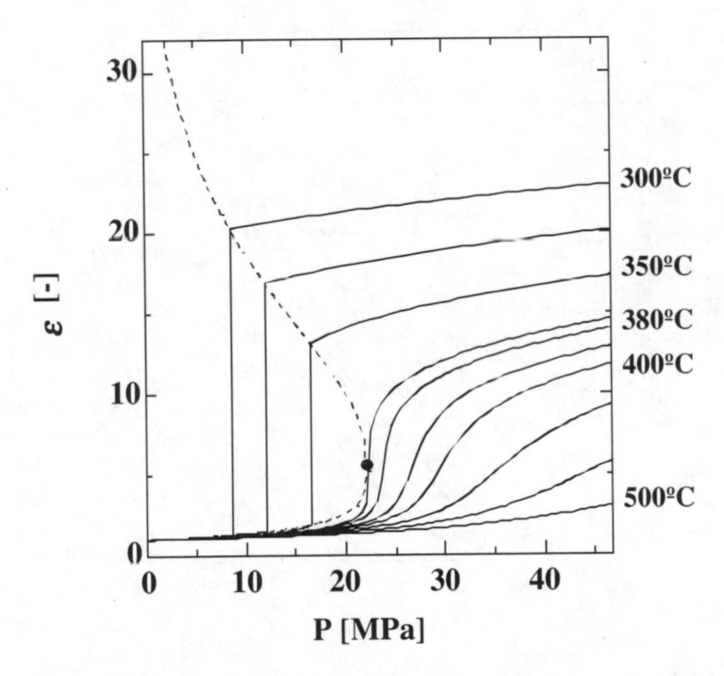

Figure 7.2 Temperature and pressure dependence of dielectric constant of water.

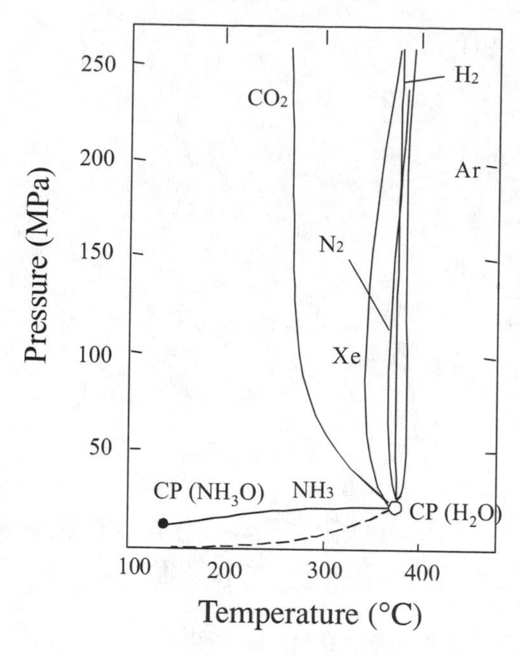

Figure 7.3 Phase behavior of a water-inorganic gas system (Redrawn from Franck, 1981).

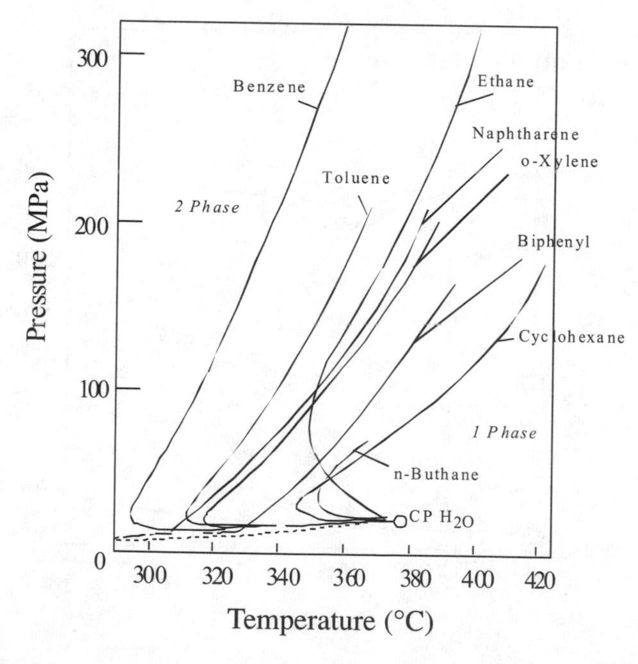

Figure 7.4 Phase behavior of a water-organic compound system (Redrawn from Schneider, 1972).

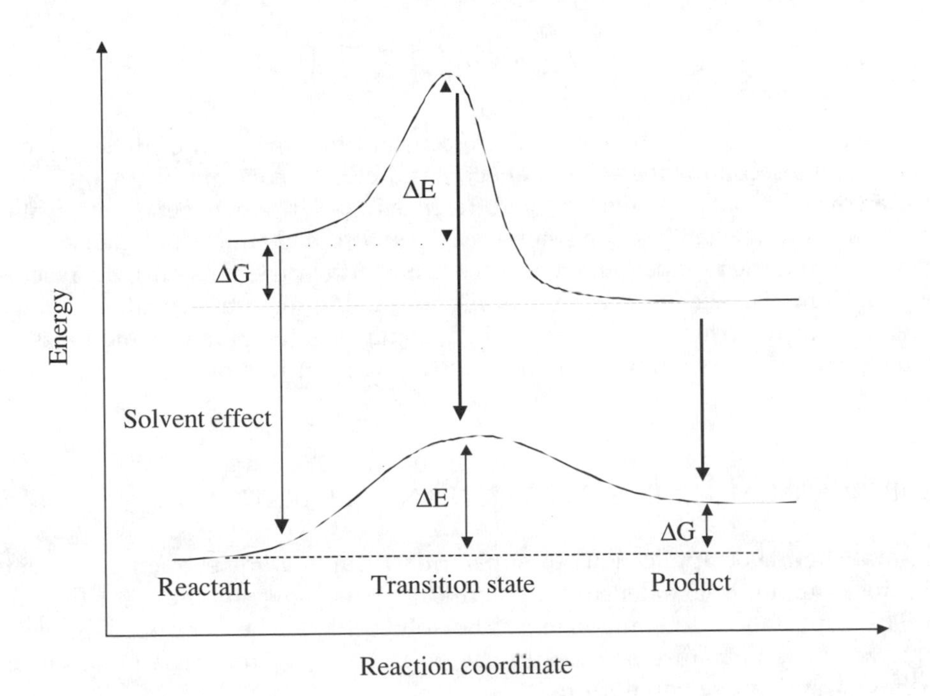

Figure 7.5 Free energy change in reaction and solvent effect.

components mix uniformly at arbitrary compositions. At room temperature, forcing a gas into water produces bubbles, but in supercritical water, which is high density steam, the gas forms a homogeneous mixture. On the other hand, water and hydrocarbons at ambient conditions do not mix and undergo phase separation, but when water is in the supercritical state, they mix uniformly. This mutual miscibility with hydrocarbons is understandable when considering that the dielectric constant of supercritical water is about 2 to 10 as shown in Figure 7.2 and approximately equivalent to that of a polar organic solvent. Also the phase equilibria varies greatly around the critical point as evident from the phase behavior.

Reaction Equilibrium and Reaction Rate

The controllability of the reaction rate and the reaction equilibrium are also key features of supercritical fluids. Consider the reaction activation barrier shown in Figure 7.5. The difference in potential between the transition state and the reactant is the activation energy. When there is a solvent involved, the state of the reactant and the molecules in the transition state will change, and thus the activation energy will also change. Solution theory (Amis and Hinton, 1973) expresses the reaction rate in a solution by the following equation:

$$\ln k = \ln k_0 - \frac{c}{RT}\left(\frac{1}{\varepsilon} - \frac{1}{\varepsilon_0}\right) \tag{1}$$

where the first term is the reaction rate constant in a medium of dielectric constant, ε_0 the second is the solvent effect, ε is dielectric constant of the fluid, and c is a constant that accounts for the difference in polarity between the transition state and the reactant. For the same reason that the chemical potentials of the reactants and the products change with solvents (dielectric constant), the reaction equilibrium also changes. As seen in Figure 7.2, the dielectric constant changes greatly near the critical point due to slight changes of temperature and pressure, which can lead to significant changes in the reaction rate or the reaction equilibrium.

SUPERCRITICAL FLUID EXTRACTION AND SEPARATION

One of the major applications of supercritical fluids is for extraction solvents. Because the liquid-like density, supercritical fluids show solubility for the nonvolatile substances to some extent. The solubility can be controlled by small changes in temperature or pressure, as was shown by the variation of phase behavior and phase equilibrium.

Beginning in the 1980s, attention focused on the controllability of the dissolution strength and phase equilibrium of supercritical fluids, and researchers have worked on the development of new extraction and separation processes (McHugh and Krukonis, 1986, Brunner 1994). Supercritical carbon dioxide, in particular, is nontoxic, odorless, and inexpensive. Since carbon dioxide has a critical temperature near a room temperature, it can be used as a solvent for the separation process operated at low temperatures, especially for the separation of thermally labile substances. It also promises high selectivity in separation, and there are few problems with post-separation of the residual solvent. For these reasons, development has proceeded on supercritical carbon dioxide extraction processes chiefly in the areas of food and pharmaceuticals.

In 1978, the first commercial plant of supercritical carbon dioxide extraction was constructed for decaffeination of coffee beans by HAG AG in Bremen, Germany. Figure 7.6 (McHugh and Krukonis, 1986) shows a photo of the commercial plant (10,000 t/day) for decaffeination of coffee beans. Table 7.2 (Saito, 1995 and 1996) shows the list of the commercial plant for supercritical carbon dioxide extraction processes. As shown here, most of the processes developed are for food and pharmaceutical industries.

However, with the imminent issue of chlorofluorocarbons, application of supercritical carbon dioxide extraction for cleaning has been intensively studied in various industries where replacement solvents need to be implemented (McHardy and Sawan, 1998). The advantage of supercritical carbon dioxide cleaning over liquid solvents is applicability for

Figure 7.6 Caffeine extraction vessel (From McHugh and Krukonis, 1986).

1) precisely, complicated or nano structured materials because of the lower surface tension of supercritical carbon dioxide,
2) thermally labile materials or chemically weak materials, and
3) materials which requires long drying time, since complete removal of the solvent can be done just by the decompression.

Thus, supercritical CO_2 solvent can be expected to be applicable to the cleaning of computer hardware, semiconductor chips, micro-machines, and other precise machines or parts.

There are some other applications of supercritical carbon dioxide extraction. In 1995, ethanol concentration of the fermented alcohol succeeded as a NEDO national project in Japan (Saito, 1995). One recently proposed process is the dye extraction

Table 7.2 Supercritical fluid extraction process (From Saito, 1996).

Year	*Company*	*City/Country*	*Object*	*Volume of extractor [L]*
1978	HAG AG	Bremen/Germany	Coffee	—
1982	SKW/Trostberg	Munchsmester/Germany	Hop	6,500 × 3
1984	Barth & Co.	Wolnzach/Germany	Hop, Coffee	500 × 1
	Natural Cane	—	Hop, Red Pepper	1,000 × 2
	SKW/Trostberg	Munchsmester/Germany	Tea, Coffee	—
	Fuji Flavor	Kawasaki/Japan	Colorant, Flavor	200 × 1
1985	Pfizer	Sydney, NB/USA	Hop	—
1986	SKW/Trostberg	Trostberg/Germany	—	200 × 2
	Fuji Flavor	Kawasaki/Japan	Colorant, Flavor	300 × 1
1987	Barth & Co.	Wolnzach/Germany	Hop	4,000 × 4
	Messer Greisheim	—	—	200 × 1
	Yasuma	Shizuoka/Japan	Paprica, Oreoresine	100 × 1
1988	SKW/Trostberg	Munchsmester/Germany	Tea	—
	Takeda Pharmacy	Hikari/Japan	Solvent from antibionics	1,200 × 1
	Maxwell (GF)	Houston, TX/USA	Coffee	(25,000 ton/year)
	CAL Pfizer	—	Aroma	100 × 4
1989	Hasegawa Koryo	—	—	300 × 2
	HAG AG	—	—	3,000 × 3
	Shigeri Seiyu	Matsuzaka/Japan	—	500 × 1
	Ensco Inc.	—	Solid Waste	2,000 × 1
	Philip Morris	Champion, VA/USA	Tobacco	7,000 × 8
	Takasago Koryo	Kanagawa/Japan	Flavor	420 × 1
1990	Jacob Suchard	—	Coffee	360 × 14
	HAG AG	Bremen/Germany	Coffee	(50,000 ton/year)
	SKW/Trostberg	Trostberg/Germany	Flavor	200 × 2
				220 × 1
	Barth & Co.	Wolnzach/Germany	Coffee	1,000 × 2
				4,000 × 2
	Raps & Co.	—	Spice	500 × 3
	Johns Maas, Inc.	Yakima, VA/USA	Hop	
	Pitt-Des & Co.	—	Hop	3,000 × 4
1991	Fuji Flavor	Kawasaki/Japan		300 × 1
	SKW/Trostberg	Venafro/Italy	Coffee	(20,000 ton/year)
	Barth & Co.	Wolnzach/Germany	—	4,000 × 2
	Texaco		Refinary Waste	2,000 × 3
1992	SKW/Trostberg	Trostberg/Germany	—	—
1993	Hasegawa Koryo	—	—	500 × 2
	Agrisana	—	Pharmaceuticals	300 × 3
1994	Barth & Co.	—	—	200 × 2
	—	Chennai/India	Spice	300 × 1
	Nan Fang	—	Aroma	350 × 4
	Fluor Mill	—	—	300 × 2

combined with dyeing by its spraying. Other applications of supercritical carbon dioxide extraction are for material processing including de-wax and drying, which will be explained later in this chapter.

SUPERCRITICAL FLUID CHROMATOGRAPHY

Chromatography is a method to separate the species according to the difference in the partition between the two phases, mobile phase and stationary phase. In a chromatography column, different compounds will partition differently, and thus they will travel at different speeds and so will separate as they move.

Diffusion in the gas phase is relatively rapid and as a result, gas chromatography gives comparatively narrow peaks and thus the separation is very efficient. However, this method cannot be used for nonvolatile species. For these separations, liquid chromatography is applicable, where a liquid solvent is employed as the mobile phase. The diffusion coefficient in supercritical carbon dioxide is intermediate between the gas and the liquid values, as shown in Table 7.1, and as many compounds can be dissolved in a supercritical fluid, supercritical chromatography provides a useful option for efficient separation of less volatile compounds. Supercritical fluid chromatography has become a field in its own, with its own special focus towards biological and environmental analyses. See Lee and Markides (1990) for further discussion of this field.

MATERIAL SYNTHESES

The controllability of supercritical fluid physical characteristics is not limited to extraction and separation solvents, but also offers promise for crystallization solvents and recently research and development are proceeding in these applications (Adschiri, 1998). Here, several material processing methods will be introduced.

Rapid Expansion of Supercritical Solutions (RESS) Method

The solvent power of supercritical fluids changes considerably by slight manipulation of temperature and pressure, which makes it possible to decrease the solubility of a solute at least several orders of magnitude. Thus, first dissolving the solute in a high density supercritical fluid, and then expanding and lowering its density to nearly atmospheric pressure by passing it through a nozzle can precipitate the solutes with a high degree of supersaturation. Consider the solubility of SiO_2 in a supercritical fluid, as shown in Figure 7.7 (Kennedy, 1950). Near the critical point, a slight decrease in pressure translates into a substantial drop in the solubility and a high degree of supersaturation is obtained. Because

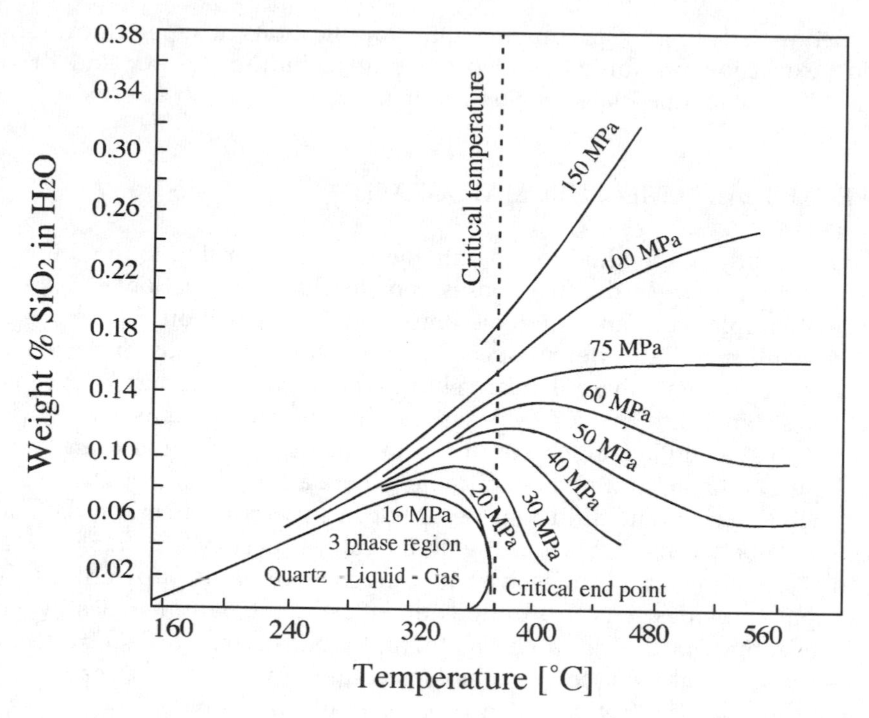

Figure 7.7 Solubility of Silica (SiO_2) in water at high temperature and pressure (Redrawn from Kennedy, 1950).

expansion occurs at the speed of sound, an extremely high degree of supersaturation is obtained in a short time of under 10^{-5} sec, which results in the formation of fine particles. This method is known as the rapid expansion of supercritical solutions (RESS).

There are many reports on the recrystallization of pharmaceuticals (Adschiri, 1998). Compound material production and encapsulation of particles has been attempted by RESS of different solutes from two nozzles. As shown in Figure 7.8, coating of the drug levostatin with lactic acid has been successful. Moreover, the research showed that by varying the conditions lactic acid coated levostatin in spherical, flat and thin film forms could be produced. Network structured materials can also be made, which can be used to prepare pharmaceuticals for drug delivery.

Gas Anti-Solvent (GAS) Method

As explained earlier, phase behavior undergoes great changes in mixed systems involving a supercritical fluid and these characteristics can be used in crystallization. Dissolving a supercritical fluid into a solvent can cause the liquid solvent

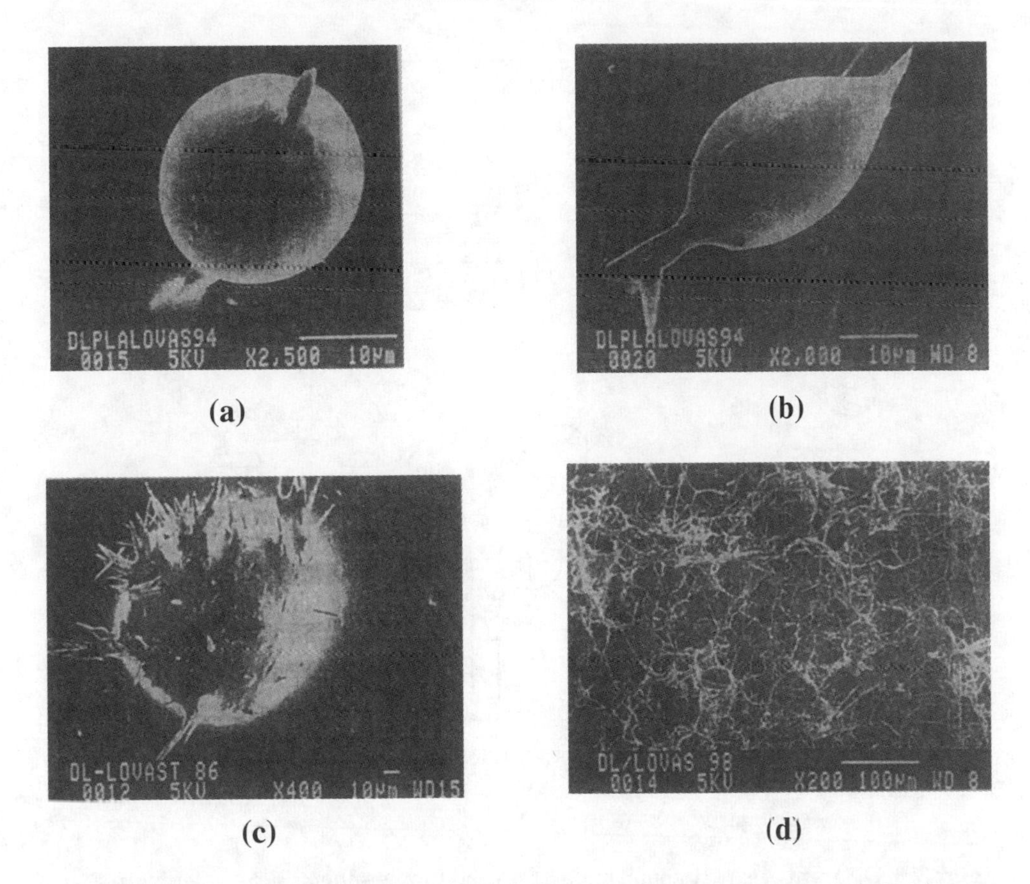

Figure 7.8 Production of the levostatin particles coated with a polymer (DL-PLA) using the RESS method. Conditions: (a) and (b) Molecular weight (DL-PLA): 5,300; extraction temperature: 550°C; extraction pressure: 22 MPa, preliminary expansion temperature: 75–80°C, (c) Molecular weight (DL-PLA): 5,300; levostatin + DL-PLA mixture extraction tank temperature: 55°C; levostatin extraction tank temperature: 35°C; extraction pressure: 20 MPa; preliminary expansion temperature: 75–80°C (when using two extraction tanks and spraying after mixing the two fluids), (d) Molecular weight (DL-PLA): 10,000; extraction temperature: 55°C; extraction pressure: 20 MPa; preliminary expansion temperature: 80°C (From Tom *et al.*, 1993).

to have poor solubility for the solute, resulting in solute precipitation. By controlling the dissolving conditions, it is also possible to synthesize fiber or needle like polymers. This is called the GAS method. GAS method with supercritical carbon dioxide can be used to recrystallize explosives or pharmaceuticals and other substances that are difficult to mechanically pulverize or that are thermally unstable.

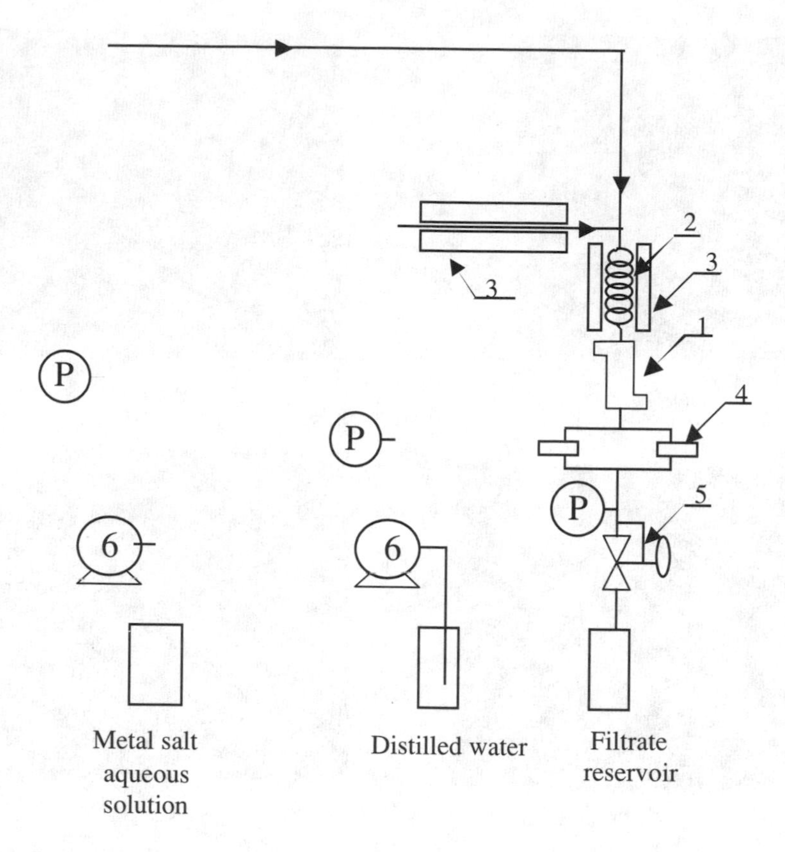

Figure 7.9 Diagram of experimental apparatus used for the supercritical water hydrothermal crystallization.
1: water jacket, 2: reactor, 3: electric furnace, 4: in-line filter (0.5 μm), 5: back pressure regulator, 6: high pressure pump.

Hydrothermal Synthesis in Supercritical Water

When a metal salt solution is heated, metal (hydro)oxides are formed through the following reactions. At a higher temperature, dehydration takes place for the metal hydroxides to form metal oxides.

$$M^{x+} + xOH^- = M(OH)_x$$

$$M(OH)_x = MO_{\frac{x}{2}} + \frac{x}{2} H_2O$$

The method to produce metal (hydro)oxides by using these reactions is the hydrothermal crystallization method. Ordinary, this method is operated at around

Table 7.3 Metal oxide fine particles produced by hydrothermal crystallization in supercritical water.

Metal salts used stating material	*Products*	*Particle size [nm]*	*Morphology*
$Al(NO_3)_3$	AlOOH	80 ~ 1,000	hexagonal plate rhombic needle-like
$Fe(NO_3)_3$	α-Fe_2O_3	~ 50	spherical
$Fe_2(SO_4)_3$	α-Fe_2O_3	~ 50	spherical
$FeCl_2$	α-Fe_2O_3	~ 50	spherical
$Fe(NH_4)2H(C_6H_5O_7)_2$	Fe_3O_4	~ 50	spherical
$Co(NO_3)_2$	Co_3O_4	~ 100	octahedral
$Ni(NO_3)_2$	NiO	~ 200	octahedral
$ZrOCl_2$	ZrO_2 (cubic)	~ 10	spherical
$Ti(SO_4)_2$	TiO_2	~ 20	spherical
$TiCl_4$	TiO_2 (anatase)	~ 20	spherical
$Ce(NO_3)_3$	CeO_2	20 ~ 300	octahedral
$Fe(NO_3)_3$, $Ba(NO_3)_2$	$BaO \cdot 6Fe_2O_3$	50 ~ 1,000	hexagonal plate
$Al(NO_3)_3$, $Y(NO_3)_3$, $TbCl_3$	$Al_5(Y+Tb)_3O_{12}$	20 ~ 600	dodecahedral

100–200°C. Adschiri *et al.* (1992a, 1992b, 1998) have developed a new processes for the hydrothermal crystallization of metal oxide fine particles in supercritical water. By using the experimental apparatus shown in Figure 7.9, an aqueous metal salt solution was fed by a HPLC pump and mixed with supercritical water at a mixing point. By this method, the metal salt solution rapidly heats up to the supercritical state and hydrothermal synthesis initiates. Results are summarized in Table 7.3. Synthesis of fine metal oxide particles with specific morphologies with this method has been demonstrated for nitrates, chlorides, and sulfates for various metals. Some specific features of the supercritical hydrothermal synthesis method that have been elucidated are: 1) nano size particles are produced due to the fast hydrolysis rate and the lower solubility in supercritical water. 2) morphology of the particles can be controlled due to the drastic change of properties of water around the critical point, as shown in Figure 7.10 (Adschiri, 1992b). 3) oxidizing or reducing atmosphere can be controlled by introducing oxygen, CO or hydrogen gases as a homogeneous reaction atmosphere can be formed. This method has been applied to produce a magnetic material (barrium hexaferrite), phosphor (YAG:Tb), and a Li ion battery material (LiCoO2) (Hakuta *et al.* 1995, 1998, Adschiri, 1999).

MATERIAL PROCESSING

When removing a liquid or solvent from materials by evaporation or drying, the presence of a gas-liquid interface exists, and surface tension can cause problems such as crack formation or fragmentation. Supercritical extraction offers a method

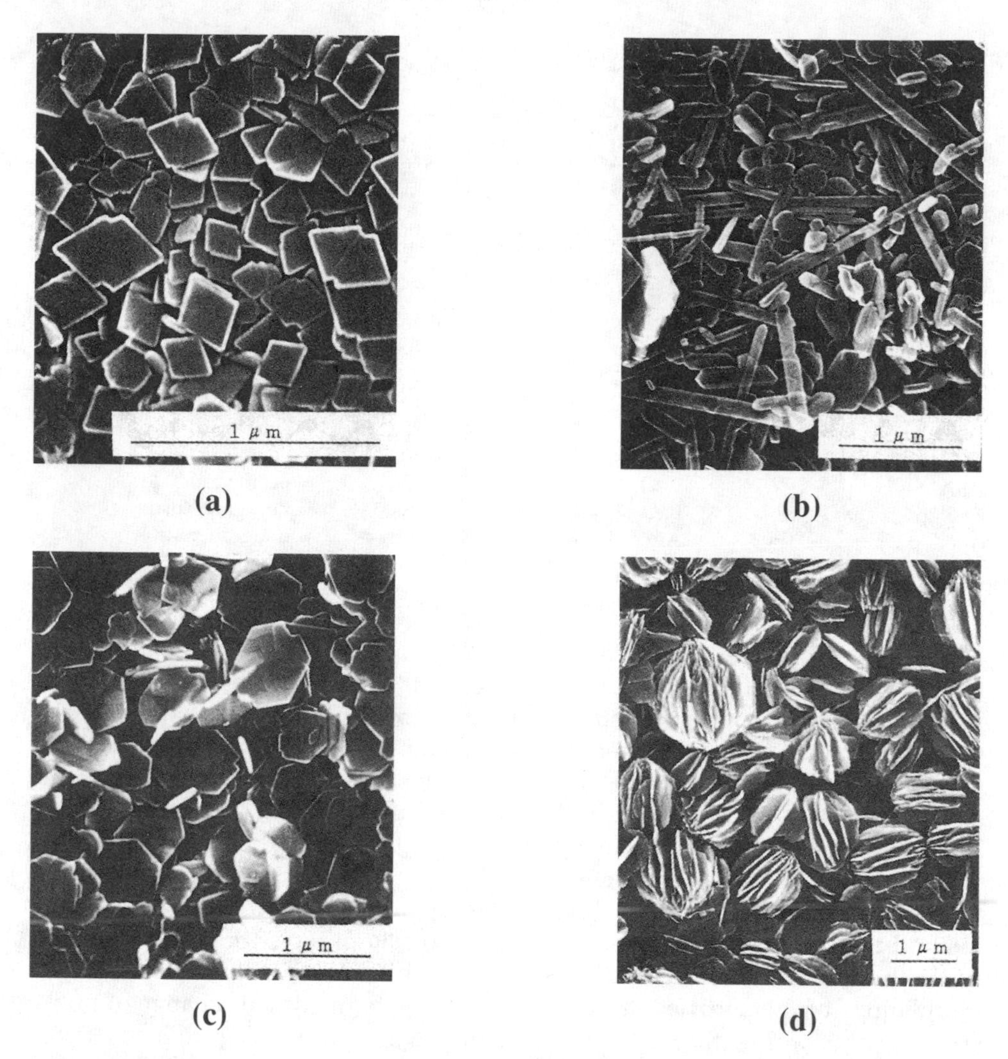

Figure 7.10 Morphology of boehmite (AlOOH). Conditions: (a) 0.006 M; 350°C; 35 MPa, (b) 0.02 M; 400°C; 35 MPa, (c) 0.05 M; 400°C; 40 MPa, (c) 0.05 M; linear temperature size from 100 to 300°C; 30 MPa (From Adschiri, 1992b).

to remove the liquid without such problems. Supercritical extraction of the liquid eliminates the gas-liquid interface, which makes it possible to obtain a dry product, while suppressing the formation of cracks and fragmentation. Aerogel made by supercritical carbon dioxide drying of gels produces a material with extremely high porosity and surface area (McHugh and Krukonis, 1986).

Supercritical drying can be used for the treatment of fine particles formed in liquid phase (Adschiri, 1998). Synthesizing fine particles by the wet method necessitates a drying process, and the drying of particles results in strong agglomeration by capillary force when the solvent vaporizes, which often has major effects on the physical properties of the final powder products. In the supercritical fluid drying method, elimination of a gas-liquid interface suppresses the strong agglomeration caused by capillary force.

ENVIRONMENTAL APPLICATIONS

Waste Treatment with Supercritical Water Oxidation

As explained previously, key features of the reactions in supercritical fluid are the controllability of reaction phases and that of reaction equilibrium and rate.

Modell (1992) proposed a new oxidation process in supercritical water. When oxygen is introduced into waste water, gas bubbles will be formed. Also, ordinarily, the organic substances form an emulsion in water. Thus, the oxidation reaction proceeds in three phases, which leads to poor mass transfer. However, at supercritical conditions, oxygen gas becomes miscible with supercritical water and organic materials can be dissolved into it, which leads to a homogeneous oxidizing atmosphere and thus extremely high oxidation rates. In 1998, Organo developed a commercial plant for the supercritical oxidation of waste water from a semiconductor factory (Suzuki, 1997).

Recycling

One reaction of great importance in high temperature water or supercritical water is non-catalytic hydrolysis. In high temperature water, hydrolysis proceeds without any acid or base catalyst. Various applications of the non-catalytic hydrolysis are expected, one of which is the chemical recycle of waste polymers to monomers. There appears to be considerable opportunity for recovering monomers from the condensation polymers by the high temperature non-catalytic hydrolysis method.

Recently, Arai and Adschiri (1999) have proposed new processes of polymer hydrolysis to recover monomers using hydrolysis in supercritical water. Cellulose can be hydrolyzed in to glucose, fructose and oligomers of glucose (cellobiose, cellotriose etc.) in 0.05 sec at 400°C and 30 MPa. Non catalytic hydrolysis of polyethylene telephthalate (PET) in supercritical water results in yields of telephthalic acid close to 100% with a purity of greater than 97% under the conditions of 673 K, 40 MPa (Adschiri *et al.* 1997). This method can be also employed for recovering monomers from the heavy distillates. It was demonstrated that phenol could be recovered from the BPA tar of the distillation residue (Adschiri *et al.* 1997). Based on these same concepts, Kobe Steel developed a commercial plant of recycling tolylenediamine by the hydrolysis of tolylenediisocyanate for

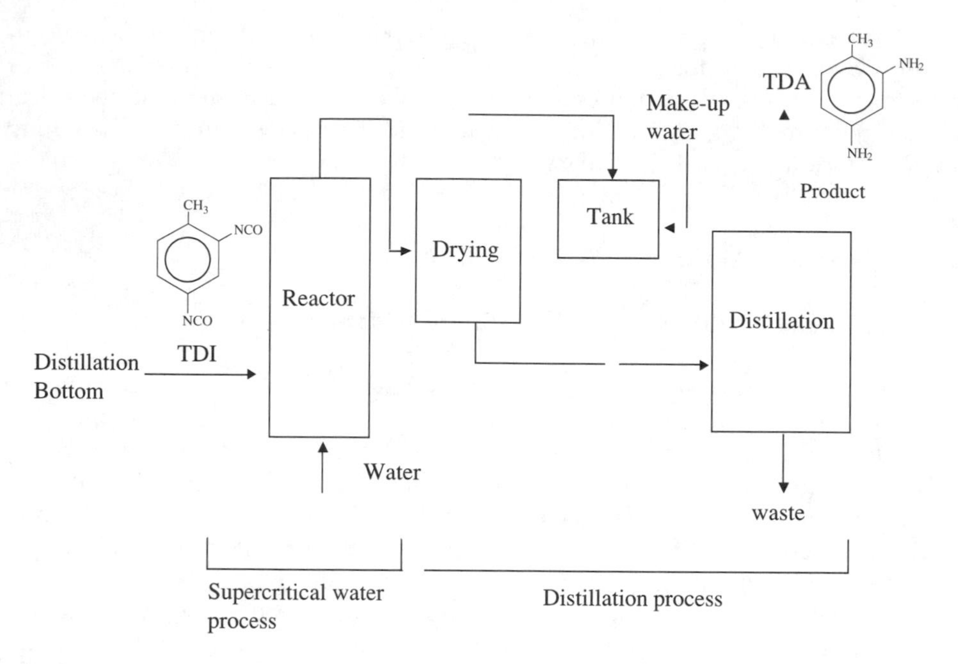

Figure 7.11 Flow chart of chemical recycling process of TDI to TDA.

a pharmaceutical company in 1998 (Fukuzato *et al.*, 1999). The process flow is shown in Figure 7.11.

ORGANIC SYNTHESES

The key features of reactions in supercritical fluid are the controllability of reaction phases and that of reaction rates and reaction equilibira. In these ten years, there have appeared a increasing number of research activities on reactions in supercritical fluids (Eckert *et al.*, 1996; Jessop and Leitner, 1999; Savage, 1999; Kendall *et al.*, 1999). Below is one example of the reaction is supercritical carbon dioxide, which shows the importance of reaction phase control is reported. Jessop *et al.* (1994, 1995) demonstrated the capability of molecular catalysts for various organic synthesis reactions, including the reaction of CO_2 and H_2. In the traditional method, an organic solvent is used to dissolve the molecular catalyst. However, since the solubility of CO_2 and H_2 gases in organic solvents is too low to obtain high reaction rate, Jessop *et al.* (1994, 1995) employed supercritical carbon dioxide as a solvent both for the catalysts and the reactants. Since the H_2 gas, and the catalyst forms a homogeneous reaction phase with supercritical carbon dioxide, the reaction proceeds on the order of 10^{+5} faster than that in the liquid solvent, as shown in Figure 7.12.

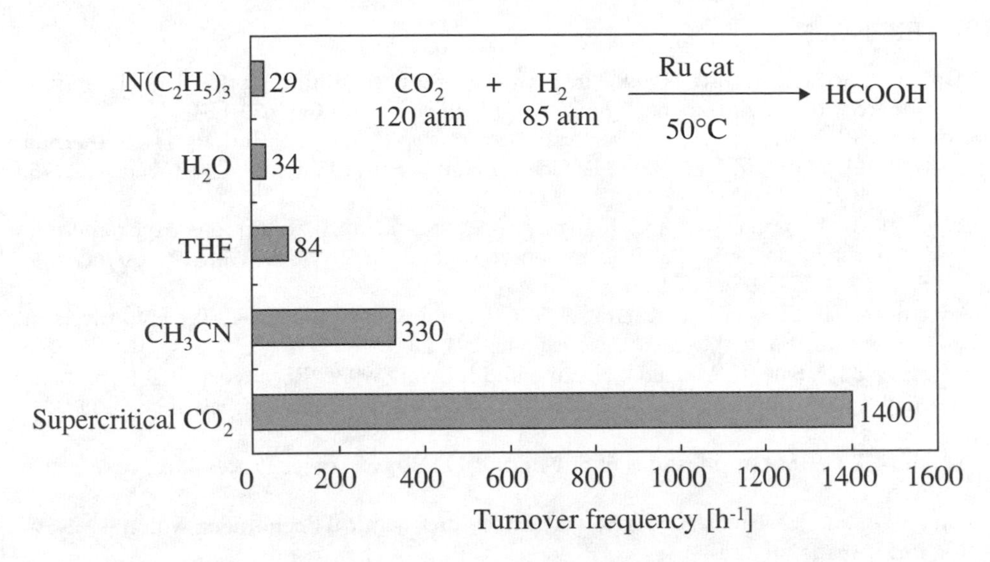

Figure 7.12 Initial turnover frequency of formic acid production by CO_2 hydrogenation with $RuH_2[P(CH_3)_3]_4$ catalyst in various solvents at 50°C (From Jessop, 1994).

CONCLUSION

In this chapter, a variety of applications of supercritical fluids have been introduced, especially for applications that use supercritical carbon dioxide and water as solvents. Supercritical fluid extraction, which was born around twenty years ago, has been grown to be a formidable industrial method. The use of supercritical fluids as reaction solvents has opened the door to broader applications. Hydrolysis without acid or base catalyst in high temperature water will be a promising technology for the chemical recycle of waste condensation polymers. Hydrothermal synthesis in supercritical water (SCW) also produces various functional materials. In the near future, I believe some new synthesis methods will be proposed from the research of reactions in supercritical fluids, which are now being studied in the fields of chemistry, physics and chemical engineering.

I think chemical engineers should have a responsibility for proposing new technologies which support the establishment of "sustainable development of the world" and supercritical fluid technology will be a promising candidate. However, supercritical fluid technology is just being established and as of today, has only been accepted in certain fields. For the further development of supercritical fluid technology and the application to wider fields, its engineering basis should be firmly established. Fundamental information on fluid properties, phase equilibrium, kinetics and mechanism of the reactions should be methodically researched and based on these results, unit operations of supercritical fluid processes can be firmly established.

REFERENCES

Arai, K. and T. Adschiri (1999) "Importance of Phase Equilibria for Understanding Supercritical Environments." *Fluid Phase Equilibria*, **158–160**, 673–684

Adschiri, T., K. Kanazawa and K. Arai (1992) "Rapid and Continuous Hydrothermal Crystallization of Metal Oxide Particles in Supercritical Water." *J. Am. Ceram. Soc.*, **75**, 1019–1022

Adschiri, T., K. Kanazawa and K. Arai (1992) "Rapid and Continuous Hydrothermal Synthesis of Boehmite Particles in Subcritical and Supercritical Water." *J. Am. Ceram. Soc.*, **75**, 2615–2618

Adschiri, T., R. Shibata and K. Arai (1997) "Phenol Recovery by BPA Tar Hydrolysis in Supercritical Water." *Sekiyu Gakkaishi*, **40**, 291–297

Adschiri, T., O. Sato, K. Machida, N. Saito and K. Arai (1997) "Recoovery of Tetraphthalic Acid by Decomposition of PET in Supercritical Water." *Kagaku Kogaku Ronbunshu*, **23**, 505–511

Adschiri, T. (1998) "Applications of Supercritical Fluids in Power Processing." *KONA*, **16**, 89–101

Amis, S. A. and J. F. Hinton (1973) Solvent Effects on Chemical Phenomena, Volume 1. New York: Academic Press

Brunner, G. (1994) Gas extraction. An introduction to Fundamentals of Supercritical Fluids and the Application to Separation Processes. New York: Springer

Eckert, C. A., B. L. Knutson and P. G. Debenedetti (1996) "Supercritical Fluids as Solvents for Chemical and Materials Processing." *Nature*, **383**, 313

Franck, E. U. (1981) *Pure & Appl. Chem.*, **53**, 1401

Fukuzato, R. (1999) "Chemical Recycle of Wastes using Supercritical Water." *ECO INDUSTRY*, **4**, 19–29

Hakuta, Y., T. Adschiri, T. Suzuki, T. Chida, K. Seino and K. Arai (1998) "Flow Method for Rapidly Producing Barium Hexaferrite Particles in Supercritical Water." *J. Am. Ceram. Soc.*, **81**, 2461–2464

Hakuta, Y., K. Seino, H. Ura, H. Takizawa, T. Adschiri and K. Arai (1998) "Production of Phosphor (YAG:Tb) Fine Particles by Supercritical Water Crystallization." Proceedings of Advanced Technologies for Particle Processing, Vol. 1, Particle Technology Forum, AIChE Annual Meeting, Miami Beach, Florida

Jessop, P. G., T. Ikariya and R. Noyori (1994) "Homogeneous Catalytic Hydrogenation of Supercritical Carbon Dioxide." *Nature*, **368**, 231–233

Jessop, P. G., T. Ikariya and R. Noyori (1995) "Homogeneous Catalysis in Supercritical Fluids." *Science*, **269**, 1065–1069

Jessop, P. G. and W. Leitner (1999) Chemical Synthesis Using Supercritical Fluids. Weinheim, Germany: Wiley-VCH

Kendall, J. L., D. A. Canelas, J. L. Young, J. M. DeSimone (1999) "Polymerizations in Supercritical Carbon Dioxide." *Chem. Rev.*, **99**, 543–563

Kennedy, G. C. (1950) "A portion of the system silica-water." *Econ. Geol.* **45**, 629

Lee, M. L. and K. E. Markides (1990) Analytical Supercritical Fluid Chromatography and Extraction. Provo, Utah: Chromatographic Conferences

McHardy, J. and S. P. Sawan (1998) Supercritical Fluid Cleaning: Fundamentals, Technology and Applications. Westwood: Noyes

McHugh, M. and V. Krukonis (1986) Supercritical Fluid Extraction: Principles and Practice. Boston: Butterworth

Modell, M., J. Larson and S. F. Sobczynski (1992) *Tappi J.*, **751**, 195–202

Savage, P. E. (1999) "Organic Chemical Reactions in Supercritical Water." *Chem. Rev.*, **99**, 603–621

Schneider, G. M. (1972) "Phase Behavior and Critical Phenomena in Fluid Mixtures under Pressure." *Ber. Bunsenges. Phys. Chem.*, **76**, 325–331

Saito, S. (1995) "Research Activities on Supercritical Fluid Science and Technology in Japan — A Review." *J. Supercritical Fluids*, **8**, 177–204

Saito, S. (1996) Science and Technology of Supercritical Fluids. Sankyo Business. Japan: Sendai

Scott, R. L. (1972) "Thermodynamics of Critical Phenomena in Fluid Mixtures." *Ber. Bunsenges. Phys. Chem.*, **76**, 296

Scott, R. L. and P. B. van Konynenburg (1970) "Static Properties of Solutions — Van der Waals and related models for hydrocarbon mixtures." *Discuss. Faraday Soc.*, **49**, 87

Suzuki, A., O. Taro, N. Anjo, H. Suzugaki and T. Nakamura (1997) "Commercialization of Supercritical Water Oxidation. Destruction of Trichloroethylene, Dimethyl Sulfoxide and Isopropyl Alcohol with Pilot-Scale Process" Proceedings of 4th International Symposium on Supercritical Fluids, pp. 895–900. Japan: Sendai

Tom, J. W., G.-B. Lim, P. G. Debenedetti and R. K. Prud'homme (1993) In *Supercritical Fluid Engineering Science: Fundamental and Applications* (eds., Kiran, E., and Brenekke, J. F.) 238–257, ACS Symposium series 514, American Chemical Society, Washington DC.

8. A PEEK INTO THE NANO-WORLD OF COLLOIDAL AND INTERFACIAL PHENOMENA

RAJ RAJAGOPALAN

Department of Chemical Engineering, University of Florida, Gainesville, Florida 32611-6005, USA

A WORLD OF NEGLECTED DIMENSIONS?

The world of colloids — a world of materials that are so small that one would need a powerful microscope to see them — was once called "the world of neglected dimensions," but not anymore! In truth, colloids and interfacial phenomena have always been an area of intense interest to us, for — as we shall see shortly — they have always been a part of many things we use or experience every day. The beautiful, iridescent opal stone on your mother's ring is nothing but a crystal-like arrangement of tiny colloidal silica particles. The Bernaise sauce you might enjoy in a fine French restaurant and the Belgium chocolate that melts so smoothly in your mouth are so appealing because of their colloidal properties. Your favorite music CD that helps you unwind after a hard day at school requires processing technologies based on interfacial phenomena — as does the printing ink used to print this book. We take many of these for granted, especially since some of them have been around for many centuries. For example, printing ink has been around at least since the time of the Pharaohs! But studies of colloids and surfaces have lately gained a drastically different perspective, thanks to some remarkable inventions in recent years.

For example, recent advances in instrumentation have made it possible not only to "see," but, indeed, to *manipulate* individual atoms on a surface. Molecular engineering of polymers, surfactants (defined later), and particles has given rise to hitherto-unknown possibilities to create new advanced materials and to produce novel drug delivery systems. Direct measurements of forces between particles and direct determination of the elasticity of single polymer chains are now commonplace. Such advances are mind-boggling when one recalls that only about a century back even the very existence of atoms was hotly debated by scientists! The new millennium holds tremendous possibilities for further taming the submicroscopic world of colloids for the benefit of mankind.

Where do chemical engineers fit in in this grand picture? Chemical engineers have traditionally focused on the processing end of the chemical industries, and surely there are plenty of challenges in the processing of colloidal materials. However, chemical engineers also have a unique combination of mathematical, scientific and engineering skills which can be put to use to innovate and design new materials and processes made possible by advances in colloid and surface science. Our goal in this chapter is to illustrate this opportunity by providing a brief look at the submicroscopic world of colloids. We shall do this through a number of snapshots taken from a variety of fields ranging from biology and medicine to advanced materials and environmental science. Some of the snapshots highlight what we have known for some time, but others will give a glimpse of what the future holds.

Colloids: A Matter of Definition?

What is the definition of colloids? Colloids are often defined simply as substances or materials whose linear dimensions (at least in one of the three directions) are somewhere between one nanometer and one micron.[1] Fogs, mists and smoke consist of colloidal-sized particles (of liquid droplets or solid particles). Many food items (milk, ice cream and whipped cream) are also colloids. Since only one of the linear dimensions needs to be small enough for a material to be classified as a colloid, soap films or foams in general also fall under the classification. Just these examples alone are sufficient to convince anyone why we should be interested in colloids, how they form and how to process them. However, there are a lot more reasons why chemical engineers (and scientists in general) are interested in colloids. As we shall see shortly, many biological materials (*e.g.*, proteins, biological cells and viruses) are of colloidal dimensions, and medicine and pharmaceutical sciences are full of examples of colloids. And, as we become more and more interested in miniaturizing computers and computer-controlled devices such as personal digital assistants or cellular phones, making thin semiconductor structures of colloidal dimensions requires engineers knowledgeable about colloids and how they behave. For that matter, just keeping microscopic dust particles away from processing environments designed for making submicroscopic devices with very fine features is a formidable task by itself!

Colloids & Surfaces: The Inseparable Twins

Colloids and surfaces are indeed inseparable twins. To appreciate this, consider a solid in the form of a cube of side 1 cm. Its surface area is 6 cm^2. Now divide it into tiny cubes of side 1 μm. There are one trillion (10^{12}) such tiny cubes in the original cube, each with a surface area of 6 μm^2, or $6 \cdot 10^{-8}$ cm^2. Therefore, the total surface area of the one trillion tiny cubes is 60000 cm^2 — four orders of magnitude larger than the original area! One can now see why surface or interfacial phenomena are so important in colloidal systems. Interfacial phenomena are any physical, chemical or biological phenomena that occur at (or across) interfaces between two materials (such as an air/water interface). Chemical engineers have always been

[1] One nanometer, denoted as 1 nm, is one-billionth of a meter, and one micron is one-millionth of a meter, denoted as 1 μm (or, 1/25000 of an inch). A typical human hair is about 200 μm in diameter.

interested in interfacial phenomena, but dealing with colloids makes interfacial phenomena all the more important. [See Chapter 10 of Ball (1997) for a first-level introduction to surfaces and surface forces in many materials and phenomena of interest to us.]

Small in Size, but Large in Scope

One thing that perhaps stands out more than anything from the above discussion is the ubiquity of colloidal and interfacial phenomena in almost all facets of life. The types of colloids are so varied and the contexts in which they arise are so numerous that it is difficult to give a balanced picture in a single chapter. What interests us, or should interest us, ranges from how colloids are made, how to adjust or "tailor" their surfaces to obtain the types of surface properties that are useful, how to transport them from one place to another, how to use them to create new materials, how to rid a material of colloids if their presence is undesirable — the list goes on and on. In fact, the field of colloidal and interfacial phenomena is truly *interdisciplinary*, for it requires the talents of chemists, physicists, biologists, and mathematicians — and engineers with training in these disciplines. This is one reason why chemical engineers — whose academic training includes a judicious combination of all these disciplines — are uniquely suited for taming the world of colloids.

All we can do in a short chapter such as the present one is to provide a collage of snapshots of the world of colloids and colloidal phenomena. In what follows we shall see how colloids rule our biological existence, assist our medical needs, contribute to the foods we eat, help us in saving the environment and enhance our material well-being through modern technology. We shall also take a moment to highlight a couple of modern tools that allow us to peer into the above submicroscopic world. In doing so, we shall barely scratch the surface (pun intended!) of what lies beneath! Many of the examples discussed here and more can be found in Hiemenz and Rajagopalan (1997).

Before we go to the examples, it is useful to review a type of classification of colloids that will help us to appreciate the similarities and differences between the various examples. In this context, it is also useful to make a note of forces that determine (and dominate) colloidal phenomena. We shall do these in the following section.

THE MANY FACES OF COLLOIDS AND COLLOIDAL FORCES

Colloids come in many varieties and can be classified in a number of different ways, but for our purpose here it is sufficient to consider some relatively simple types and use a classification based on the constituents of the colloidal materials. This classification leaves a lot out, but helps us to focus on some of the most essential concepts and to appreciate the examples discussed later.

Association Colloids

Association colloids are so called because they are formed from the association of certain types of molecules known as *surfactants*. Surfactants are typically molecules that contain two parts with differing liking (affinity) to water. One part is *hydrophilic* ("water loving") and the other *hydrophobic* ("water hating"). They are a little like Siamese twins, for they are physically attached to each other, but the analogy does not go very far, as they have differing tastes when it comes to solvents. Soap molecules are an example of surfactants. When mixed with water, they tend to migrate to the surface, with the hydrophilic part straddling the water surface and the hydrophobic part sticking out into the air. When more surfactant molecules are "dissolved" in the water, they start to "associate" with each other and form little clusters with all hydrophobic parts facing each other and all the hydrophilic parts forming an envelope around the hydrophobic segments. Such "self-assembled" entities form spontaneously for thermodynamic reasons (*i.e.*, such a formation lowers the thermodynamic energies of the system) and are called *micelles*. The micelles are typically spherical at low surfactant concentrations, but can assume other shapes depending on the relative "bulkiness" or size of the two segments. As we shall see shortly, such association colloids appear in numerous contexts in nature (*e.g.*, biological cells in our bodies) and are used in many consumer products (*e.g.*, ice cream, hand lotion, and shampoo). You can find an excellent discussion of surfactant solutions and similar "soft" materials in a highly readable book by Ball (1994).

"Conventional" Colloids

There is actually no such thing as a "conventional" colloid, but we shall use the term here to refer to solid particles of microscopic dimensions in liquids, or tiny liquid droplets in another, immiscible liquid (or particles in a gas). We come across such colloids in common products such as latex paints or inks. In contrast to association colloids, which are structurally stable, particles in a latex paint or inks tend to agglomerate. To avoid permanent agglomeration (try painting a wall with a paint that has agglomerated and formed lumpy clusters!), one would have to "stabilize" them with additives, such as polymer molecules. The polymer molecules adsorb on the surfaces of the particles and form a thin "cushion" that allows us to redisperse the particles if they form aggregates. Some of the examples we shall consider in this chapter will be concerned with such colloids (*e.g.*, see the discussion below on high-strength ceramics).[2]

---

[2] A primitive recipe for ink is a dispersion of fine particles of carbon (carbon black) in water. Such an ink, however, will not be stable, as the carbon particles will eventually settle down. However, if a very small amount of Arabic gum (a polymer) is added to the dispersion, the ink becomes stable, for up to a year, because the particles acquire a thin layer of polymer chains on the surface which form a cushion against aggregation. A technique dating back to the ancient Egyptians, such a process is still used today for the preparation of watercolors. A fascinating account of examples such as this is presented by the 1991 Physics Nobel laureate Pierre-Gilles de Gennes in a book consisting of talks to French high school students (de Gennes and Badoz, 1996).

The Forces that Shape Colloidal Phenomena

It is probably evident from some of the above descriptions that it is difficult to understand colloids and colloidal phenomena without understanding the intermolecular and other forces that cause the colloids to form and behave as they do. One reason for this is that, as we mentioned above, the small sizes of the particles lead to enormous surface areas for a given amount of material. The exposed surfaces interact with each other through intermolecular forces. We shall mention only a few types of forces of interest here, so that the examples below can be understood.

1) One of the most important forces is a class of forces known as *van der Waals forces*, which are usually attractive and make the particles stick to each other or make soap films collapse. (The strength or weakness of van der Waals forces is also the reason why some substances exist as liquids at room temperature while others remain as gases.)
2) Another type of force that is important in colloidal systems arises from electrostatic (*Coulombic*) interactions caused by accumulation of ions around the particles. Often colloidal particles acquire charges at their surfaces when dispersed in water because of adsorption of salt ions in the water or dissociation of chemical species on the surface. If the particles are, say, negatively charged as a result, positive ions from any salts present in water accumulate around the particles. When two such particles come close to each other, a repulsion or attraction results (depending on the charges on the particles).
3) Another important type of force arises from the affinity (or lack thereof) of surface molecules to polar substances (such as water) or nonpolar substances (such as oils). For example, such hydrophilic or hydrophobic interactions determine the formation of association colloids mentioned above.
4) In addition, one can also have forces induced by any polymers added to the dispersions (such as the polymers that are used in latex paints to prevent the paint from forming permanent clumps). In the simplest case, the polymers impart a repulsive, "steric" force arising from the fact that the regions occupied by the polymer chains are not physically accessible to other particles. However, polymer-induced forces can be attractive also. What happens depends on the mutual likes and dislikes between the individual segments of a polymer chain relative to their affinity to the solvent.

Equipped with the above information, we are now ready to sample examples of colloidal phenomena that are of interest to scientists and engineers. In what follows we will take as wide a perspective as space permits. We will consider colloidal and interfacial phenomena in life sciences (the problems of interest to biochemical and biomedical engineers, for example), food processing, advanced materials, and environmental issues (which are of ever-increasing interest to chemical engineers). We shall also take a look at two modern developments in instrumentation for probing colloids and surfaces at microscopic (or nanoscopic) scales, for such developments increase our ability to understand the phenomena

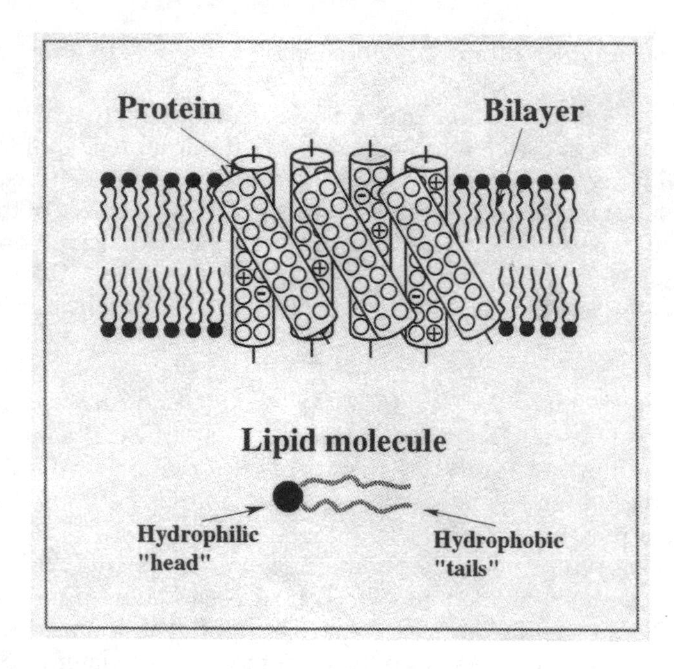

Figure 8.1 The colloidal bilayer structure of a biological membrane formed by self-assembly of surfactants.
The figure shows a simplified sketch of a hypothetical protein molecule embedded in a bilayer (a biological membrane). The bilayer shown is a two-dimensional cross-section of a membrane. The bundle of cylinders shown represents the "helices" of a protein. The cylinders are part of the same protein and are joined together by other segments of the protein protruding out of the bilayer on either side [from Hiemenz and Rajagopalan (1997)].

of interest from first principles and expand the scope of what scientists and engineers can do.

COLLOIDS RULE BIOLOGY AND MEDICINE

Let us begin close to home — with a few examples of colloids and colloidal forces and phenomena that are central to our very being.

Colloidal Basis of Biological Membranes and Cells

Life sciences provide a fascinating array of examples in which colloid and surface science plays a vital (pun intended!) role in maintaining and promoting supramolecular structures and processes that sustain life. A specific example is the phospholipid "bilayers" that form the "walls" of biological cells and separate the interior of the cells from the rest of the environment (see Figure 8.1)[3]. These bilayers arise from self-assembly of component molecules, each of which consists of a hydrophilic "head" group and hydrophobic "tails," as we saw in our introduction earlier to association colloids. The nature of the forces that create such layers (or membranes) and maintain their structure and functionality is a major topic of research today and is pursued by cell biologists, physiologists, biophysicists, biochemists and *chemical engineers*, among others. However, a bilayer by itself is insufficient to create and maintain a "living" cell. The cell membrane is actually a mosaic of a number of functional units that include protein molecules embedded in the membrane. The protein molecules themselves contain hydrophobic and hydrophilic components and provide the pathways for life-sustaining ions and polar molecules to move across the cell membrane. The arrangements of the hydrophilic and hydrophobic parts of a protein molecule are far more complex than those in a simpler structure such as a bilayer. The configuration of the hydrophilic and hydrophobic parts of a protein and how a protein is embedded in a bilayer are important for the molecular events that contribute to the functions of a biological cell (see Figure 8.1). As shown in the figure, the hydrophobic parts of the protein molecules form the outer surface of the cylinders shown and hold the molecules in place, away from water, in the hydrocarbon part of the bilayer. The inner surfaces of the cylinders are hydrophilic and allow the transfer of ions and polar molecules from one side of the bilayer to the other. These pathways remain closed until an appropriate internal or external stimulus triggers them to open to allow transport across the membrane. It therefore comes as no surprise that colloid and surface science plays a central role in such phenomena in life sciences.

In addition to illustrating the importance of colloid and surface science *and* the opportunities for chemical engineers in biological and life sciences, this example highlights the importance of phenomena peculiar to surfactant systems, which explains why chemical engineers are interested in the behavior of surfactants in solutions and the tendency of the surfactants to self-assemble when dissolved in water or in water-oil mixtures. We shall have more to say about the use of surfactants in our discussion later on their applications in environmental science and chemical processing.

Colloidal Cargo Carriers: From Cosmetics to Medicine and Genetic Engineering

The geometric structure of micelles also suggests that we might be able to package appropriate doses of medicine in physiologically friendly surfactant capsules,

[3] A relatively nontechnical and very engaging discussion of bilayers and their implications to biology and biochemistry is presented by Tanford, a pioneer of the so-called *hydrophobic effect* (Tanford 1989). This book also presents a very interesting historical account of Benjamin Franklin's now-famous experiments on stilling water waves with oil.

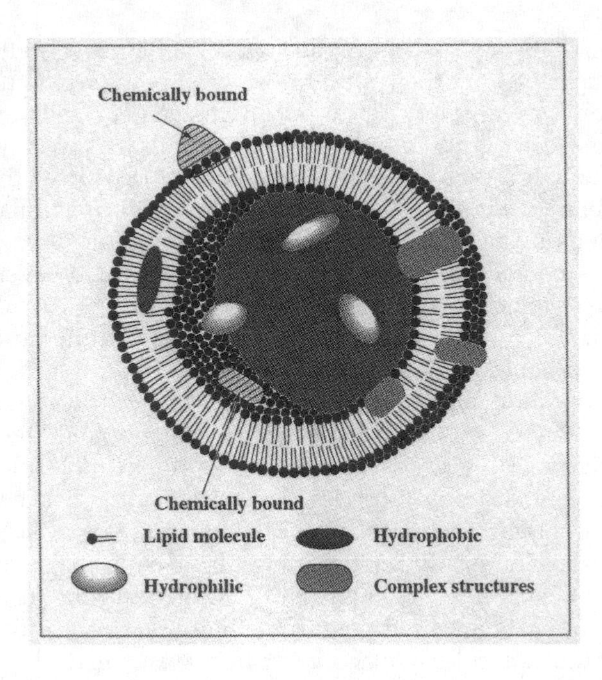

Figure 8.2 A liposome as a carrier of molecular cargo.
The cargo molecules are carried in different parts of the liposome depending on their chemical nature. Hydrophobic molecules are carried inside the hydrophobic part of the bilayer, whereas hydrophilic molecules reside in the interior. More complex molecules are wholly or partly imbedded in the bilayer or chemically bound to the interior or exterior surface [from Hiemenz and Rajagopalan (1997)].

which can then deliver the medicine specifically to the organs that need it while at the same time keeping it away from areas that may find the medicine toxic. The same idea has worked well in a more mundane application; namely, using surfactant "vessels" to trap perfumes in our skins so that the fragrance lasts longer. These were hardly the questions that engaged the attention of Alec Bangham, a British scientist who discovered surfactant capsules known as *liposomes* in the early 1960s while studying the effect of lipid molecules on the clotting of blood.

Liposomes, illustrated in Figure 8.2, are colloidal-sized containers made of lipid bilayers. As we already saw in Figure 8.1, a lipid molecule consists of a polar, hydrophilic "head" that is attached to (one or two) hydrophobic, hydrocarbon "tails." At appropriate concentrations, the lipid molecules in water self-assemble to form bilayers since the hydrophobic tails like to avoid contact with the water. When such bilayers are broken up into small pieces, the fragments wrap themselves into closed structures known as liposomes and encapsulate some of the water inside. The potential applications of liposomes in cosmetics, pharmaceutical and medical technology, and genetic engineering (for studying basic properties of genes by isolating them inside a liposome or for developing schemes for gene- or protein-replacement therapies) are numerous.

Why should we be interested in using liposomes for drug delivery? In many cases, drugs administered in "free form" cause "side effects" because of their toxicity to areas of the body that are not affected by the disease or disorder. It appears possible to improve the effectiveness of drugs and minimize their toxicity by encapsulating the drugs in liposomes and delivering them efficiently and specifically to the affected organs. It is also possible to design liposomes that avoid detection[4] by "hunters" such as macrophages in the body so that the drugs can be delivered to the cells that need treatment. Excellent descriptions of these and related issues in liposome technology are presented by Lasic in a lucidly written review in the *American Scientist* (Lasic 1992).

There are a number of problems of interest to chemical engineers in this context, ranging from, for example, synthesis of new (short-chain as well as polymeric) surfactants for developing special membranes, artificial skins, pharmaceutical lotions and (of course) drug delivery systems. Formation of surfactant micelles and the relation between the molecular architecture of the surfactants and the shapes of the self-assembled structures that result from the surfactants are areas that hold considerable promise. Moreover, additional opportunities are afforded by the microenvironments inside surfactant micelles and liposomes for studying catalysis and material synthesis.

Colloidal Forces in Molecular Recognition

Molecular recognition and specificity are the stuff of life. How are biological macromolecules able to recognize each other or recognize a membrane or a substrate and form specific associations? What do we need to know about the interplay between charged surface groups on interacting macromolecules such as proteins in order to design better drugs or to understand and modify enzyme catalysis? What determines the encounter between an influenza virus and its host site (a sialic acid residue) on a cell before the virus binds and enters the cell (by what is known as endocytosis)? These are the types of questions whose answers are central to the functioning of biological systems, design of drugs, development of artificial biomaterials, design of specific chromatographic techniques, and the like.

Although the answers to questions such as these depend on a complex array of factors ranging from the structure of the relevant molecules to their environment and the chemical activity of the medium containing the molecules, intermolecular (guest/host) interactions play a central role in determining the rate and

[4] These are known by their registered tradename *stealth liposomes*. The "stealth" effect is the result of the steric layer (recall the earlier discussion on forces) surrounding the liposomes, and these liposomes are therefore also known as sterically stabilized liposomes (see Lasic 1992).

the efficiency of the ultimate result. And a major component of the many possible intermolecular forces is the electrostatic interaction, particularly because of the long-range nature of the Coulombic forces and the inevitable influence of the ionic atmospheres that surround the macromolecules and substrates (recall our brief introduction to these in the section on colloidal forces).

The electrostatic interaction works in tandem with the diffusional (translational as well as rotational) motion of the macromolecules as the macromolecules find their way to their destination for that crucial first encounter. Moreover, the eventual mutual recognition of the guest and the host and the stability of the resulting binding depend on the changes in the (thermodynamic) free energy due to the encounter, contact and binding. Such free energy changes can be sensitive to the details of the electrostatic contributions, and it is necessary to understand the electrostatic interactions between a target (*i.e.*, a host such as a cell surface or an ion-exchange resin) and a guest (say, a protein) as a function of the ionic strength, separation distance, orientation and details of the structure of the materials involved (*e.g.*, both the guest and the host may have a mosaic of charged patches; see Figure 8.3). In many cases, a rather involved analysis of diffusional encounters (Brownian motion) in an ionic environment may be necessary, and problems such as these are particularly attractive to chemical engineers as their training often bridges the gap between chemistry and physics.

COLLOIDS AND THE CONSUMMATE GOURMET

Having started with examples from biology, physiology and medicine, it is only natural to move on to foods — the stuff that sustains our physical well-being. Food science is a rich source of colloidal problems and the food industry is a good home for chemical engineers.

Structure and Processing of Food Products

Most food products and food preparations are colloids. They are typically multicomponent and multiphase systems, consisting of colloidal species of different kinds, shapes and sizes and different phases. Ice cream, for example, is a combination of emulsions, foams, particles and gels since it consists of a frozen aqueous phase containing fat droplets, ice crystals, and very small air pockets (microvoids). Salad dressing, special sauce, and the like are complicated emulsions and may contain small surfactant clusters known as micelles. The dimensions of the "particles" in these entities usually cover a rather broad spectrum, ranging from nanometers (typical micellar units) to micrometers (emulsion droplets) or millimeters (foams). Food products may also contain macromolecules (such as proteins) and gels formed from other food particles aggregated by adsorbed protein molecules. The texture (how a food feels to touch or in the mouth) depends on the structure of the food.

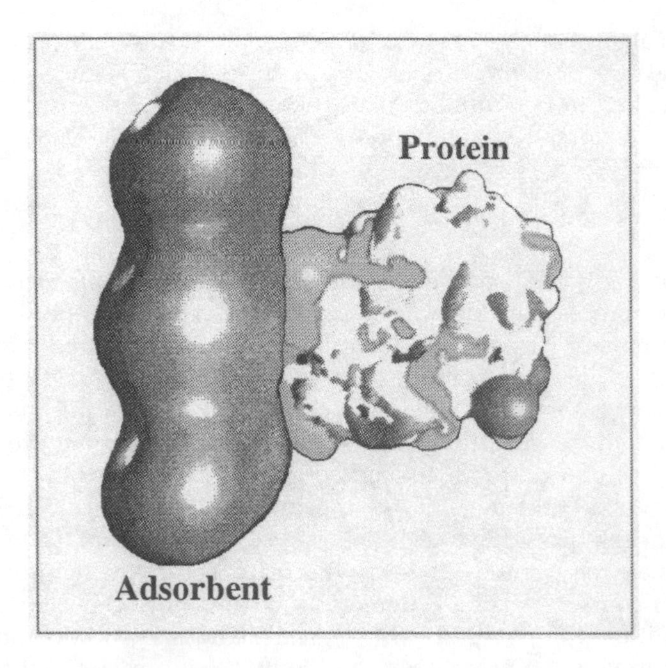

Figure 8.3 The role of electrostatic forces in the "recognition" of a guest molecule by a substrate ("guest/host recognition").
The figure illustrates the preferred orientation (dictated by electrostatic interactions) of a rat cytochrome b5 (a protein) near a simulated anion-exchange resin. The white regions and the regions at different gray scales represent regions of constant electrostatic potentials of different strengths according to the following scheme. The molecular surfaces of the protein and the resin are in an off-white color. The positively-charged, equipotential surfaces (on the protein as well as on the resin) are represented by the dark regions. The negative equipotential surface on the protein is in gray. [See Hiemenz and Rajagopalan (1997) for more details.]

One of the major concerns of an engineer concerned with processing of food products is the physical stability of food colloids with time. Requirements change with the type of food and may depend upon nutritional requirements, governmental regulations or economy of production. For example, products such as margarine, cream, and spreads require specific textural and flow characteristics. Beers may require additives in order to produce froth on dispensing. Sauces and gravies need to maintain their structure at fairly high temperatures. In practical terms, these imply that one is concerned with the microstructure of the constituents, how it varies with chemical additives, temperature, etc., and how to produce products with appropriate "shelf-life" and structural integrity. In general terms these problems are not unlike those encountered in the manufacture of other colloidal products such as paints, face creams, printing inks and toners, for example. The difference, of course, is that the latter are hardly appetizing and one seldom cares about how they taste! (Of course, physical stability alone does not determine the taste or the worthiness of a food product. Chemical reactions play a role in taste and in determining the structure, and other considerations such as nutritional value and esthetic appeal may also apply.)

What Makes a Good Chocolate

Flow behavior of materials, which is technically known as *rheology*, is of major importance in processing operations in industry. The rheological behavior of process streams and dispersions determines the pumping and transportation costs, the ease of mixing operations in reactors and the "quality" of the final product in many cases. What we do not often realize is that it is also one of the factors that determine the esthetic and sensual appeal of certain products, especially in the case of food products and cosmetics. Differences in the "feel" of a face lotion or a skin crème or the "consistency" of ketchup can make or break the market for the product!

Take chocolate for example. The way chocolate feels in the mouth (in addition to its taste, of course) makes a difference in its appeal to chocolate enthusiasts, and much of it has to do with the rheological changes due to the melting of the fat in the mouth. From a processing standpoint, the intricacies of the molds and coating patterns require that the chocolate be free-flowing at high shear stresses but have negligible flow at low stresses; that is, it should be "shear-thinning."

Why are colloids and the rheology of colloids relevant to chocolate? Melted chocolate is a complex, multiphase fluid consisting of solid non-fat particles (about 70% by volume; mostly sugar granules and the rest from crushed cocoa bean) and cocoa butter. (Milk chocolate includes, in addition, some milk solids and milk fat. In both cases, the particle size range is broad and the shapes are irregular.) Melted chocolate is a non-Newtonian fluid that is shear-thinning and has a yield stress. The viscosity could be adjusted by adding more cocoa butter, but that is an expensive proposition. A much less expensive and quite effective way is to add lecithin or a suitable emulsifier made up of small molecules. The rheology of melted chocolate is very sensitive to lecithin, which first reduces the viscosity and yield stress but increases them at larger concentrations. The optimum amount needed is less than one-tenth of the cocoa butter that would be needed for similar results. The current theory is that lecithin adsorbs onto the sugar particles and prevents them from forming agglomerates, thereby influencing the rheological behavior. This is the domain of the colloid scientist, who is more likely to excel in the needed task if he or she is also a chocolate enthusiast!

COLLOIDS FOR MATERIAL CONVENIENCE

Colloidal and macromolecular materials also make possible the development of materials with special properties, often known simply as "advanced materials."

For example, it is possible to develop materials that are very light, like aluminum, but extremely strong, like steel. Such developments depend on our ability to control the microstructure of the new materials, or to generate intricate microstructure, by clever processing techniques. Let us consider a couple of examples next.

High-Strength Ceramics to Replace Metals

Considerable recent activity in the area of ceramic processing is aimed toward the formulation of materials with large strengths, comparable to the room-temperature strength of metal alloys, at high temperatures (of the order of 2000 K). The impetus comes from the significant gains made in the last twenty years with materials formed from submicron powders of silicon nitride and silicon carbide and the promise of similar improvements in the near future. The problems with the ceramics lie as much in reproducibility as in absolute strength. With brittle materials, microscopic flaws concentrate stress and cause failure at stresses, which vary according to the classical crack theory as $\delta^{-1/2}$, where δ is the characteristic dimension of the flaw. For materials formed from powders, the flaws represent structural imperfections, introduced in the initial forming stage, which survive the subsequent processing steps. Hence variability in the initial powder or incomplete control of the fabrication steps leads to unreliable performance.

Of the many methods that are explored, one class of techniques relies on what is known as the colloidal processing route. Here, one starts with a dispersion of precursor powders and densifies the dispersion to an appropriate consistency. The resulting slurry may be slip cast to produce a so-called "green" specimen in the required shape of the product. Slip casting employs a porous mold, which imbibes the fluid but excludes the particles. The type of operation used may employ drain, pressure, or vacuum casting and is determined by the geometry desired and the time constraints. Regardless of the choice of casting method used, a pressure difference draws the fluid into the mold and leaves a dense solid casting adjacent to the wall and a fluid slip in the interior. The physical mechanism bears a strong resemblance to filtration and sedimentation. A broad sketch of the above process is illustrated in Figure 8.4. The structure of the green body depends on the processing technique and conditions as well as the underlying colloid science. For example, as we already saw, the dispersions are often stabilized using polymer additives so that the van der Waals attractive forces between the particles are masked by polymer "brushes" adsorbed on the particle surfaces. This prevents the formation of loose, fragile aggregates that give rise to large voids and defects in the green body upon consolidation.

The questions of interest to an engineer in this case are: How do the initial concentration, the particle size, and the nature of the interparticle potential affect the structure of the dispersion, the structure of the final specimen, and the processing time? How long does the process take? What kinds of chemical additives are suitable? The permeability and the capillary suction in the mold determine the

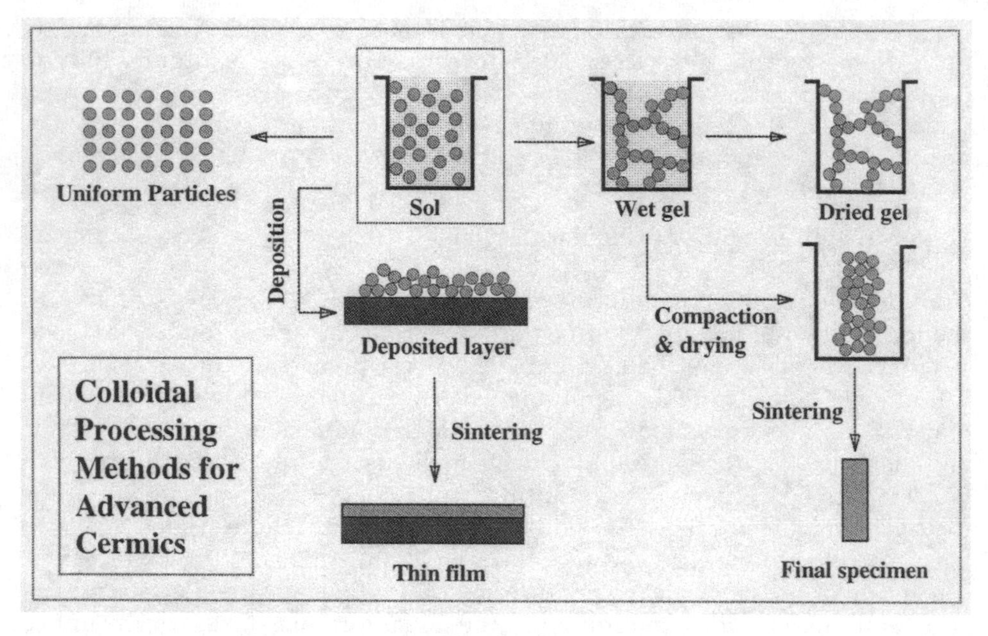

Figure 8.4 A schematic diagram of colloidal processing methods for fabricating ceramic materials with desired properties.
The figure illustrates some of the ways in which a dispersion is densified and transformed into porous or compact films or bulk objects [from Hiemenz and Rajagopalan (1997)].

rate of production of the specimens. How does one adjust the two to optimize production? These questions are likely to engage the attention of chemical and materials engineers in the years to come.

Synthetic Materials with Intricate Microstructure

What works against the formation of controlled and uniform microstructure in the previous example is the aggregation of the particles. And the trick is to devise additives and processing methods to control the *kinetics* of aggregation. Similar things also happen when two immiscible liquids are mixed. All it takes is a simple demonstration with a salad dressing to show that oil and vinegar do not mix. What we have in this case is an *emulsion*. However, if we add a surfactant with molecules containing both oil-liking and vinegar-liking parts to the mixture, the surfactant molecules will preferentially position themselves at oil/vinegar interfaces, very much like the way a surfactant with hydrophilic and hydrophobic parts will behave in an oil/water mixture. One can then have a stable mixture consisting of very small oil droplets dispersed in vinegar or vice versa, depending on the proportion of oil to vinegar. Such *micro*emulsions can exist in very com-

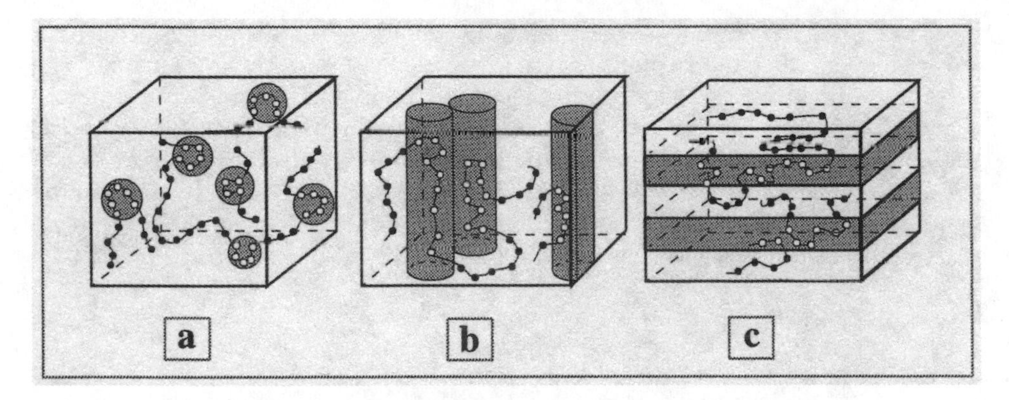

Figure 8.5 The intricate microstructures that can be generated spontaneously in a polymer solution containing polymer chains with two types of polymeric species.
The figure illustrates the spherical, cylindrical and lamellar structures (among others) that are possible in such solutions. Each diblock polymer chain consists of strings of white beads (representing one type of homopolymer) and strings of black beads (representing the second type of homopolymer) [from Hiemenz and Rajagopalan (1997)].

plicated and intricate structures, which can be put to good use in practice. The formation of micelles mentioned in the section on association colloids occurs for similar reasons, and the structures formed are dictated by *thermodynamics* (rather than *kinetics*) and are in general easier to control and reproduce reliably. And, it is always very nice to be able to coax thermodynamics to make materials with intricate structures with special properties!

We have such a situation in a blend of two mutually immiscible polymers [*e.g.*, polymethylbutene (PMB), polyethyl-butene (PEB)]. When mixed, such homopolymers form coarse blends that are nonequilibrium structures (*i.e.*, only kinetically stable, although the time scale for phase separation is extremely large). If we add the corresponding (PEB-PMB) diblock copolymer (*i.e.*, a polymer that has a chain of PEB attached to a chain of PMB) to the mixture, we can produce a rich variety of microstructures of colloidal dimensions. Theoretical predictions show that cylindrical, lamellar and bicontinuous microstructures can be achieved by manipulating the molecular architecture of block-copolymer additives. In fact, even in pure block-copolymer (say, diblock copolymer) solutions the self-association behavior of blocks of each type leads to very useful microstructures (Figure 8.5), analogous to association colloids formed by short-chain surfactants. The optical, electrical and mechanical properties of such composites can be significantly different from those of conventional polymer blends (usually simple spherical dispersions). Conventional blends are formed by quenching processes and result in coarse composites; in contrast, the above materials result from equilibrium structures and reversible phase transitions, and therefore could lead to "smart materials" capable of responding to suitable external stimuli.

As we noted earlier, one advantage of the colloidal structures described above is that they are dictated by thermodynamics (*i.e., not* by processing history). The resulting structures are therefore thermodynamically stable and can usually be obtained reproducibly, without any extra effort that may be needed if a specific sequence of time-dependent processing is required. Chemical engineering has a significant role in the development of such materials through involvement in the synthesis of polymers with the required functional groups, in the large-scale production of the polymers and in the processing of the composites.

COLLOIDS ADVANCE TECHNOLOGY

We discussed the flow characteristics of chocolate in an earlier section. The "particles" in chocolate do not carry any significant amount of charges. But if we are faced with charge-carrying particles, the situation can be quite different and even more interesting. The flow characteristics of charged particles open up a whole class of new phenomena known as *electrokinetic phenomena*. Let us take a look at a couple of applications of such phenomena in modern technology.

Electrophotography

We all know what a Xerox machine is. Xerox machines use what are known as photoconductive insulating surfaces (*i.e.*, surfaces that acquire charges when exposed to light) to produce electrostatic images of the document copied — a process that was suggested over fifty years back. The images thus produced are "latent"; *i.e.*, they are formed by a collection of charges in the form of the images. You cannot see them until they are "developed" using toner particles of colloidal size that stick to the charged images and are subsequently transferred to paper. This is the basis of *xerography*, which is a "dry" copying technique. (A modification of this process in which paper with a coating of a photoconductive layer containing a ZnO binder is used directly is known as *electrofax*. It is seldom used today.)

There are a number of advantages for using *liquid*-developing processes instead of "dry" processes: (i) the possibility of high resolution using fine-grained suspensions, (ii) minimizing "edge effect" (namely, the partially higher image density caused by stronger electric field in the edge regions of the latent image) through the control of the conductivity of the developer, and (iii) the possibility of compact equipment design. The liquid-development process was first proposed in the 1950s and seeks to develop the latent image by "electrophoretic" particle deposition from a colloidal dispersion, *i.e.*, moving the charged particles using an electric field and depositing them on suitable surfaces.

Let us focus here on some of the general issues that we should be concerned with in this context: (i) The toner must have stable polarity and stable charges on the particles and must be stable against settling; (ii) The liquid must have low ion concentration and a low dielectric constant to avoid "electrical leakage" of the electrostatic image; and (iii) Color, concentration and distribution of the pigment particles inside the toner particles must be such that the quality of the developed image is high. In order to guarantee these, information on physicochemical properties of the liquid developer (*e.g.*, viscosity, dielectric constant, *etc.*), the mechanisms responsible for the charges on the toner particles, and the interaction forces responsible for stabilization (usually steric hindrance and polymer-mediated forces arising from polymer additives, as described in the section on colloidal forces) must be examined. These factors also determine the electrophoretic mobility of the particles and are, therefore, important since the electrophoretic mobility, in turn, determines the image density and the developing time. Colloids and colloidal phenomena are thus central to imaging science and technology of the kind described here as well as others.

Colloid-Based Imaging Devices

In a sense, the application discussed in the previous section is almost "classical." Colloids also hold a considerable potential for applications that are unusual in the classical sense. Most of us are familiar with imaging devices such as the picture tube in a television set. These tubes are bulky and consume large amounts of electrical power. There is, therefore, a large incentive to develop compact imaging devices, known as flat-panel devices, that are easily portable and have lower power requirements. (Displays based on liquid-crystal technology fall into this class.)

Another possible flat-panel device is one known as electrophoretic image display, or EPID. EPIDs contain submicron-sized particles of pigments dispersed in a liquid along with a dye that provides contrast. When an electrical potential is applied to the system, pigment particles are driven to the interface between the suspending liquid and a viewing plate, usually made of glass. There they can be seen under normal illumination. EPIDs have the potential of providing an image that has extremely high optical contrast under normal lighting, that is legible over a wide range of viewing angles, that is inherently retained on the display (as opposed to an image needing constant refreshing as on a TV picture tube), and that requires low voltage and power.

The EPID concept can be combined with other developments in imaging technology to produce optical devices such as light valves and X-ray imagers. An example of an electrophoretic X-ray imager is illustrated in Figure 8.6. One of the most crucial aspects of this "chemical" display technology is the liquid dispersion of pigment particles and dye. The dispersion must be stable, even when the particles are compressed (to concentrations as much as ten times higher than in the bulk) to form an image at the viewing plate. The pigment particles must be able to retain their charges after numerous switching operations. The fluid must allow fast response time. Unwanted migration of particles when an electrical field is applied and other electrohydrodynamic effects must be controlled.

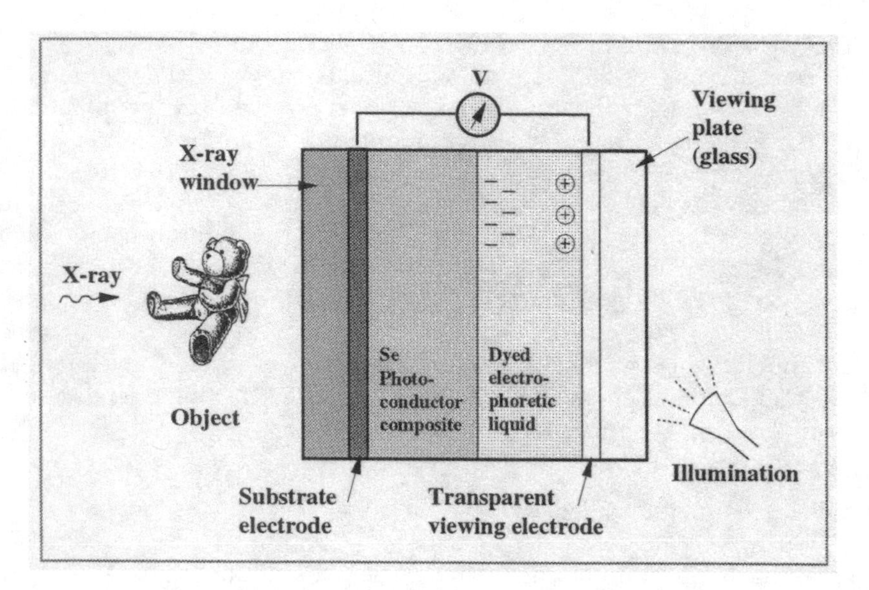

Figure 8.6 A large-area electrophoretic image display (based on colloidal dispersions) with a solid-state X-ray receptor.
When a voltage is applied across the image cell, pigment particles and counterions in the liquid separate. Most of the voltage drop occurs across the selinium (Se) layer. X-ray exposure under this condition leads to the creation of a charge-image at the photoconductor-composite/liquid interface due to the generation of X-ray-induced charges in the Se. After the X-ray exposure, the applied voltage is reduced to zero, and the pigment particles are driven to the viewing plate. The image becomes visible upon illumination [from Hiemenz and Rajagopalan (1997)].

Many of the crucial problems for research and development in this area are the same as those encountered in other areas of colloid and interfacial phenomena. The questions that need to be addressed are: How do particles interact when they are repeatedly and forcefully packed and unpacked in electrical fields? How are particles, counterions, polymers, and surfactants transported across an EPID cell? How does fluid motion in an EPID device affect this overall transport? How fast do structures formed by charged particles dissipate in applied fields? Formally and conceptually, these questions are not unlike many of the problems of interest in chemical engineering.

COLLOIDS TO MANAGE THE ENVIRONMENT

Preserving the environment requires two important steps: One is to develop treatment techniques for wastes that are produced by our processing industries. The other, where the chemical engineers have a special role to play, is to eliminate or minimize production of waste in the processing step itself, so that post-production cleaning operations (known often as "downstream processing") are unnecessary. Eliminating or minimizing production of waste in the processing stage has a catchy name now and is called "environmentally friendly" (or "environmentally benign") processing. We shall take a look at both types of environmental treatments, using examples that highlight the use of colloidal phenomena. Before we go to these examples, let us look at one case that illustrates how colloids play a role in transporting contaminants.

Colloid-Enhanced Transport in the Subsurface

Contaminated bed sediments exist at numerous locations in the United States and around the world. These sediments result mainly from past indiscriminate pollution of our aquatic environments and consist of freshwater and marine bodies including streams, lakes, wetlands and estuaries. The bed sediments contain many hydrophobic organic compounds and metal ions that in the course of time act as sources of pollutants to the overlying aqueous phase. There are a number of transport pathways by which pollutants are transferred to the aqueous phase from contaminated sediments. One of the lesser-known, but potentially important, modes of transport of pollutants from bed sediments is by diffusion and advection of contaminants associated with colloidal-sized dissolved macromolecules in pure water. These colloids are measured in the aqueous phase as Dissolved Organic Compounds (DOCs). (These are defined operationally as particles with diameter smaller than 0.45 micrometer.)

The facilitated transport of compounds by colloids, illustrated schematically in Figure 8.7, is important in several areas and especially in the study of the fate and transport processes of hydrophobic organic compounds and metal ions in the environment. This also has implications in other areas where colloid diffusion through porous media is important, such as those processes utilizing porous sorbents to clean up water and *in situ* flushing of subsurface soil with surfactant solutions for both oil recovery and cleanup of soils contaminated with oil. The understanding of facilitated transport of hydrophobic compounds from unconsolidated media such as bed sediments and soils is therefore very important and requires a knowledge of colloid and surface science of pollutant/particle interactions and transport. In general, colloids are known to have a large capacity to bind hydrophobic compounds and have been reported to transport contaminants to very large distances in ground water. It has been shown using simple mathematical models that orders of magnitude larger flux of contaminants is possible in the presence of colloids as compared to simple molecular diffusion of contaminants in the absence of any association with colloids. It is also known that contaminant binding to colloids is a fast, equilibrium process with partition constants larger than that between the bed sediment and water. In the literature there exists some information on the possible effects of colloids on the transport

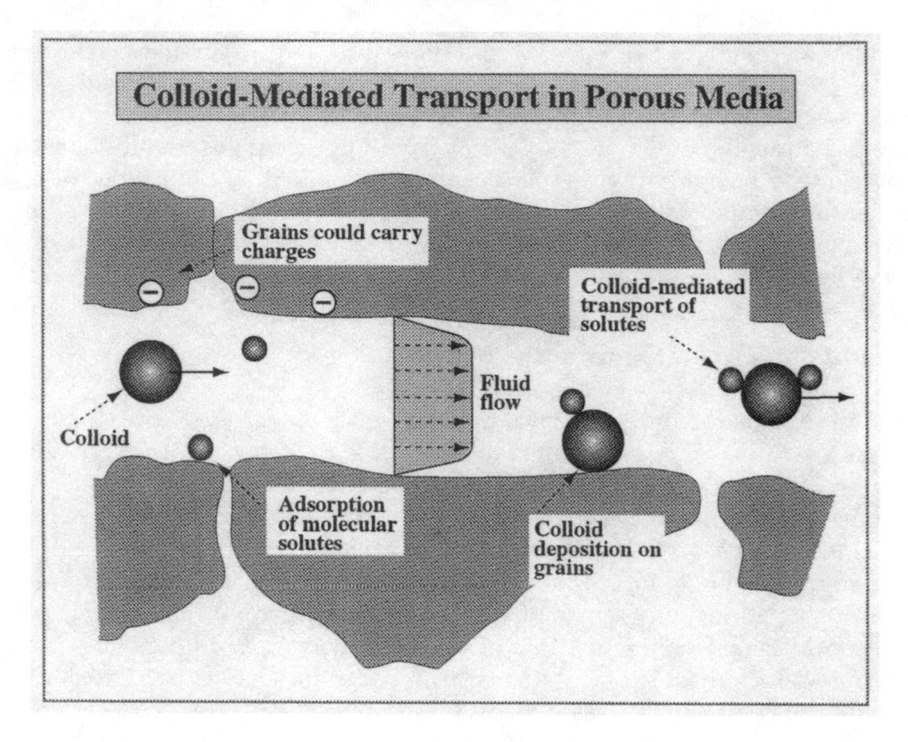

Figure 8.7 A schematic illustration of colloid-mediated transport in porous media. The sketch illustrates the transport of molecular solutes by colloidal particles. The extent of such transport and its importance are determined by a number of factors such as the extent of adsorption of molecular solutes on the colloids and on the grains, the deposition and retention of colloids in the pores, the influence of charges on the colloids and on the pore walls, *etc.* [from Hiemenz and Rajagopalan (1997)].

of pollutants from bed sediments. However, detailed studies on the flux of contaminants by colloids under controlled laboratory conditions are not available. This example, in fact, captures many of the concepts and phenomena a chemical engineer comes across routinely.

Micellar-Enhanced Ultrafiltration

The above example highlights the fact that removal of dissolved organics and polyvalent cations from contaminated groundwater, industrial wastewater and oil-field-produced water in a cost-effective manner is likely to be one of the most important engineering challenges of the future. Hazardous-waste sites contain groundwater polluted by leaking underground organic storage tanks, and the leaching of the hazardous materials and accidental waste releases contribute further to the water pollution problems. The use of water in chemical industries in separation operations is common, and the water thus used picks up at least small amounts of dissolved organics and other hazardous chemicals.

The current technologies popular for removing such wastes include adsorption on activated carbon, ion-exchange membranes, and bioremediation. Activated-carbon adsorption is effective but expensive since regeneration of the carbon is often difficult and expensive. Ion-exchange membranes are restricted to specific ions. An alternative technology that is currently investigated relies on the ability of surfactants to self-assemble in water as micelles and trap hydrocarbon contaminants. We have already seen an example of encapsulation of chemical species by surfactant aggregates known as liposomes in an earlier section. Micellar encapsulation works on similar principles and may turn out to be simultaneously effective for both dissolved organics and polyvalent ions, since an ionic micelle with an appropriate headgroup can adsorb ionic contaminants.

"Biologically friendly" ionic surfactants can be added to the wastewater at concentrations above the threshold value beyond which the surfactants self-assemble to form micelles. The resulting micelles can trap the hydrocarbon wastes since the hydrocarbon solutes prefer the hydrocarbon interior of the micelle over the aqueous environment outside. In addition, ionic wastes in the water adsorb to the polar heads of the surfactants (see Figure 8.8). The resulting waste-laden micelles can then be removed more easily using ultrafiltration methods. Such a process, known as micellar-enhanced ultrafiltration, can be made continuous, scalable, cost-effective and environmentally friendly (through the use of biodegradable surfactants).

Environmentally Friendly Processing Methods

Perhaps the term steric stabilization normally brings to mind dispersions such as paints and food colloids, which are usually stabilized against coagulation by a layer (sometimes a "brush") of polymer chains that mask out the van der Waals attraction at short inter-particle separations (recall the discussion in the section on colloidal forces). However, there are other applications where steric stabilization (and the steric effect, in general) provides interesting and exciting possibilities. We have already come across the role of the steric effect in the case of the so-called stealth liposomes as drug delivery vehicles. Here, let us look at another, namely, the use of specially designed polymeric surfactants ("Designer" surfactants?) for manufacturing polymer particles in an environmentally friendly fashion!

First, let us look at the classical manufacturing route to the large-scale production of many polymers of commercial importance [*e.g.*, polystyrene, poly(vinyl chlorine), poly(acrylic acid), *etc.*]. These polymers are synthesized typically with water or an organic solvent as the dispersing medium (depending on whether the polymer is water-insoluble or water-soluble, respectively). This is a heterogeneous polymerization process that has two or more phases with the monomer

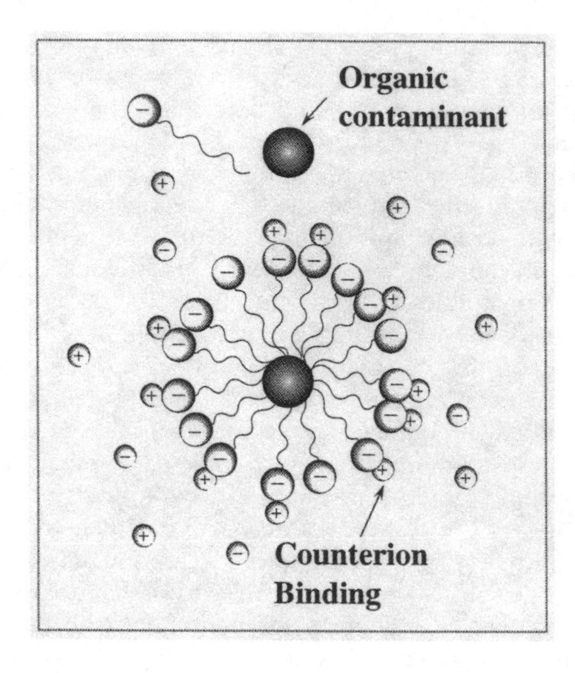

Figure 8.8 Use of micelles to trap hydrocarbon and ionic wastes.
Such "encapsulation" of wastes in micelles can be used to remove the waste materials (*e.g.*, hydrocarbons or ionic wastes) from water by filtering out the waste-laden micelles [from Hiemenz and Rajagopalan (1997)].

or the polymer (or both) in a finely divided form. The final product, in the form of a powder or a dispersion, is then used in subsequent fabrication steps to produce the end products in desired shapes through compaction, pressing, *etc.* The particle sizes in the polymerization step are usually controlled by the addition of suitable surfactants that stabilize the particles against coagulation to prevent a large size distribution. To begin with, this is already an example of steric stabilization in action! But what is wrong with it? The problem is that this processing route produces a large amount of hazardous waste, namely, either contaminated water or organic solvents (such as chlorinated hydrocarbons and toluene), or both.

How can we minimize the damage to the environment? Common sense says that prevention is better than cure. If one can use an environmentally friendly solvent in the processing step, the discharge of hazardous chemicals and the need for environmental post-treatment can be avoided. This is an approach that is currently being explored in polymer processing as well as in other manufacturing technologies. In the case of the polymerization operations, one possibility is to use supercritical CO_2, which resembles a liquid but has low viscosity, as the solvent. However, most polymers, with the exception of fluoropolymers, do not

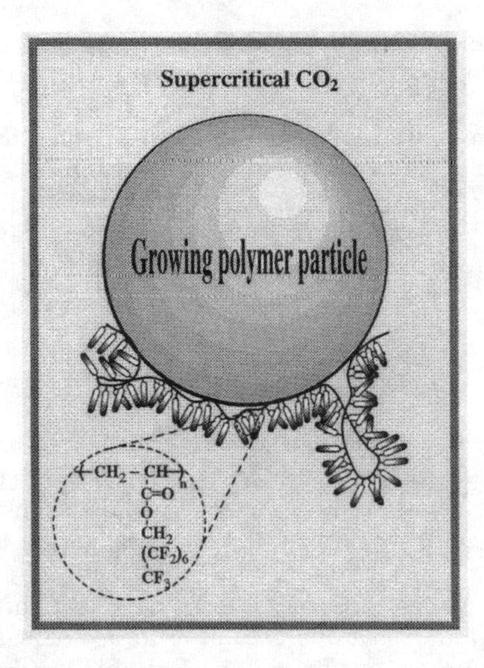

Figure 8.9 An example of an environmentally benign polymer processing technique. A polymer particle in supercritical CO_2 is protected against coagulation by the steric action of a specially designed polymer chain with fluorine-containing tails on the polymer backbone [see Hiemenz and Rajagopalan (1997) for more details].

dissolve in CO_2. Here is where specially designed surfactants come in. What one needs is a molecularly engineered polymer surfactant which has a backbone that can adsorb on the growing polymer particle but with side chains or "tentacles" that are CO_2-loving so that the growing polymer particles can stay dispersed in the solvent (see Figure 8.9). In addition, adsorption of the surfactant backbone anchors the polymer on the surface of the particles and the loops and the tails of the polymer provide a steric barrier against coagulation as desired. Recent research shows that such approaches are not just feasible but are likely to open up promising new avenues for other processing operations as well.

PEERING INTO NANO-WORLDS: NEW TOOLS TO SEE ATOMS AND TO MEASURE ATOMIC FORCES

We noted in the introduction to this chapter that one exciting aspect of modern colloid science is the development of an ever-expanding array of instruments that allow us to probe materials at atomic scales. The present chapter will not be complete if we do not take a look at, at least one or two examples of the above development. Let us, therefore, look briefly at two typical examples: (i) a class of techniques known as *scanning-probe microscopes* that can be used to image surfaces of all kinds and to measure forces down to the atomic levels, and (ii) an apparatus known as the *surface force apparatus*, which has led to a level of understanding of static and dynamic surface forces in colloidal and biological systems that was previously unimaginable. These techniques have also been extended to the study of frictional forces and rheology of thin films; these extensions and a brief history of such measurements are highlighted in a recent book (Meyer *et al.* 1998)

Scanning-Probe Microscopes: Probing Surfaces at the Atomic Scale

Speculations about the existence of atoms are often traced to the ancients, but a fierce (and sometimes vitriolic) debate was being waged in scientific circles about the reality of atoms even as recently as the dawn of the twentieth century. In fact, Ludwig Boltzmann's depression that drove him to suicide in 1906 is partly attributed to the criticism he faced from his detractors for his staunch advocacy of the atomic theory of matter. Who, then, would have thought that before the end of the century we would have the capability to not only "see" atoms but to manipulate and move them one by one on a surface? Yet that is what has been made possible by the ingenuity of Gerd Binnig and Heinrich Rohrer, two physicists at the IBM Zurich Research Laboratory, who invented what is known as the *scanning tunneling microscope* (STM) in 1982 — an invention for which they shared the 1986 Nobel Prize in Physics (with Ernst Ruska of Germany, the inventor of the electron microscope).

Scanning Tunneling Microscopy

When a probe, usually a metal tip down to the size of an atom, is brought to within a nanometer from the surface of a conductor or a semiconductor (see Figure 8.10), electrons can be made to jump from the surface to the tip by applying what is known as a bias potential; this is called the *tunneling* effect, and the resulting current is known as the tunneling current. By keeping the tunneling current constant as one moves the probe along a surface, one can follow the surface contour with atomic scale precision! In the constant-current mode of operation, for example, the vertical displacement of the probe can be mapped with Angstrom-level precision using the deflections of the piezo-electric ceramic piece on which the metal tip is mounted. The data can be converted to images to allow one to "see" the surface (see Figure 8.10, inset). In fact, what is mapped is not the topography of the surface but contours of constant electron densities; that is, the results are sensitive to the type of atom "seen" on the surface by the probe. This implies that the STM can serve as a powerful and highly specific and sensitive local probe to study a surface (and materials adsorbed on a surface) with high precision.

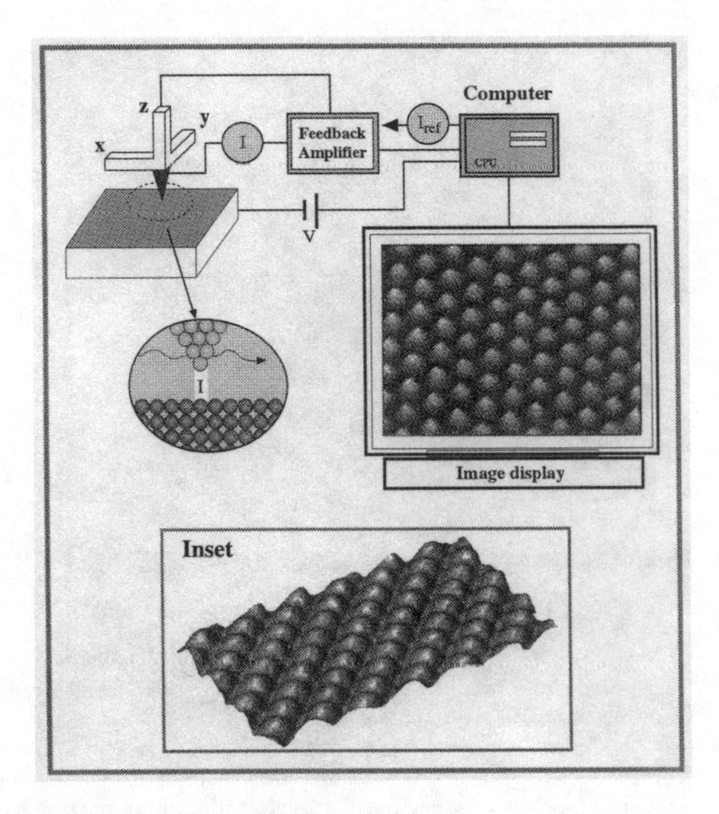

Figure 8.10 Scanning tunneling microscopy.
The figure illustrates one type of operation of a scanning tunneling microscope (STM). A tunneling current I flows between the sharp tip of the probe and the surface when a bias voltage V is applied to the sample. A computer monitors the tunneling current and adjusts the distance between the probe and the surface such that a constant tunneling current, I_{ref}, is maintained. The resulting changes in the position of the tip are then recorded and converted to an image, such as the one shown on the monitor or the one shown in the inset. The image shown in the inset is that of an atomically smooth nickel surface. The periodic arrangement of the atoms on the surface can be seen clearly in the STM image [see Hiemenz and Rajagopalan (1997) for more details].

Atomic Force Microscopy

Vacuum tunneling of electrons is not the only option, however. The invention of STM has triggered the development of a whole family of techniques known collectively as scanning-probe microscopy (SPM) that are not restricted to tunneling as the mechanism or to only conducting or semiconducting surfaces. One type of SPM, known as *atomic force microscopy* (AFM), can be used to measure forces such as van der Waals forces, ion-ion interactions, and hydration forces, on polymer-coated surfaces, biologically relevant bilayers, *etc.*

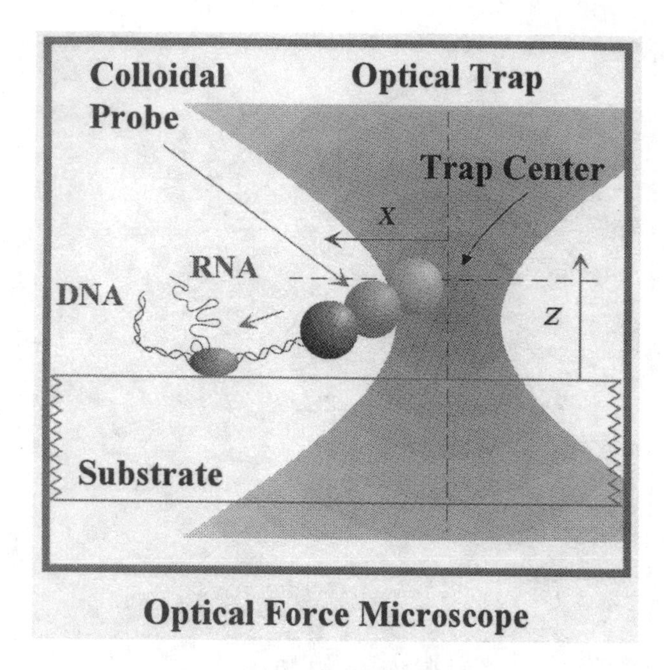

Figure 8.11 A schematic representation of an optically held probe particle near a substrate, as used in a scanning optical force microscope.
The figure illustrates the probe particle being dragged away from the trap center by the copying action of ribonucleic acid (RNA), attached to the substrate, on a deoxyribonucleic acid (DNA) chain. The mechanical force needed to stop the copying action can be determined by measuring the departure of the colloidal probe from the trap center. Such information is of value to scientists in understanding biochemical phenomena.

In the force-measuring mode, an SPM is often referred to as an *SFM – scanning force* microscopy, but the label SPM signifies the more general capability of these devices. For example, atoms on surfaces can be manipulated and rearranged individually using STM and SPM, and conformation of polymers and biological macromolecules on surfaces can be probed. In addition, the STM can be combined with suitable spectroscopic techniques to design powerful local probes of a surface to a precision not possible through other techniques; these are known as scanning-probe *spectroscopy*. Because of its precision, the STM/SPM family can provide details that are often missed by other techniques. The development of SPM's has significantly changed the "landscape" of the instrumentation available to colloid and surface scientists, and the possibilities are endless![5]

[5] A short volume by Milburn (1997), on "Schrodinger's Machines," has an excellent introduction to scanning tunneling microscopy and its potential applications to building "machines" at the nanoscale and is highly recommended.

Optical Force Microscopy

As should be clear by now, members of the family of scanning probe microscopes differ from each other with respect to the type of probe used and the type of substrate/probe interaction that is used as the basis of the measurement. The sensitivity of the force measurement and the resolution of the images obtained depend on the characteristics of the probe and the probe/substrate interactions, among others. The stiffness of a typical AFM cantilever is typically relatively large, and, therefore, an AFM is not always suitable for studying very "soft" surfaces. For example, when one is interested in understanding weak adsorption of a single polymer chain, one needs probes that are much less stiff than the cantilever of an AFM. One exciting possibility that has now become possible for this is the use of a submicron particle held by the radiation pressure of a laser beam as a probe (see Figure 8.11), thanks to the optical manipulation of particles developed by Arthur Ashkin of AT&T (see Ashkin 1997).[6] The resulting device is known as *optical force microscopy* (OFM). In the OFM, the probe (typically a spherical particle) is virtually a prisoner in the laser beam, but is free to move to a certain extent within the beam as it interacts with the substrate. The focused laser beam functions essentially as an invisible spring holding the particle, with a spring stiffness that can be four orders of magnitude lower than the stiffness of an AFM cantilever. The beam can be moved laterally to move the particle; hence the arrangement shown is known as an "optical tweezer" or a "tweezer trap." Such optical traps can be used to measure femtoNewton-level (10^{-15} N) forces (forces smaller than the weight of a tiny droplet of water in the morning mist). OFM and related optical tweezer-based techniques are leading a quiet revolution in the way we can probe colloidal forces and biopolymers and biophysical processes (see, for example, Sheetz 1998).

Surface Force Apparatus

The surface force apparatus (SFA) was originally developed in the late 1960s for measuring the van der Waals forces between molecularly smooth mica surfaces in air or vacuum. The apparatus, which has since been modified for making measurements in liquids, consists of two curved surfaces of mica, the interaction forces between which are measured using highly sensitive force-measuring springs (see Figure 8.12).[7] The separation between the two surfaces can be measured by use of an optical technique (which employs multiple beam interference fringes)

[6] Ashkin's pioneering work was subsequently developed further by Steven Chu and William D. Phillips of USA and Claude Cohen-Tannoudji of France for capturing and trapping *atoms* using lasers — an achievement that was recognized through the 1997 Nobel Prize in Physics.

[7] Ball (1997) contains an excellent introduction to the role of surface forces in many phenomena of interest to us (see Chapter 10). The book also describes the surface force apparatus and its use.

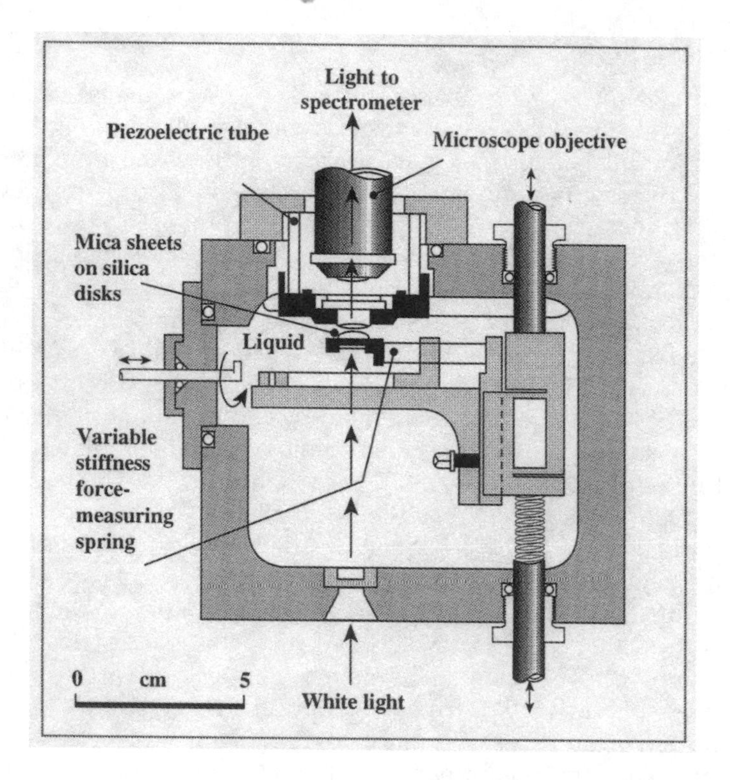

Figure 8.12 A sketch of the surface force apparatus used for measuring colloidal and surface forces directly.
The surface-force apparatus can be used for direct measurements of the forces between surfaces in liquids or vapors. As shown in the sketch, two atomically smooth surfaces immersed in a liquid are brought towards each other, with the surface separation being controlled to 0.1 nm. The distance between the surfaces is measured using an optical technique down to 1 Angstrom (0.1 nm). The force between the two surfaces is determined from the deflection of the lower surface using highly sensitive springs. The surfaces can also be moved laterally past each other to measure shear forces (Israelachvili 1991).

from microns down to molecular contact, usually to better than 0.1 nm. In addition, the exact shapes of the two surfaces and the refractive index of the liquid (or material) between them can also be measured. The ability to measure the refractive index allows one to determine the quantity of material (*e.g.*, lipid or polymer) deposited or adsorbed on the surfaces. The force is measured through the expansion or contraction of the piezoelectric crystal by known amounts and measuring optically how much the two surfaces have actually moved. The difference in the two values is then multiplied by the stiffness of the force-measuring spring to obtain the force difference between the initial and final positions. Thus, both repulsive and attractive forces can be measured as functions of the distance between interacting surfaces, with a sensitivity of the order of 10^{-8} N. In the past mica has been the material of choice for the interacting surfaces because of the ease of handling, and since molecularly smooth surfaces can be fabricated, mica surface coated with a thin film of other materials (*e.g.*, lipid monolayers or bilayers, metal films, polymer films, or other macromolecules such as proteins) can also be used. The use of alternative materials such as molecularly smooth sapphire and silica sheets and carbon and metal oxide surfaces is also being explored.

The invention and refinement of the SFA has been one of the most significant advances in experimental colloid science and has allowed researchers to identify and quantify most of the fundamental interactions occurring between surfaces in aqueous solutions as well as nonaqueous liquids. Attractive van der Waals and repulsive electrostatic double-layer forces, oscillatory (solvation or structural) forces, repulsive hydration forces, attractive hydrophobic forces, steric interactions involving polymeric systems, and capillary and adhesion forces have all been measured using this apparatus or variations of it. Moreover, the scope of phenomena that can be studied using the SFA has also been extended to measurements of dynamic interactions and time-dependent effects such as the viscosity of liquids in very thin films, shear and frictional forces, and the fusion of lipid bilayers.

The impact of the SFA and similar direct measurement of forces goes beyond simple testing of theories of intermolecular forces. For instance, such measurements are also useful for explaining more complex phenomena such as the unexpected stability of certain colloidal dispersions in high salt, the crucial role of hydration and ion-correlation forces in clay swelling and ceramic processing, and the deformed shapes of adhering particles and vesicles (see Israelachvili 1991). Further, both static (*i.e.*, equilibrium) and dynamic forces can now be studied with remarkable precision and accuracy for obtaining information on the structure of liquids adjacent to surfaces and related interfacial phenomena. Such studies demonstrate that properties of ultra-thin films are profoundly different from those of bulk liquids.

CONCLUDING REMARKS

We have seen in this chapter a number of examples of colloids and colloidal phenomena that are important to us in every facet of our lives. Each example poses innumerable questions for scientists and engineers to solve. For a student just beginning to contemplate pursuing a science or engineering degree, it may not be easy to follow all the technical aspects of the examples. However, the examples provide a "feel" for the variety, importance and intellectual challenges of colloid and interfacial phenomena and, we hope, whet the appetite of the aspiring scientists and engineers. The examples also illustrate the inevitable need for an interdisciplinary orientation for those wishing to contribute in emerging areas of science and engineering.

Hendrik Casimir, a prominent Dutch physicist, is said to have remarked (de Gennes and Badoz, 1996):

> "Almost everything we do is written in sand and fades away in the wind. Once in a while, we are fortunate enough to be given a metal tablet on which to inscribe a more permanent message."

There is a lot we know that are established facts in science, but a lot of what we think we know (and a lot of what we have described in this chapter) are "written in sand." And it is left to you — the future generation of scientists — to rewrite what needs to be rewritten!

REFERENCES

Ashkin, A. (1997) *Proc. Natl. Acad. Sci. USA*, **94**, 4853
[*Advanced level.* An overview of optical trapping of atoms and neutral particles by the pioneer who developed the technique. Describes developments in optical manipulation of particles and polymers. Has numerous references to applications in biology, atom trapping and other more recent developments.]

Ball, P. (1994) *Designing the Molecular World: Chemistry at the Frontier.* Princeton, NJ: Princeton Univ. Pr.
[*High school and undergraduate level.* A highly readable, *Scientific American*-style, popular introduction to the many facets of molecular design in chemistry. Chapter 7 is an excellent introduction to self-organization of surfactant molecules into micelles, vesicles, Langmuir layers, and biological cells. This chapter also discusses the formation of polymer gels.]

Ball, P. (1997) *Made to Measure: New Materials for the 21st Century.* Princeton, NJ: Princeton Univ. Pr
[*High school and undergraduate level.* Another highly readable, popular introduction to materials science. Chapter 10 presents a good introduction to surfaces and interfaces and also contains easy-to-understand descriptions of scanning-probe microscopy and the surface force apparatus.]

de Gennes, P.-G. and Badoz, J. (1996) *Fragile Objects: Soft Matter, Hard Science, and the Thrill of Discovery.* New York: Springer-Verlag
[*High school level.* A delightfully personal account of research on polymers and colloids delivered as a series of talks to high school students by the 1991 Physics Nobel laureate Pierre-Gilles de Gennes. In addition to simple, readable descriptions of the fascinating science behind many common examples of polymers and colloids, the book also discusses how science is done, the role of individual and teamwork and the interplay of conscience and knowledge.]

Hiemenz, P. C. and Rajagopalan, R. (1997) *Principles of Colloid and Surface Chemistry*, 3rd edn. New York: Marcel Dekker
[*Undergraduate level.* An introductory textbook on colloid and surface chemistry. Contains numerous other examples of the type discussed in the present chapter.]

Israelachvili, J. N. (1991) *Intermolecular and Surface Forces*, 2nd edn. New York: Academic Pr

[*Undergraduate and graduate level*. The best reference available currently on the topic. Contains numerous examples of the application of surface forces in biological systems. The links between molecular forces and surface forces and the relation between molecular forces and bulk properties of materials are discussed in a manner accessible to advanced undergraduate students.]

Lasic, D. (1992) *American Scientist*, **80**, 20

[*High school and undergraduate level*. A lucid overview of the use of liposomes for medical applications.]

Meyer, E., Overney, R.M., Dransfeld, K. and Gyalog, T. (1998) *Nanoscience: Friction and Rheology at the Nanometer Scale*. Singapore: World Scientific

[*Advanced level*. A historical and technical introduction to the use of scanning-probe microscopes for studying friction and rheology of thin films.]

Milburn, G. J. (1997) *Schrodinger's Machines: The Quantum Technology Reshaping Everyday Life*. New York: W. H. Freeman

[*Undergraduate level*. This popular introduction to new technologies based on quantum mechanics has lucid descriptions of scanning-probe techniques and their use in "atomic calligraphy." There is also a chapter on laser traps for particles and atoms.]

Sheetz, M. P., Editor (1998) *Laser Tweezers in Cell Biology*. San Diego: Academic Pr

[*Advanced level*. A manual on the use of optical tweezers in biology and for imaging.]

Tanford, C. (1989) *Benjamin Franklin Stilled the Waves*. Durham, NC: Duke Univ. Pr

[*High school and undergraduate level*. Provides a nontechnical, popular introduction to surfactants, surfactant films and related self-assembled structures, intertwined with a historic perspective.]

9. BUBBLE COLUMN REACTORS

D. J. LEE and L.-S. FAN

Department of Chemical Engineering, The Ohio State University, 140 West Nineteenth Avenue, Columbus, Ohio 43210, USA

Bubble columns are simple contactors in which a gas or a mixture of gases is introduced at the bottom and rises up in the form of a dispersed phase of bubbles through a continuous liquid phase to provide efficient contact between gas and liquid phases. Various types of bubble column reactors and their modifications are shown in Figure 9.1. The simplest form of a bubble column reactor is a vertical cylinder as shown in Figure 9.1(a). The gas enters at the bottom through a gas distributor. The liquid phase can be operated in batch or continuous form, and may move co-currently or counter-currently with the flow direction of the gas phase. Bubble columns can also be single-staged or multi-staged as shown in Figures 9.1(b) and (c). Often, reactive or catalytic solid particles are suspended or fluidized in the system, *e.g.*, slurry bubble columns and gas-liquid-solid fluidized beds. There are many variations of bubble columns, which offer directional fluid circulation. For example, inserting a tube or divider to stabilize the circulation pattern often forms loop reactors as shown in Figures 9.1(e) to (h). These loop reactors permit the processing of large amounts of gas and provide a homogeneous flow zone.

Some of the advantages offered by bubble columns are the absence of moving parts which eliminates the need for seals, minimal maintenance, small floor space, excellent heat transfer and mixing characteristics, ability to handle solids, low cost, and high values of effective interfacial area and overall mass transfer rate. Bubble columns are thus especially suited for reactions involving high gas flow rate as well as for high-pressure and/or high-temperature processes. On the other hand, a considerable amount of backmixing in the liquid phase is present in a bubble column, which can have a significant effect on the effective reactor volume. When high conversion is desired, a relatively larger reactor volume is necessary. Also, bubble columns may not be feasible when the gas supply is at atmospheric pressure, but bubble columns can be operated efficiently at elevated pressures.

Bubble columns are widely used in chemical, biochemical, and petroleum industries as mass transfer and reaction devices (Fan, 1989) that have been modified in many ways to suit particular applications over the past century, *e.g.*, loop reactors are used in fermentation. For certain reactions such as air oxidation of organic and inorganic compounds, where a high gas component conversion is not necessary and a low pressure-drop is required, a bubble column with a horizontal gas distribution can be used. When complete gas phase conversion and hence a long gas residence time is required, downflow bubble columns, as shown in Figure 9.1(i), can be used. For biotechnological processes where low backmixing is desired to achieve high conversion, a sectionalized bubble column can be used.

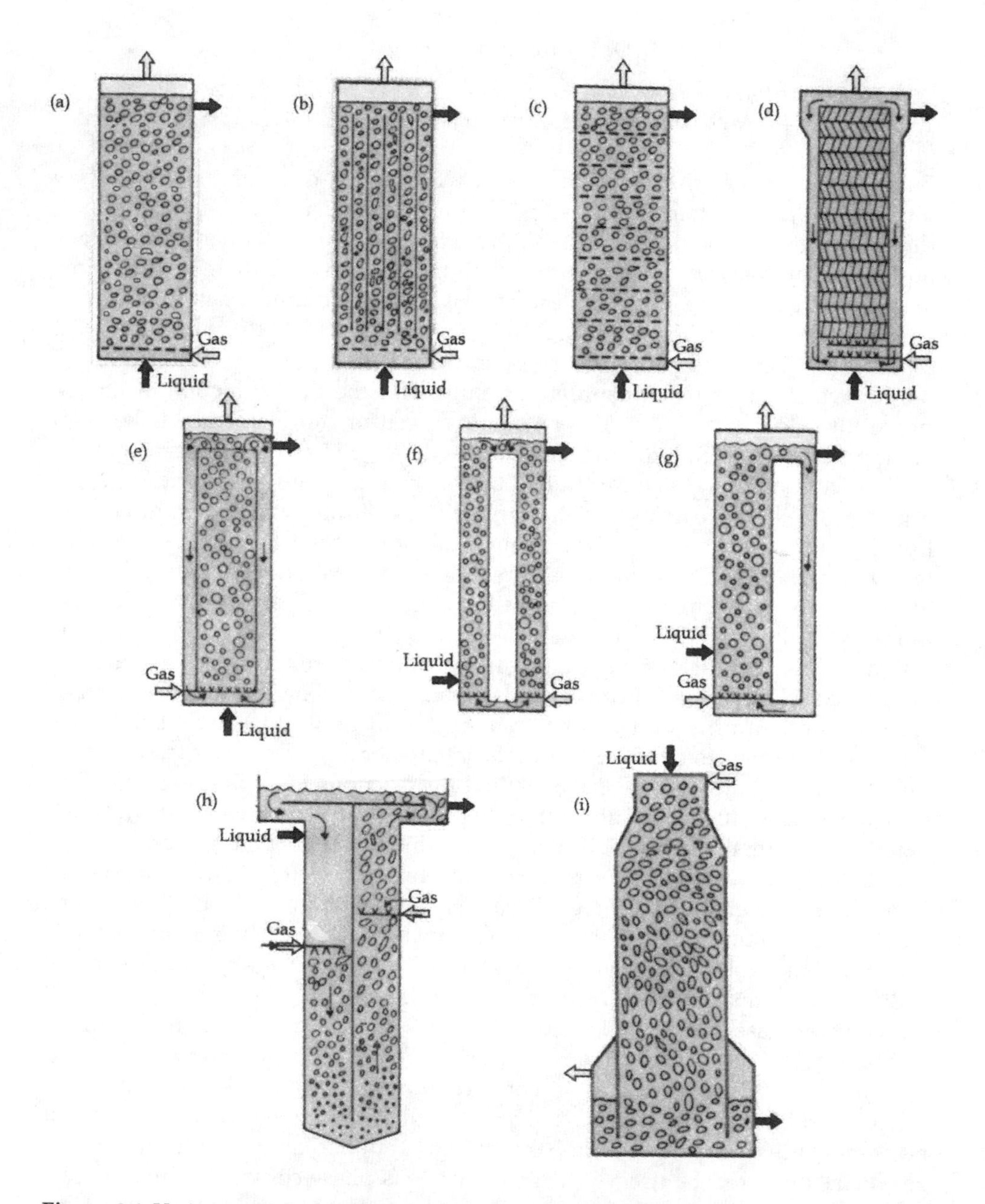

Figure 9.1 Various types of bubble columns and their modifications: (a) Simple bubble column, (b) & (c) Multi-staged bubble column, (d) Bubble column with static mixers, (e), (f), (g) & (h) Loop bubble columns; (i) Downward-flow bubble column.

Table 9.1 Some applications of bubble columns.

Process	*Methods and/or reactants*
Acetone	Oxidation of cumene
Acetic acid	Oxidation of acetaldehyde
	Oxidation of sec-butanol
	Carbonylation of methanol
Acetic anhydride	Oxidation of acetaldehyde
Acetaldehyde	Partial oxidation of ethylene
Acetophenone	Oxidation of ethylbenzene
Barium chloride	Barium sulfide and chlorine
Benzoic acid	Oxidation of toluene
Bleaching powder	Aqueous calcium oxide and chlorine
Bromine	Aqueous sodium bromide and chlorine
Butene	Absorption in aqueous solutions of sulfuric acid
Carbon dioxide	Absorption in ammoniated brine
Carbon tetrachloride	Carbon disulphide and chlorine
Copper oxychloride	Oxidation of cuprous chloride
Cumene	Oxidation of phenol
Cupric chloride	Copper and cupric acid or hydrochloric acid
Dichlorination	Oxychlorination of ethylene
Ethyl benzene	Benzene and ethylene
Hexachlorobenzene	Benzene and chlorine
Hydrogen peroxide	Oxidation of hydroquinone
Isobutylene	Absorption in aqueous solutions of sulfuric acid
Phthalic acid	Oxidation of xylene
Phenol	Oxidation of cumene
Potassium bicarbonate	Aqueous potassium carbonate
Sodium bicarbonate	Aqueous sodium carbonate
Sodium metabisulphite	Carbon dioxide, aqueous sodium carbonate, and sulfur dioxide
Thiuram disulphides	Dithiocarbamates, chlorine, and air
Vinyl acetate	Oxidation of ethylene in acetic acid solutions
Water	Wet oxidation of waste water

Vertically sparged bubble columns are most commonly utilized in the industry. These vertically sparged columns are frequently used as absorbers, strippers, and reactors in various chemical and petrochemical processes. Examples of bubble column uses in the petrochemical industry are hydrocracking of heavy petroleum fractions, coal hydrogenation, gasification of coal, methanol synthesis, and Fischer-Tropsch synthesis. Table 9.1 summarizes some of the applications of bubble columns in organic and inorganic chemical processes for gas-liquid systems.

In this chapter, some fundamental bubble dynamics in liquid media are introduced. Understanding bubble dynamics is critical for bubble column design since the hydrodynamic behaviors are induced and dictated by the gas phase in the form of bubbles. Furthermore, important design parameters for bubble column reactors, including flow structures and flow regimes, phase holdup, and mass and heat transfer characteristics, are discussed. The ultimate design of multiphase

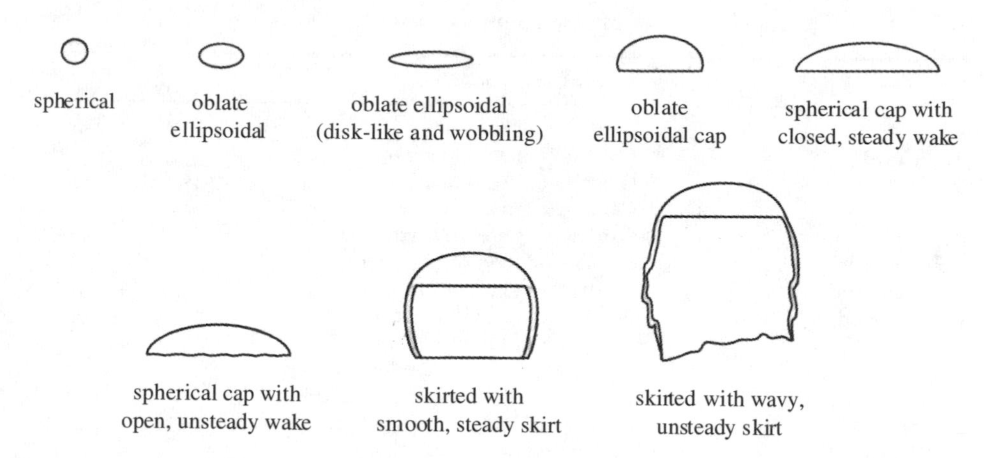

Figure 9.2 Sketches of various bubble shapes.

systems for industrial applications can be strengthened with comprehensive knowledge of the above-mentioned fluid dynamic/transport phenomena.

BUBBLE DYNAMICS

Complex hydrodynamic behavior in bubble columns is closely related to the characteristics of rising bubbles and bubble wake structures. A rising bubble can be described in terms of the rise velocity, shape and motion of the bubble. These rise characteristics are closely associated with the flow and physical properties (mainly viscosity and presence/absence of solid particles) of the surrounding medium, as well as the interfacial properties (*i.e.* presence/absence of surfactant) of the bubble surface.

Bubble Geometry and Motion

The interaction between a rising bubble and the surrounding liquid medium produces the bubble shape and determines the extent of the disturbance in the surrounding flow field. Bubbles in motion are generally classified by shape as spherical, oblate ellipsoidal, and spherical-cap (see Figure 9.2) with the actual shape depending upon the dominating forces, such as surface tension, viscous, or inertial force acting on a bubble.

For small bubbles, *i.e.* diameter of volume-equivalent sphere (d_b) less than 0.1 cm in water, the bubble shape is approximately a sphere, the surface tension dominates, and the bubble rises steadily in a rectilinear path. For intermediate size bubbles, effects of both surface tension and inertia of liquid flowing around

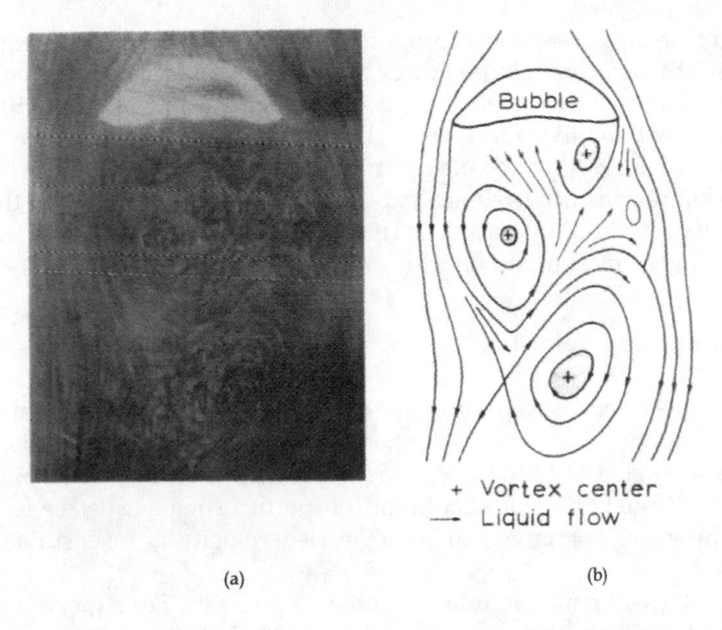

Figure 9.3 Bubble wake phenomena: (a) Photograph of a circular-cap bubble rising in stationary water and its wake visualized through hydrogen bubble tracers and (b) Schematic interpretation of the wake flow (Tsuchiya and Fan, 1988, reproduced with permission).

the bubble are important and liquid viscosity and the presence of surface-active contaminants influence the bubble dynamics. The motion of intermediate size bubbles is extremely complex; ellipsoidal bubbles exhibit secondary motion characteristics as they rise, such as zigzag or spiral trajectories and shape dilations or oscillations. For large bubbles, *i.e.* $d_b > 2$ cm, the inertial effect of flows surrounding the bubble predominates, and the effects of surface tension, viscosity, and impurities of the liquid medium are negligible. Large bubbles have spherical-cap shape and follow a rectilinear path with some rocking and/or base oscillations.

Bubble Wake

In gas-liquid contacting systems, the velocity distribution in the wake region controls the diffusion/dispersion of gas from bubbles into liquid media. Further, bubbles induce liquid mixing by the process of entrapping liquid into and shedding it from their wakes. Figure 9.3(a) shows a photograph of a relatively large two-dimensional nitrogen bubble rising in water. The schematic diagram shown in Figure 9.3(b) indicates two regions. The primary wake region includes two vortices; on the left-hand side is well-established circulatory motion and on the right-hand side the vortex is just forming. Outside the primary wake region, there exists a slightly deformed, large vortex, which is isolated by streams of external flow across the wake from right to left. As seen in the figures, the solids concentration varies within the wake; lower solids concentration regions were observed immediately beneath the bubble base and around the vortex center, while higher concentration regions occurred around the vortices and especially in the regions where the two vortices interact. The dynamic behavior of bubble wake is thoroughly discussed in Fan and Tsuchiya (1990).

Single-bubble Rise Velocity

The bubble rise velocity, u_b, is the single most critical parameter in characterizing the hydrodynamics and transport phenomena of bubbles in liquids. The rise velocity of a single gas bubble depends first on its size. For small bubbles, the rise velocity strongly depends on liquid properties such as surface tension and viscosity; however, for large bubbles, the rise velocity is insensitive to liquid properties.

Pressure plays a significant role in bubble rise velocity. For a given bubble size, u_b tends to decrease with increasing pressure. The effects of pressure and temperature, or more directly, the effects of physical properties of the gas and liquid phases on the variation of u_b could be represented or predicted most generally by the following three predictive equations. One is the Fan-Tsuchiya equation (Fan and Tsuchiya, 1990). The other two are the modified Mendelson's wave-analogy equation (Mendelson, 1967) by Maneri (1995) and a correlation proposed by Tomiyama *et al.* (1995).

The Fan-Tsuchiya equation can be written in a dimensionless form:

$$u_b' = u_b\left(\frac{\rho_l}{\sigma g}\right)^{1/4} = \left\{\left[\frac{Mo^{-1/4}}{K_b}\left(\frac{\Delta\rho}{\rho_l}\right)^{5/4} d_b'^2\right]^{-n} + \left[\frac{2c}{d_b'} + \left(\frac{\Delta\rho}{\rho_l}\right)\frac{d_b'}{2}\right]^{-n/2}\right\}^{-1/n} \tag{9.1}$$

where the dimensionless bubble diameter is given by

$$d_b' = d_b(\rho_l g / \sigma)^{1/2} \tag{9.2}$$

and d_b is the volume-equivalent spherical diameter of the bubble. Three empirical parameters, n, c, and K_b, in Equation 9.1 reflect three specific factors governing the rate of bubble rise. They relate to the contamination level of the liquid phase, the varying dynamic effects of the surface tension, and the viscous nature of the surrounding medium. The suggested values of these parameters are:

$$n = \begin{cases} 0.8 & \text{for contaminated liquids} \\ 1.6 & \text{for purified liquids} \end{cases} \tag{9.3a}$$

$$c = \begin{cases} 1.2 & \text{for monocomponent liquids} \\ 1.4 & \text{for multicomponent liquids} \end{cases} \tag{9.3b}$$

and

$$K_b = \max\left(K_{b0} Mo^{-0.038}, 12\right) \tag{9.3c}$$

where

$$K_{b0} = \begin{cases} 14.7 & \text{for aqueous solutions} \\ 10.2 & \text{for organic solvents / mixtures} \end{cases} \tag{9.3d}$$

The modified Mendelson's equation (Maneri, 1995) is a special form of the Fan-Tsuchiya equation where the viscous term, *i.e.* the first term from the right side of Equation 9.1, is omitted. Equally as general as the Fan-Tsuchiya equation for bubbles in liquids, the correlation by Tomiyama *et al.* (1995), which is given in terms of drag coefficient,

$$C_D = \frac{4}{3} \frac{g \Delta\rho \, d_b}{\rho_l u_b^2} \tag{9.4}$$

consists of three equations used for different system purities:

$$C_D = \max\left\{ \min\left[\frac{16}{Re}\left(1 + 0.15 Re^{0.687}\right), \frac{48}{Re} \right], \frac{8}{3} \frac{Eo}{Eo + 4} \right\} \tag{9.5a}$$

for purified systems;

$$C_D = \max\left\{ \min\left[\frac{24}{Re}\left(1 + 0.15 Re^{0.687}\right), \frac{72}{Re} \right], \frac{8}{3} \frac{Eo}{Eo + 4} \right\} \tag{9.5b}$$

for partially contaminated systems; and

$$C_D = \max\left[\frac{24}{Re}\left(1 + 0.15 \mathrm{Re}^{0.687}\right), \frac{8}{3} \frac{Eo}{Eo + 4} \right] \tag{9.5c}$$

for sufficiently contaminated systems. In Equations 9.1 and 9.5, the dimensionless groups are defined as

$$Mo = \frac{g \Delta\rho \, \mu_l^4}{\rho_l^2 \sigma^3} \tag{9.6a}$$

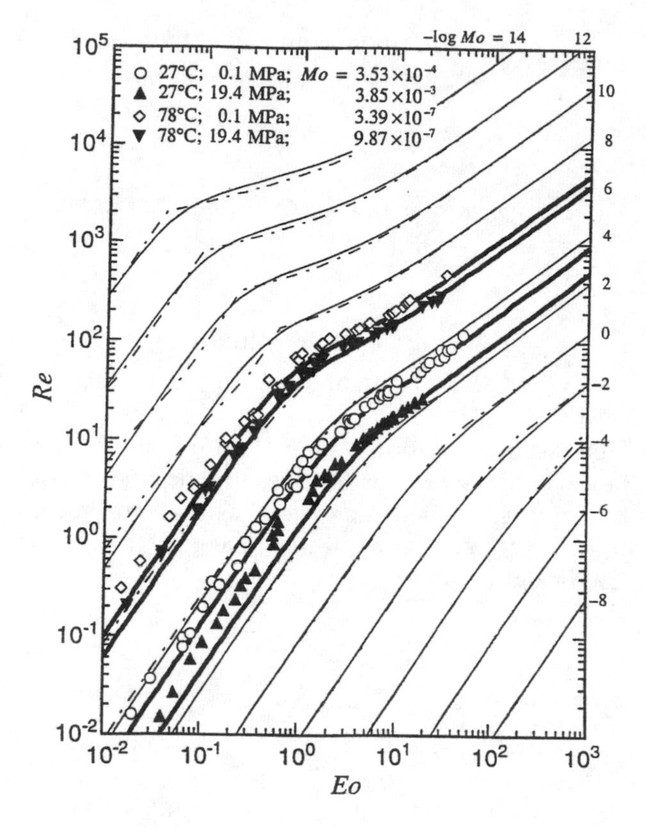

Figure 9.4 Comparisons of measured and calculated *Re* of single bubbles in Paratherm NF heat transfer fluid under varied pressure and temperature conditions. The Fan-Tsuchiya (1990) and Tomiyama *et al.* (1995) correlations are plotted (——— and – – –, respectively) at regular intervals of *Mo* values. The Fan-Tsuchiya correlation at measured *Mo* values for comparison with measured *Re-Eo* data (—) (Fan *et al.*, 1999, reproduced with permission).

$$Re = \frac{d_b u_b \rho_l}{\mu_l} \tag{9.6b}$$

and

$$Eo = \frac{g\,\Delta\rho\; d_b^2}{\sigma} \tag{9.6c}$$

where $\Delta\rho = \rho_l - \rho_g$. The bubble rise velocity, u_b, can be obtained explicitly from Equation 9.1 for a given d_b as well as given gas and liquid physical properties.

Figure 9.4 shows the *Re–Eo* relationship often used to represent the general rise characteristics of single bubbles in liquids (Clift *et al.*, 1978). The thin background lines are plotted with constant intervals of log *Mo* and signify the general quantitative trends for the rise velocity of single bubbles in purified Newtonian liquids under ambient conditions. The experimental results under various pressure and temperature conditions (Lin *et al.*, 1998) are also plotted in the figure. By employing accurate values for physical properties of the liquid phase and the gas density at given pressures and temperatures, the experimental data can be successfully represented over the entire *Eo* range, *i.e.* bubble size range, by Equation 9.1. The prediction is proven to represent experimental data for various liquids under ambient conditions, within minor deviation. Furthermore, the single bubble rise velocity at high pressures can be reasonably estimated by incorporating the physical properties of the gas and liquid under the operating conditions.

Bubble Coalescence

There are three steps in the bubble coalescence process: (1) approach of two bubbles to form a thin liquid film between them; (2) thinning of the film by the drainage of the liquid under the influence of gravity and suction due to capillary forces; and (3) rupture of the film at a critically small thickness. The second step is the rate-controlling step in the coalescence process, and the bubble coalescence rate can be approximated by the film-thinning rate. The film thinning velocity can be expressed as (Sagert and Quinn, 1978)

$$-\frac{\mathrm{d}l}{\mathrm{d}t}=\frac{32l^3\sigma}{3\phi R_d^2\mu_l d_b} \tag{9.7}$$

where the parameter ϕ is a measure of the surface drag or velocity gradient at the surface due to the adsorbed layer of the gas.

For gas-liquid systems, the experimental results available in the literature indicate that an increase of pressure retards the bubble coalescence. As pressure increases, surface tension decreases and liquid viscosity increases. In addition, ϕ increases with pressure. As implied from Equation 9.7, all these variations contribute to the reduction of the film thinning velocity, and hence, the bubble coalescence rate, as pressure increases. As a result, the time required for two bubbles to coalesce is longer and the rate of overall bubble coalescence in the bed is reduced at high pressures. Moreover, the frequency of bubble collision decreases with increasing pressure. An important mechanism causing bubble collision is bubble wake effect. When the distributions of bubble size and bubble rise velocity are narrower at high pressures, the likelihood of small bubbles being caught and trapped by the wakes of large bubbles decreases. Therefore, bubble coalescence is suppressed by the increase in pressure due to both the longer bubble coalescence time and the smaller bubble collision frequency.

Bubble Breakup and Maximum Stable Bubble Size

In gas-liquid systems, the upper limit of the bubble size is set by the maximum stable bubble size, D_{max}, above which the bubble is subject to breakup and is hence unstable. Several mechanisms have been proposed for the bubble breakup phenomenon. Based on these mechanisms, theories have been established to predict the maximum bubble size in bubble column systems.

Hinze *et al.* (1955) proposed that the bubble breakup is caused by the dynamic pressure and the shear stresses on the bubble surface induced by different liquid flow patterns, *e.g.*, shear flow and turbulence. When the maximum hydrodynamic force in the liquid is larger than the surface tension force, the bubble disintegrates into smaller bubbles. This mechanism can be quantified by the liquid Weber number. When the Weber number is larger than a critical value, the bubble is not stable and disintegrates. This theory was adopted to predict the breakup of bubbles in gas-liquid systems (Walter and Blanch, 1986).

A maximum stable bubble size exists for bubbles rising freely in a stagnant liquid without external stresses, *e.g.*, rapid acceleration, shear stress, and/or turbulence fluctuations. The Rayleigh-Taylor instability has been regarded as the mechanism for bubble breakup under such conditions. A horizontal interface between two stationary fluids is unstable to disturbances with wavelengths exceeding a critical value if the upper fluid has a higher density than the lower one (Bellman and Pennington, 1954):

$$\lambda_c = 2\pi\sqrt{\frac{\sigma}{g(\rho_l - \rho_g)}} \tag{9.8}$$

Grace *et al.* (1978) modified the Rayleigh-Taylor instability theory by considering the time available for the disturbance to grow and the time required for the disturbance to grow to adequate amplitude. Batchelor (1987) pointed out that the observed size of air bubbles in water was considerably larger than that predicted by the model of Grace *et al.* (1978), and further took into account the stabilizing effects of the liquid acceleration along the bubble surface and the non-constant growth rate of the disturbance. In Batchelor's model, the magnitude of the disturbance is required for the prediction of the maximum bubble size; however, this information is not known. The models based on the Rayleigh-Taylor instability predict an almost negligible pressure effect on the maximum bubble size; in fact, Equation 9.8 implies that the bubble is more stable when the gas density is higher.

The Kelvin-Helmholtz instability is similar to the Rayleigh-Taylor instability except that the former allows a relative velocity between the fluids, u_r. Using the same concept as Grace *et al.* (1978), Kitscha and Kocamustafaogullari (1989) applied the Kelvin-Helmholtz instability theory to model the breakup of large bubbles in liquids. Wilkinson and Van Dierendonck (1990) applied the critical wavelength to explain the maximum stable bubble size in high-pressure bubble columns:

$$\lambda_c = \frac{2\pi\sqrt{\dfrac{\sigma}{g(\rho_l - \rho_g)}}}{\dfrac{\rho_l}{\rho_l + \rho_g}\dfrac{\rho_g u_r^2}{2\sqrt{\sigma g(\rho_l - \rho_g)}} + \sqrt{1 + \dfrac{\rho_l^2 \rho_g^2 u_r^4}{4(\rho_l + \rho_g)^2 \sigma g(\rho_l - \rho_g)}}}. \qquad (9.9)$$

Disturbances in the liquid with a wavelength larger than the critical wavelength can break up a bubble. Equation 9.9 indicates that the critical wavelength decreases with an increase in pressure, and therefore bubbles are easier to disintegrate by disturbances at higher pressures. However, the critical wavelength is not equivalent to the maximum stable bubble size, and Equation 9.9 alone cannot quantify the effect of pressure on bubble size.

None of the models mentioned above accounts for the internal circulation of the gas. The internal circulation velocity is of the same order of magnitude as the bubble rise velocity. A centrifugal force is induced by this circulation, pointing outwards toward the bubble surface. This force can suppress the disturbances at the gas-liquid interface and thereby stabilize the interface. On the other hand, the centrifugal force can also disintegrate the bubble, as it increases with an increase in bubble size. The bubble breaks up when the centrifugal force exceeds the surface tension force, especially at high pressures when gas density is high. Levich (1962) assumed the centrifugal force to be equal to the dynamic pressure induced by the gas moving at the bubble rise velocity, *i.e.* $k_f \rho_g u_b^2 / 2$ ($k_f \approx 0.5$), and proposed a simple equation to calculate the maximum stable bubble size:

$$D_{\max} \approx \frac{3.63\sigma}{u_b^2 \sqrt[3]{\rho_l^2 \rho_g}}, \qquad (9.10)$$

Equation 9.10 severely underpredicts the maximum bubble size in air-water systems, but shows the significant effect of pressure on the maximum bubble size. Considering all the theories proposed in the literature, the mechanism for bubble breakup at high pressures is still unknown.

An analytical criterion for the bubble breakup is derived by considering a single large bubble rising in a stagnant liquid or slurry at a velocity of u_b without any disturbances on the gas-liquid interface. The bubble is subject to breakup when its size exceeds the maximum stable bubble size due to the circulation-induced centrifugal force (Luo *et al.*, 1999). The circulation of gas inside the bubble can be described by Hill's vortex (Hill, 1894). To model the bubble breakup, it is necessary to evaluate the x-component of the centrifugal force, F_x, induced by the

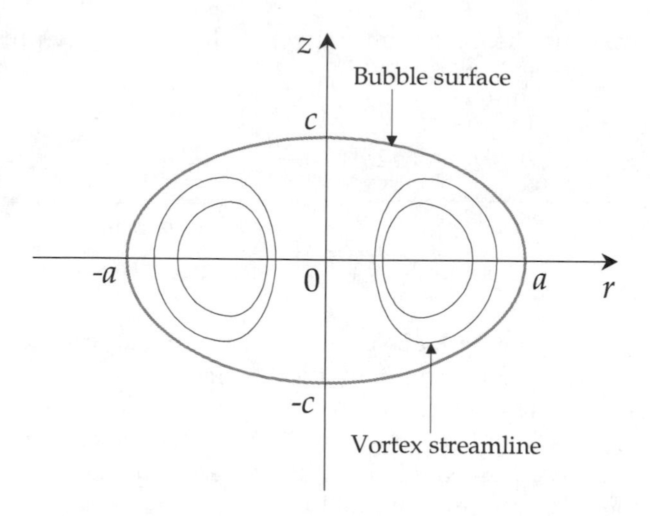

Figure 9.5 Schematic of the internal circulation model for bubble breakup.

circulation on the entire bubble surface as shown in Figure 9.5. A rigorous theoretical derivation from Hill's vortex yields the expression for F_x:

$$F_x = \frac{9\pi \rho_g u_b^2 a^2}{64\sqrt{2}\alpha}. \tag{9.11}$$

The surface tension force is the product of the surface tension and the circumference of the bubble,

$$F_\sigma = \sigma L = \sigma \int_{\text{ellipse}} \sqrt{(\delta r_c)^2 + (\delta z)^2} = 4\sigma a \, \mathrm{E}(\sqrt{1-\alpha^2}). \tag{9.12}$$

Also, the volume equivalent bubble diameter, d_b, is related to a and the aspect ratio, α, by

$$a = \frac{d_b}{\sqrt[3]{8\alpha}}. \tag{9.13}$$

Note that the centrifugal force is affected significantly by the gas density, the aspect ratio of the bubble, the bubble size, and the bubble rise velocity. The bubble is not stable if F_x is larger than F_σ, *i.e.*

$$u_b^2 d_b \geq \frac{8\alpha^{4/3}\mathrm{E}(\sqrt{1-\alpha^2})}{0.312} \frac{\sigma}{\rho_g}. \tag{9.14}$$

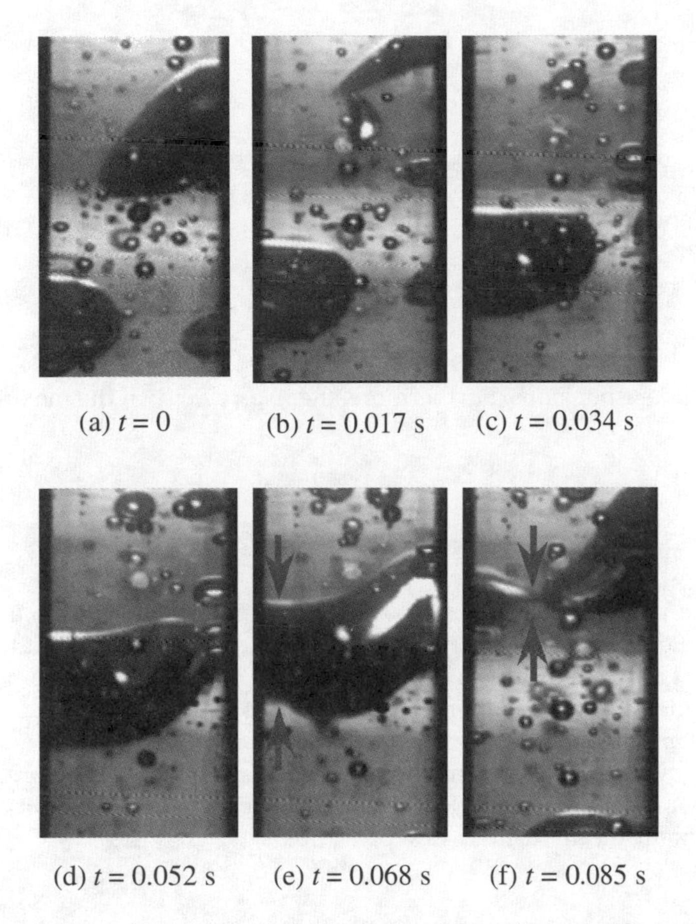

Figure 9.6 A sequence of bubble images showing the process of bubble breakup at P = 3.5 MPa (Fan *et al.*, 1999, reproduced with permission).

When the centrifugal force is larger than the surface tension force, the bubble will be stretched in the x direction. During the stretching, α becomes smaller, while d_b and u_b are assumed to remain constant. As a result, the centrifugal force increases, the surface tension force decreases, and the bubble stretching becomes an irreversible process. In Figure 9.6, the sequence of bubble images obtained at a pressure of 3.5 MPa shows the proposed mechanism of bubble breakup. The figure shows the bubble is stretched in the x direction before the breakup. Using the Davies-Taylor equation (Davies and Taylor, 1950) for the bubble rise velocity, the maximum stable bubble size is

$$D_{\max} \approx 7.16\alpha^{2/3}\mathrm{E}(\sqrt{1-\alpha^2})^{1/2}\sqrt{\frac{\sigma}{g\rho_g}}. \tag{9.15}$$

The simplified forms are (Luo *et al.*, 1999):

$$D_{max} \approx 2.53\sqrt{\frac{\sigma}{g\rho_g}} \quad (\text{for } \alpha = 0.21) \tag{9.16a}$$

in liquids, and

$$D_{max} \approx 3.27\sqrt{\frac{\sigma}{g\rho_g}} \quad (\text{for } \alpha = 0.3) \tag{9.16b}$$

in liquid-solid suspensions. Furthermore, the rise velocity of the maximum stable bubble is

$$u_{max} = C\left(\frac{\sigma g}{\rho_g}\right)^{1/4} \tag{9.17}$$

where C is a constant. The comparison between experimental data and the predictions of Equations 9.16(a) and 9.16(b) indicates that this internal circulation model can explain the observed effect of pressure on the bubble size. The internal circulation model captures the intrinsic physics of bubble breakup at high pressures. The comparisons of predictions by different models indicate that the bubble breakup is governed by the internal circulation mechanism at high pressures over 10 atm, whereas the Rayleigh-Taylor instability or Kelvin-Helmholtz instability is the dominant mechanism at low pressures.

FLOW STRUCTURES AND FLOW REGIMES

In bubble column reactors, the hydrodynamics, transport, and mixing properties, such as pressure drop, holdup (volume fraction) of various phases, gas-liquid interfacial areas, and mass and heat transfers, depend strongly on the prevailing flow regime. Three flow regimes can be identified based on the bubble flow behavior in order of increasing gas flow rate.

(1) Bubbly flow regime, dispersed bubble regime, or homogeneous bubble regime: No bubble coalescence occurs, and almost uniformly sized bubbles with equal radial distribution characterize this regime. This regime occurs at very low superficial gas velocities, typically lower than 1 cm/s in an air-water system. The liquid is carried up by the bubble driven motion in the vicinity of the ascending bubble streams and falls downward between bubble streams.

(2) Churn-turbulent, coalesced bubble, or heterogeneous bubble regime: At higher gas velocities the homogeneous dispersion cannot be maintained and an unsteady flow pattern occurs. Large bubbles moving up with high velocities in

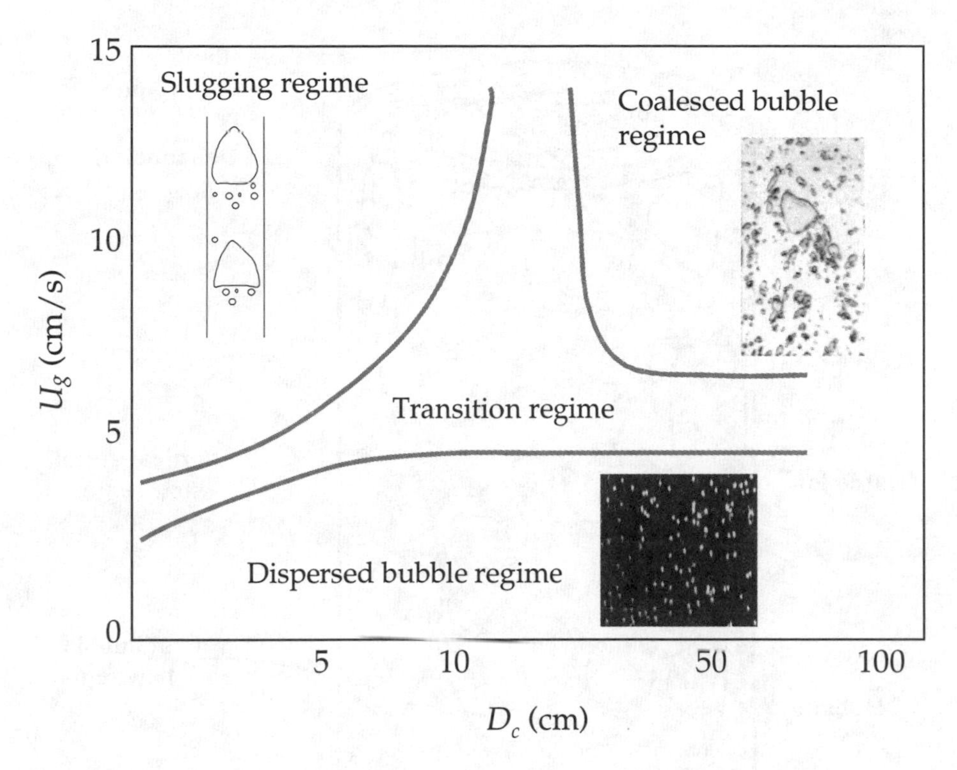

Figure 9.7 Flow regime map for a liquid-batch bubble column of a low viscosity liquid phase.

the presence of small bubbles characterize this heterogeneous bubble regime. Due to the migration motion, the bubbles start forming one or more bubble streams, which move in a rocking spiral manner. The liquid phase is carried upward by the spiral motion of the bubble streams and flows downward in the same spiral or vortical manner between the bubble stream and column wall. A detailed flow structure in this regime is described below.

(3) Slug flow regime: At high gas flow rates in small columns only (smaller than 15 cm in diameter), large bubbles are stabilized by the column wall, leading to the formation of bubble slugs. In columns of large diameter, the churn-turbulent flow regime always occurs even at high gas velocities. These flow regimes are schematically represented by Figure 9.7. It must be noted that many factors, *e.g.*, distributor type, physical properties (*e.g.*, viscosity and density) of liquid, and surface tension, play a role in affecting the transition between flow regimes. For example, the coalesced bubble flow regime can be observed at very low gas velocity for highly viscous fluids, and a small amount of contaminants in an air-water system (tap water compared to distilled water) can delay the flow regime transition.

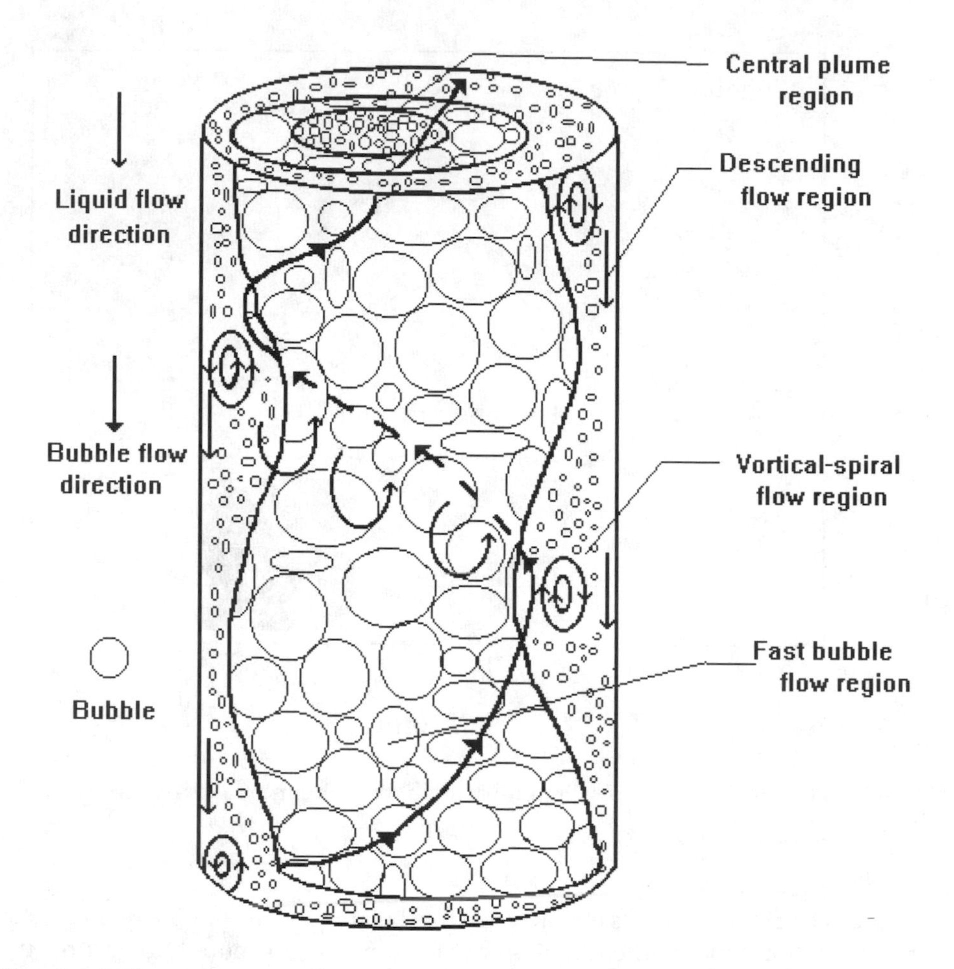

Figure 9.8 Flow structure in the coalesced bubble regime in a bubble column (Chen *et al.*, 1994, reproduced with permission).

Coherent Flow Structure in Coalesced Bubble Regime

A generalized macroscopic flow structure of a cylindrical bubble column in the coalesced bubble regime is shown in Figure 9.8. In the figure, four flow regions, namely, descending flow region, vortical-spiral flow region, fast bubble flow region, and central plume region, are indicated. The descending flow region, located adjacent to the column wall, is characterized by downward liquid streams moving in either straight or spiral manner depending on the gas and liquid velocities. This region is free of bubbles at low gas velocities. However, at relatively high gas velocities, tiny bubbles can be captured and dragged downward in the strong downward liquid flow.

The vortical-spiral flow region, located between the descending flow region and the central bubble stream, is characterized by the existence of spiral-downward liquid. The spiral-descending velocity varies with gas and liquid velocities. The entire region swings laterally back and forth, closely related to the swinging motion of the neighboring fast bubble flow region. Vortices are rather dynamic due to disturbances caused by small bubbles trapped in this region and by large bubbles rising in the fast bubble flow region. The vortical structures can lead to enhancement of the inter-phase mixing characteristics, yielding regions of increased chemical reaction.

Adjacent to the vortical-spiral flow region is the fast bubble flow region, also referred to as the central bubble stream, where significant bubble coalescence and breakup takes place. In this region, clusters of bubbles or coalesced bubbles move upward in a spiral manner with high velocity. There could be more than one spiral bubble stream simultaneously existing, and the spiral direction will change dynamically. These spiral bubble streams swing laterally back and forth over almost the entire bed diameter at high gas velocities. The fast bubble flow region dictates the macroscopic flow structure of the system in the coalesced bubble regime.

The central plume region can be observed in a relatively large bubble column. This region is characterized by a relatively uniform bubble size distribution and less bubble-bubble interaction.

Flow Regime Transition

In industrial applications of bubble columns, the churn-turbulent regime is most commonly encountered. Knowledge of the transition between the homogeneous bubble flow and the churn-turbulent flow regime is important for the design and operation of industrial reactors because the achievable conversion depends strongly on the flow regime. The transition velocity depends on gas distributor design, physical properties of the phases, operating conditions, and column sizes. The liquid-phase turbulence plays a critical role during the regime transition. Furthermore, as mentioned earlier, operations of bubble column systems under high-pressure conditions are common in industrial applications. A higher gas density has a stabilizing effect on the flow, and the gas fraction at the instability point increases with gas density, while the gas velocity at the instability point only slightly increases with gas density. Increasing pressure or temperature delays the regime transition.

Wilkinson *et al.* (1992) proposed a correlation to estimate the gas holdup and gas velocity at the transition point under high-pressure conditions. This predictive scheme incorporates the concept of bimodal bubble size distribution, *i.e.* the churn-turbulent regime is characterized by a bimodal bubble size distribution, consisting of fast rising large bubbles (>5 cm in diameter) and small bubbles (typically, <5 mm in diameter). Wilkinson *et al.* (1992) found that the transition velocity depends on the liquid properties and can be estimated by the following correlations:

$$\varepsilon_{g,tran} = \frac{U_{g,tran}}{u_{small}} = 0.5\exp(-193\rho_g^{-0.61}\mu_l^{0.5}\sigma^{0.11}) \tag{9.18}$$

and

$$u_{small} = 2.25\left(\frac{\sigma}{\mu_l}\right)\left(\frac{\sigma^3\rho_l}{g\mu_l^4}\right)^{-0.273}\left(\frac{\rho_l}{\rho_g}\right)^{0.03} \tag{9.19}$$

where u_{small} is the rise velocity of small bubbles. It must be noted that the *in-situ* physical properties of the fluids at a given temperature and pressure must be used in the correlation to predict the properties at the flow regime transition.

The effect of particles on the regime transition in bubble columns shows, in general, that the addition of fine particles to the liquid phase promotes bubble coalescence, which accelerates the transition to the churn-turbulent regime. However, the regime transition at high-pressure conditions in slurry bubble columns is still not fully understood, and further studies are needed to examine the effect of solids concentration on the transition velocity, develop an accurate correlation, and explore the transition mechanism.

In general, the pressure effect on the flow regime transition is a result of the variation in bubble characteristics, such as bubble size and bubble size distribution. The bubble size and distribution are closely associated with factors such as initial bubble size, bubble coalescence and breakup rates. Under high-pressure conditions, bubble coalescence is suppressed, bubble breakup is enhanced, and the distributor tends to generate smaller bubbles. All these factors contribute to small bubble sizes, narrow bubble size distributions, and, consequently, delay the flow regime transition in high-pressure bubble columns and slurry bubble columns.

BUBBLE SIZE AND BUBBLE SIZE DISTRIBUTION

In bubble columns, bubbles are formed at the distributor orifices or nozzles. As the bubbles rise, however, they may grow in size through coalescence or split into smaller bubbles by turbulent shear in the liquid phase. Bubbles formed at the distributor are almost uniform in size for given conditions, and the initial bubble size is affected only by the orifice diameter and the gas velocity through the orifice. The effect of gas velocity on bubble size in a bubble column depends on the flow regime. In the dispersed bubble regime, bubble size is small and increases only slightly with gas velocity. In the coalesced bubble regime, bubble size distribution is broader and the maximum bubble size increases more rapidly with gas velocity.

An estimation of the bubble size is important because it has a direct relation to the mass or heat transfer coefficient and transfer area. In general, at low gas velocities, *i.e.* a superficial gas velocity less than 0.5 cm/s, the bubble diameter will be a strong function of the orifice diameter and a weak function of the gas velocity in the orifice. At moderate gas flow rates, these functionalities are reversed and bubble diameter becomes a stronger function of the gas velocity in the orifice. At higher gas velocities, *i.e.* the superficial gas velocity higher than 10 cm/s, however, both the orifice diameter and the gas velocity will have a lesser effect on the bubble size.

The effect of electrolytic solutions also plays an important role on the bubble size. Because of surface tension and the electrostatic potential of the resultant ions at the gas-liquid interface, smaller bubbles are formed in the presence of electrolytes. The bubble size is likely to depend upon the electrolyte concentration as well as the type of electrolytes used.

In multi-bubble systems, a mean bubble size is usually used to describe the system. The mean bubble size is commonly expressed through the Sauter, or volume-surface, mean. For a group of bubbles with measured diameters, the Sauter mean is

$$d_{vs} = \frac{\sum n_i d_{bi}^3}{\sum n_i d_{bi}^2} \tag{9.20}$$

where n_i is the number of bubbles in the class i with its volume equivalent size d_{bi}.

Pressure has a significant effect on mean bubble diameter. In general, the mean bubble diameter decreases with increasing pressure; however, above a certain pressure, the bubble size reduction is not significant. The effect of pressure on the mean bubble size is due to the change of bubble size distribution with pressure. At atmospheric pressure, the bubble size distribution is broad; while under high pressure, the bubble size distribution becomes narrower and is in smaller size ranges. The bubble size is affected by bubble formation at the gas distributor, bubble coalescence and bubble breakup. When the pressure is increased, the bubble size at the distributor is reduced, bubble coalescence is suppressed, and large bubbles tend to breakup, *i.e.* the maximum stable bubble size is reduced. The combination of these three factors causes the decrease of mean bubble size with increasing pressure.

The bubble size distribution can normally be approximated by a log-normal distribution with its upper limit at the maximum stable bubble size. The contribution of bubbles of different sizes can be examined by analyzing the relationship between overall gas holdup and bubble size distribution. In slurry bubble columns, the gas holdup can be related to the superficial gas velocity, U_g, and the average bubble rise velocity, $\overline{u}_b$, (based on bubble volume) by a simple equation:

$$U_g = \varepsilon_g \bar{u}_b. \tag{9.21}$$

When the distributions of bubble size and bubble rise velocity are taken into account, $\bar{u}_b$ can be expressed as

$$\bar{u}_b = \frac{\int_{d_{b,\min}}^{d_{b,\max}} V_b(d_b) f(d_b) u_b(d_b) \mathrm{d}d_b}{\int_{d_{b,\min}}^{d_{b,\max}} V_b(d_b) f(d_b) \mathrm{d}d_b}. \tag{9.22}$$

The outcome of Equation 9.22 and the gas holdup strongly depend on the existence of large bubbles, because of the large volume and high rise velocity of such large bubbles. In the coalesced bubble regime, more than half of the small bubbles are entrained by the wakes of large bubbles and consequently have a velocity close to that of large bubbles. The large bubbles have a dominant effect on the overall hydrodynamics of bubble columns due to their large volume, their high rise velocity, and the wakes associated with the large bubbles.

GAS HOLDUP

Gas holdup is defined as the percentage of volume of the gas phase in the total volume of all phases. Gas holdup is one of the most important parameters characterizing the hydrodynamics of bubble columns. The gas holdup in a bubble column characterizes the retention of the bubbles under the liquid. The information of gas holdup in the column is important. The gas holdup is indicative of the residence time of the gas, effective interfacial area, *etc.* Gas holdup can be measured directly from the difference between the aerated liquid level and the static liquid level or by the pressure gradient method

$$\frac{\Delta p}{\Delta z} = \varepsilon_g (\rho_l - \rho_g) g \tag{9.23}$$

where ε_g denotes overall gas holdup.

The gas holdup depends on gas and liquid velocities, gas distributor design, column geometry (diameter and height), physical properties of the gas and liquid, particle concentration, and physical properties of the particles. The gas holdup generally increases with gas velocity, with a larger rate of increase in the dispersed bubble regime than in the churn-turbulent regime. Such distributors as perforated plate, nozzle injector, and distributor affect the gas holdup significantly only at low gas velocities.

In bubble columns, the effect of column size on gas holdup is negligible when the column diameter is larger than 0.1 to 0.15 m. The influence of the column height is insignificant if the height is above 1 to 3 m and the ratio of the column height to the diameter is larger than 5.

Gas holdup decreases as liquid viscosity and/or gas-liquid surface tension increase; however, the effect of liquid density is not clear. The addition of particles into a bubble column leads to larger bubble sizes and thus a decreased gas holdup, especially for low particle concentrations. Increasing solids concentration increases the *pseudo-viscosity* of the suspension, which in turn promotes bubble coalescence, resulting in an increase in bubble size and a decrease in gas holdup.

Liquid interfacial properties play a very significant role in gas holdup. For instance, adding a small amount of surface-active material to water, such as a short-chain alcohol, produces significantly higher gas holdup. The presence of electrolytes or impurities in water can also lead to higher gas holdup. In both cases, the higher gas holdup can be attributed, in part, to a decrease in or suppression of coalescing tendencies. Foaming properties of the liquid are also important. In liquid-surfactant systems, a transition to a foaming system is observed under some conditions and can be detrimental to system operability; however, these systems provide important data for industrial bubble columns, which experience small bubble sizes and large gas holdup.

A large number of empirical correlations are available to predict the overall gas holdup in bubble columns (Fan, 1989). At low superficial liquid velocities, gas holdup is independent of the liquid velocity and coincides with that in a batch-liquid system. At moderate superficial liquid velocities, the liquid velocity affects gas holdup; an increase in liquid velocity may increase the bubble rise velocity, decrease the initial bubble sized formed at the distributor, or change the coalescence/breakup tendencies.

In industrial applications, often the bubble columns are operated in high-pressure and high-temperature conditions. Numerous studies have been conducted to investigate the effect of pressure on the gas holdup of bubble columns (Fan *et al.*, 1999). Further, empirical correlations have been proposed for gas holdup in bubble columns operated at elevated pressure and temperature. Elevated pressures generally lead to a higher gas holdup in both bubble columns, except in those systems which are operated with porous plate distributors and at low gas velocities. The increased gas holdup is directly related to the smaller bubble size and, to a lesser extent, to the slower bubble rise velocity at higher pressures. Figure 9.9 shows bubbles in a bubble column of Paratherm NF heat transfer fluid and nitrogen over a wide range of operating conditions. As shown in the figure, the bubble size is drastically reduced as pressure increases. The most fundamental reason for the bubble size reduction can be attributed to the variation in physical properties of the gas and liquid with pressure.

A significant pressure effect on the gas holdup exists in slurry bubble columns; however, little is reported concerning such an effect. Studies show that the gas holdup increases with pressure. However, in general, no viable model is available

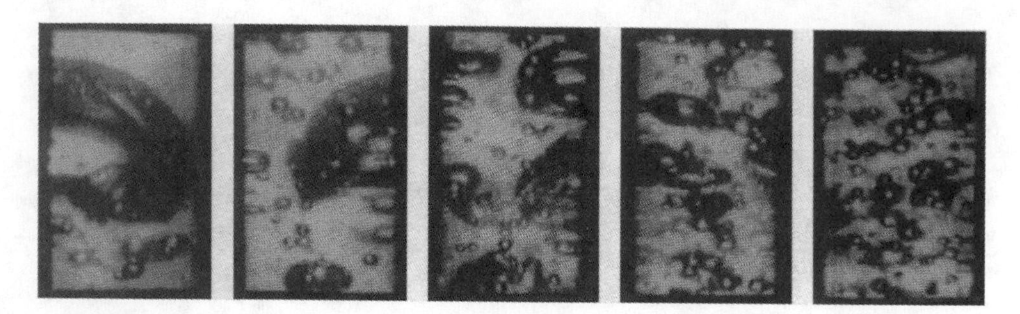

Figure 9.9 Visualization of bubbles in a bubble column system (From left: $P = 0.1, 3.5, 6.8, 11.6$, and 17.4 MPa).

to predict the gas holdup in high-pressure slurry bubble columns. The gas holdup behavior in high-pressure slurry bubble columns is not well understood, especially at high gas velocities.

Recently, Luo *et al.* (1999) obtained an empirical correlation to estimate the gas holdup in high-pressure bubble columns and slurry bubble columns as

$$\frac{\varepsilon_g}{1-\varepsilon_g} = \frac{2.9\left(\dfrac{U_g^4 \rho_g}{\sigma g}\right)^{\alpha}\left(\dfrac{\rho_g}{\rho_m}\right)^{\beta}}{[\cosh(Mo_m^{0.054})]^{4.1}} \tag{9.24}$$

where Mo_m is the modified Morton number for the slurry phase, $g(\rho_m - \rho_g)(\xi \mu_l)^4 / \rho_m^2 \sigma^3$, and

$$\alpha = 0.21 Mo_m^{0.0079} \tag{9.25a}$$

$$\beta = 0.096 Mo_m^{-0.011} \tag{9.25b}$$

A correction factor ξ accounts for the effect of particles on the slurry viscosity:

$$\ln \xi = 4.6\varepsilon_s\{5.7\varepsilon_s^{0.58} \sinh[-0.71\exp(-5.8\varepsilon_s)\ln(Mo)^{0.22}] + 1\}\,. \tag{9.26}$$

The applicable ranges of the correlation are summarized in Table 9.2.

The physical meaning of the dimensionless group of $U_g^4 \rho_g / \sigma g$ in Equation 9.24 can be shown by substituting Equation 9.17 into the group:

$$\frac{U_g^4 \rho_g}{\sigma g} \propto \left(\frac{U_g}{u_{\max}}\right)^4 \tag{9.27}$$

Table 9.2 Applicable ranges for the gas holdup correlation.

Parameter (units)	*Range*
ρ_l (kg/m^3)	668–2965
μ_l (mPa-s)	0.29–30
σ (N/m)	0.019–0.073
ρ_g (kg/m^3)	0.2–90
ε_s (–)	0–0.4
d_p (μm)	20–143
ρ_s (kg/m^3)	2200–5730
U_g (*m/s*)	0.05–0.69
U_l (*m/s*)	0 (batch liquid)
D_c (m)	0.1–0.61
H/D_c (–)	> 5
Distributor types	Perforated plate, sparger, and bubble cap

Clearly, the dimensionless group represents the contribution of large bubbles to the overall gas holdup, which is the major reason why the correlation can cover such wide ranges of experimental conditions.

For high-pressure bubble columns operated under the wide range of conditions outlined in Table 9.2, hydrodynamic similarity requires the following dimensionless groups to be the same: U_g/u_{max}, Mo_m, and ρ_g/ρ_m. To simulate the hydrodynamics of industrial reactors, cold models with milder pressure and temperature conditions may be chosen as long as the three groups are similar to those in the industrial reactor. However, the similarity rule needs to be tested in industrial reactors.

GAS-LIQUID MASS TRANSFER

The mass transfer of gas bubbles is strongly influenced by the bubble and wake flow behavior (see Figure 9.3). The solute is carried by the flow on the roof of the bubble along the boundary of the wake and is separated into two regions, within the primary wake region by the wake vortex and outside the wake region by the shedding vortex. The solute that flows into the wake is carried back to the bubble base. The shed vortex carrying the solute generates an external concentration vortex and eventually diffuses into the bulk flow. In addition to the convective diffusion by the liquid-solid flow, there is slow molecular diffusion of solute from the vortex sheet into the vortex center in the wake and from the wake surface into the bulk flow, but these contributions are negligible compared to convective diffusion.

The variations of the mass transfer patterns around bubbles with respect to time are given in Figure 9.10. The figure shows the circular-cap ozone-oxygen bubble and its wake rising in a starch-iodine-water and 0.46 mm glass bead fluidized

Figure 9.10 Ozone bubble rising in a KI-starch solution.

bed. As the bubble begins to rise, the reacted ozone molecules are carried from the edge of the bubble by the vortex sheet, and the wake underneath the circular-cap bubble is gradually saturated with ozone molecules. The shape of the bubble plus the wake is roughly circular. The zigzag trail behind the bubble is formed in the bulk of the liquid phase as a result of the vortex shedding. There is no trace of ozone molecules on the surface of the bubble roof since the reacted molecules are swept by the liquid flow. As a consequence, the convective diffusion induced by potential flow plays an important role in the mass transfer mechanism on the bubble roof. As the bubble rises further, the wake starts to shed vortices. As shown in the figure, alternate sheddings are observed in low bed expansion conditions of the liquid-solid fluidized media. The shed vortices elongate their shape and the gas molecules begin to diffuse out from the center of the vortices into the bulk liquid by molecular diffusion, especially in the case of high bed expansion.

HEAT TRANSFER CHARACTERISTICS

Since heat transfer behavior is closely associated with macroscopic flow structures and microscopic flow characteristics, a variation in pressure, which alters the physical properties of the gas and liquid, and also affects the hydrodynamics, would affect heat transfer behavior in the system in a complex manner.

Table 9.3 Variations of various parameters with pressure and the effects on heat transfer coefficient.

Parameters	*Effect of the parameter on heat transfer coefficient*	*Effect of pressure increase on parametric value*	*Parametric effect on heat transfer coefficient with increase in pressure*
μ_l	–	+	–
ρ_l	+	+ (small)	+ (small)
k_l	+	+ (small)	+ (small)
C_{pl}	+	+ (small)	+ (small)
σ	No direct effect	–	No direct effect
d_b	+	–	–
ε_g	+	+	+
ε_s	+	–	–

+ : increase, –: decrease

Liquid viscosity has a negative effect on heat transfer, and since liquid viscosity increases with pressure, pressure would have a negative effect on heat transfer. Other physical properties of liquid which are less affected by pressure include density, ρ_l, thermal conductivity, k_l, and heat capacity, C_{pl}.

Studies on instantaneous heat transfer behaviors in liquids and liquid-solid systems, which involve the injection of single bubbles, revealed the importance of bubble wakes to heat transfer behavior (Kumar *et al.*, 1992). The bubble-induced enhancement of heat transfer increases with bubble size due to the increased wake size and wake vortical intensity. When the pressure increases, the bubble size decreases, and hence the wake contribution to the heat transfer by single bubbles is reduced. In chain bubbling systems, Kumar and Fan (1994) reported that the time-averaged heat transfer coefficient increases with bubbling frequency due to the intense bubble-wake, bubble-bubble, and bubble-surface interactions. The effect of pressure on heat transfer due to the variations in liquid properties and hydrodynamic parameters is summarized in Table 9.3. The overall effect of pressure on heat transfer behavior depends on the outcome of the counteracting effects of each individual factor.

Pressure has a significant effect on the heat transfer characteristics in a slurry bubble column. The heat transfer coefficient decreases appreciably with increasing pressure except under very high pressures. The variation of the heat transfer coefficient with pressure is attributed to the counteracting effects of the variations of liquid viscosity, bubble size, and bubbling frequency with pressure. When pressure increases, bubble size decreases; however, the bubbling frequency increases, which augments the rate of heat transfer (Kumar and Fan, 1994). The counteracting effects of the above two factors give rise to the overall effect of pressure on the heat transfer rate. In a slurry bubble column, pressure reduces bubble size significantly at pressures lower than 4 MPa, which results in the

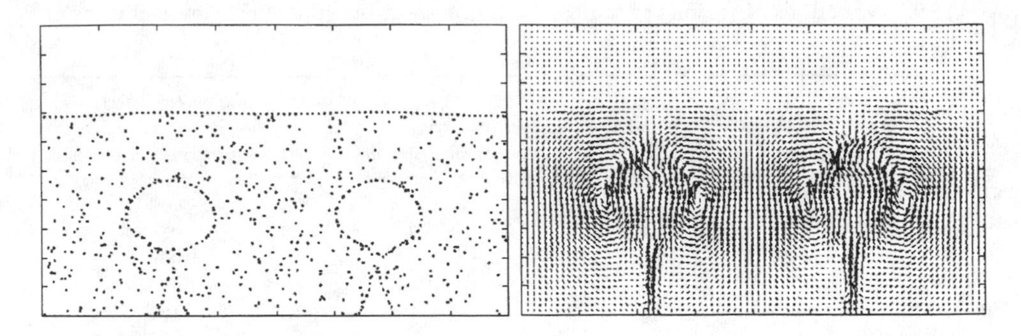

Figure 9.11 Simulation results of two-bubble formation in liquid-solid suspension with symmetric orifices at P = 19.4 MPa.

decrease of the heat transfer coefficient. When the pressure is further increased, the bubble size reduction is relatively smaller, and the increase in bubbling frequency contributes to an increase in the heat transfer coefficient. However, in a three-phase fluidized bed, due to the large particle size, the bubble size reduction becomes a less important factor in affecting the heat transfer coefficient, and the heat transfer coefficient increases with the increase of bubbling frequency.

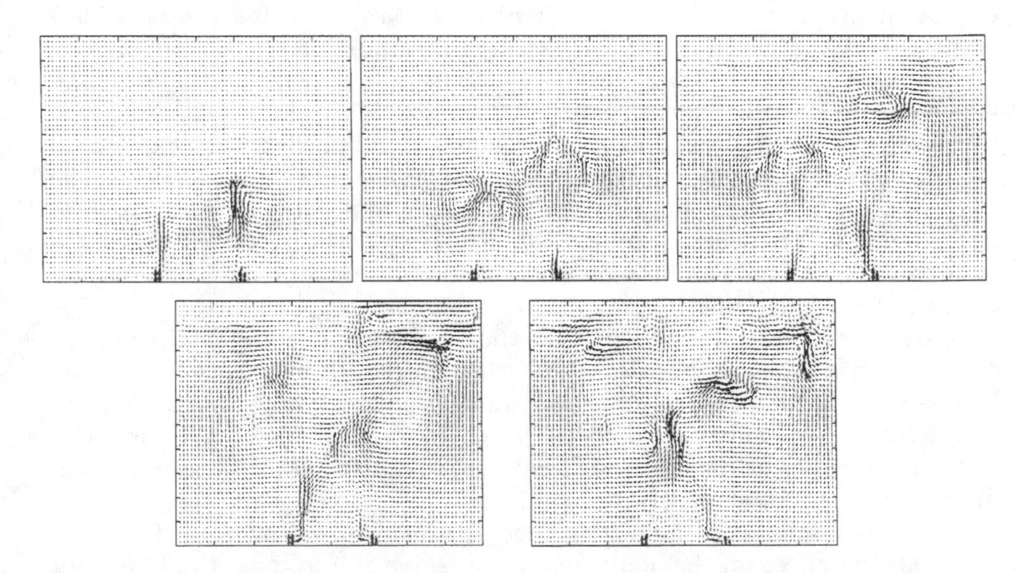

Figure 9.12 Velocity vector field of gas and liquid phases during two-bubble formation in liquid-solid suspension with asymmetric orifices at P = 19.4MPa.

CONCLUDING REMARKS

Bubble column reactors have been used in numerous applications and have enormous potential as a multiphase reactor. There have been many contributions toward understanding hydrodynamics and transport phenomena of such multiphase reactors for more efficient operation and application developments. One of the future directions of this research is computational fluid dynamics (CFD). For example, as shown in Figures 9.11 and 9.12, the flow fields of different phases in a high-pressure and high-temperature multiphase reactor can be visualized by development of the equation and computational code for discrete gas-liquid-solid flow simulation (Zhang *et al.*, 1999). With the development of more sophisticated CFD codes and innovative approaches by chemical engineers, it will be possible not only to design and develop more efficient multiphase reactors, but also to comprehend all unanswered physical and chemical behaviors in multiphase flows.

REFERENCES

Batchelor, G. K. (1987) The stability of a large gas bubble rising through liquid. *J. Fluid Mech.*, **184**, 399

Bellman, R. and Pennington, R. H. (1954) Effect of surface tension and viscosity on Taylor instability. *Q. Appl. Math.*, **51**, 151

Chen, R. C., Reese, J. and Fan, L.-S. (1994) Flow structure in a three-dimensional bubble column and three-phase fluidized bed. *AIChE J.*, **40**, 1093

Clift, R., Grace, J. R. and Weber, M. E. (1978) *Bubbles, Drops, and Particles*. New York: Academic Press

Davies, R. M. and Taylor, G. I. (1950) The mechanics of large bubbles rising through extended liquids and through liquids in tubes. *Proc. Roy. Soc. London*, **A200**, 375

Fan, L.-S. (1989) *Gas-Liquid-Solid Fluidization Engineering*. Stoneham, MA: Butterworths

Fan, L.-S., Yang, G. Q. Lee, D. J. Tsuchiya, K. and Luo, X. (1999) Some aspects of high-pressure phenomena of bubbles in liquids and liquid-solid suspensions. *Chem. Eng. Sci.*, **54**, 4681

Fan, L.-S. and Tsuchiya, K. (1990) *Bubble Wake Dynamics in Liquids and Liquid-Solid Suspensions*. Stoneham, MA: Butterworth-Heinemann

Grace, J. R., Wairegi, T. and Brophy, J. (1978) Break-up of drops and bubbles in stagnant media. *Can. J. Chem. Eng.*, **56**, 3

Hill, M. J. M. (1894) On a spherical vortex. *Phil. Trans. Roy. Soc. London*, **185**, 213

Hinze, J. O. (1955) Fundamentals of the hydrodynamic mechanism of splitting in dispersion processes. *AIChE J.*, **1**, 289

Kitscha, J. and Kocamustafaogullari, G. (1989) Breakup criteria for fluid particles. *Int. J. Multiphase Flow*, **15**, 573

Kumar, S., Kusakabe, K., Raghunathan, K. and Fan, L.-S. (1992) Mechanism of heat transfer in bubbly liquid and liquid-solid systems: single bubble injection. *AIChE J.*, **38**, 733

Kumar, S. and Fan, L.-S. (1994) Heat-transfer characteristics in viscous gas-liquid and gas-liquid-solid systems. *AIChE J.*, **40**, 745

Levich, V. G. (1962) *Physiochemical Hydrodynamics*. Englewood Cliffs, NJ: Prentice Hall
Lin, T.-J., Tsuchiya, K. and Fan, L.-S. (1998) Bubble flow characteristics in bubble columns at elevated pressure and temperature. *AIChE J.*, **44**, 545
Luo, X., Lee, D. J., Lau, R., Yang, G. Q. and Fan, L.-S. (1999) Maximum stable bubble size and gas holdup in high pressure slurry bubble columns. *AIChE J.*, **45**, 664
Maneri, C. C. (1995) New look at wave analogy for prediction of bubble terminal velocities. *AIChE J.*, **41**, 481
Mendelson, H. D. (1967) The motion of an air bubble rising in water. *AIChE J.*, **13**, 250
Sagert, N. H. and Quinn, M. J. (1978) Surface viscosities at high pressure gas-liquid interfaces. *J. Colloid Int. Sci.*, **65**, 415
Tomiyama, A., Kataoka, I. and Sakaguchi, T. (1995) Drag coefficients of bubbles (1st report, drag coefficients of a single bubble in a stagnant liquid). *Nippon Kikai Gakkai Ronbunshu B Hen*, **61**(587), 2357
Walter, J. F. and Blanch, H. W. (1986) Bubble break-up in gas-liquid bioreactors: break-up in turbulent flows. *Chem. Eng. J.*, **32**, B7
Wilkinson, P. M., Sper, A. P. and Van Dierendonck, L. L. (1992) Design parameters estimation for scale-up of high-pressure bubble columns. *AIChE J.*, **38**, 544
Wilkinson, P. M. and Van Dierendonck, L. L. (1990) Pressure and gas density effects on bubble break-up and gas hold-up in bubble columns. *Chem. Eng. Sci.*, **45**, 2309
Yu, Y. H. and Kim, S. D. (1991) Bubble properties and local liquid velocity in the radial direction of cocurrent gas-liquid flow. *Chem. Eng. Sci.*, **46**, 313

10. AN OVERVIEW OF PARTICLE TECHNOLOGY

ANTHONY D. ROSATO*

Mechanical Engineering Department, New Jersey Institute of Technology, University Heights, Newark, NJ 07102, USA

WHAT IS PARTICLE TECHNOLOGY?

The word "technology" refers to an organized system of knowledge and its application to industrial processes. Hence, "particle technology" broadly pertains to the system of knowledge and processes related to particles. More specifically, it is concerned with the characterization, production, modification, flow, handling, and utilization of bulk granular solids and powders, both dry and in slurries†. The individual particles that constitute the bulk material can possess a very wide range of sizes. This can be seen from Table 10.1, which presents a classification system (Brown and Richards 1970) including several examples of each category. It would be difficult if not impossible indeed to conceive of a branch of the physical sciences or engineering that does not involve some aspect of the behavior of particles. Market driven demand for increased production and the development of new products has spurred the need for advances in systems and operations needed to create, handle and process granular or bulk solids. Concurrent with this are major environmental issues related to the fact that the release of particulates into the air and waterways is the major contributor to pollution problems here on Earth. It is because of this reason and the universality of particulates in a host of industries (*i.e.*, agriculture, ceramics, chemicals, energy,

Table 10.1 Particle size categories and examples (Brown and Richards 1970).

Category[a]	*Size range*	*Examples*[b]
Ultra-fine powders	$0.1\mu - 1.0\mu$	paint pigments, smog
Super-fine powders	$1.0\mu - 10\mu$	milled flour, silt
Granular powders	$10\mu - 100\mu$	cement dust, pulverized coal, human hair
Granular solids	$100\mu - 3000\mu$	beach sand, gravel
Broken solids	$> 3000\mu$ (3mm)	coal, pebbles, almonds

[a] It should be noted that the term "nanoparticles" is used today to denote the class of powders that are smaller than 0.1μ.
[b] Beddow, J. K. (1980) Particulate Science and Technology. New York: Chemical Publishing Co.

*Phone/Fax: (973) 596-5829; rosato@njit.edu
† A slurry is a mixture of particles and fluid.

geological systems, manufacturing, minerals and ores, pharmaceuticals, plastics, pollution control systems, powder metallurgy, *etc.*), that the field of particle technology cuts across the disciplines of chemical, civil, industrial and mechanical engineering, as well as mathematics, physics, geology and chemistry. This being said, one may be amazed to discover that the science dealing with the behavior of bulk solids is really in its infancy; that as a branch of science, it lags well behind traditional fluids and solids which have been understood for many years. This will be discussed further on in this chapter.

It is not an exaggeration to state that particulates are universally found as constituents of most commonly used products. Consider, for example, plastic based materials, which begin their lives as small pellets or powders that are melted and formed into products in everyday usage (Charrier 1991). This includes packaging for consumable goods, children's toys, components of automobiles and other vehicles, building and construction materials, furniture, computers, audio and video media, and electronics, to name only a few. In 1994, the estimated US polymer consumption was approximately 38×10^6 tons, which is nearly a three-fold increase from reported 1974 levels (Rodriguez 1996).

Even more pervasive than plastics are the products and materials created from mining, which is the foundation of the building/construction, defense, transportation and manufacturing industries. The raw materials, which are essentially particulate or granular in form, include ores to produce aluminum, lead, iron, zinc, copper, silver, tungsten, and gold as well as fire clay, lime, gypsum, diatomite, coal, barite, cement, crushed stone, salt, sand, sulfur, gravel and silica. To quote from the Nevada Mining Association (1999), "Everything harvested, transported or published requires materials that come from mining." The few examples which follow will serve to provide an estimate of the magnitudes involved.

- In order to produce a single (average) automobile containing 3050 pounds of fabricated components, approximately 10 950 pounds of raw materials from mining operations are needed (Mining Industry Council of Missouri, 1999) as demonstrated in Table 10.2.

Table 10.2 Materials needed to produce a typical automobile.

Refined material (lbs)	*Mined material (lbs)*
Iron/Steel: 2245	Ore: 4960
Copper: 26	Copper Ore: 2600
Lead: 24	Lead Ore: 960
Zinc: 18	Zinc Ore: 720
Rubber/Plastic: 512	Crude Oil: 980
Glass: 85	Silica Sand: 170
Aluminum: 140	Barite: 560

Table 10.3 Comparative agriculture production statistics of select items for 1999 in metric tons (Food and Agriculture Organization of the United Nations 1999).

Item	*World*	*United States*	*US %-age*
Almonds	1 575 709	627 000	39.8
Barley	133 567 421	6 136 000	4.6
Buckwheat	2 775 192	40 000	1.44
Cereals	2 073 181 400	338 601 740	16.3
Lentils	3 008 630	108 270	3.6
Maize	604 400 398	242 254 000	40.08
Millet	28 621 561	200 000	0.70
Oats	26 156 823	2 126 500	8.13
Rice	588 766 882	9 603 000	1.63
Rye	20 354 473	279 249	1.37
Sorghum	68 074 625	15 151 000	22.25
Sugar Cane	1 276 911 480	34 106 500	2.67
TOTAL	4 845 370 839	649 233 259	13.4

- In 1998, the State of Nevada produced gold, silver, barite, diatomite, gypsum and lime for a total worth of $3.126 billion.
- During the first six months of 1999, 1.827×10^9 metric tons[‡] of crushed sand and gravel were mined for consumption (Pit and Quarry, 1999) in the United States.
- The annual world coal production amounts to approximately 4.6×10^9 tons, of which 1.09×10^9 tons (or 24%) is mined in the United States. It is noteworthy that coal, the most abundant fossil fuel on the Earth, has estimated reserves of 10^{12} tons that are predicted to last nearly 200 years at the current consumption rate.
- The mining industry in the United Kingdom produces 165.6×10^6 tons of granular materials each year (Bates 1997).
- Mining and related industries encompass a broad segment of the work force, including chemists, engineers, geologists, technicians, contractors, laborers and transporters.

Perhaps the granular materials most familiar to us are the foods and products found in farmer's markets and on the shelves of grocery store chains. All of these find their origin in the agriculture industry, where enormous quantities of raw materials in granular form are handled, processed, and stored for subsequent packaging and distribution to the public. For instance, Table 10.3 (Food and Agricultural Organization of the United Nations 1999) lists the 1999 world production of various agricultural commodities totaling 4.845×10^9 metric tons, of

[‡] 1 metric ton = 2204.6 pounds = 1.102 tons = 1000 kilograms.

which approximately 13.4% originated from the United States. Data compiled from "Statistical Highlights of U.S. Agriculture 1998–99" indicates that the cumulative field crop[•] value in 1988 was worth 65.9 billion dollars.

There are many other industries (*e.g.*, chemicals, electronics, pharmaceuticals, glass, ceramics, powder metallurgy, and paints, *etc.*) heavily involved with particulates. The handling of these materials in large processing plants requires a number of different, interdependent systems and operations in order to obtain an end product with desired properties.

PROCESSING AND HANDLING OF BULK SOLIDS

In today's bulk solids industries, productivity and profitability depend on the efficiency and reliability of systems within plants, that handle slurries and dry materials. There are essentially five basic objectives of the operations which take place in solids processing industries: the creation of particles, size reduction, size enlargement, separation and mixing (Ennis *et al.* 1994; see also Savage *et al.* 1996). Typical operations include mechanical and pneumatic conveying, screening, mixing, dust collection, sampling/analysis, communition, agglomeration, drying, storage, weighing, and bagging. The design and selection of the equipment components that carry out these operations impacts the capacity of the plant to output its product. Most important in the performance of operations are the material flow properties, which are related to particle characterization and classification. While it is not possible to discuss in detail all operations (which is well beyond the scope of this chapter), the following descriptions have been selected for illustrative purposes, and to provide a snapshot that highlights some of the complexities involved.

Screening

A screener is a mechanical device whose primary purpose is to sort or grade granular solids by size through the application of vibrations, which causes the bulk solid to flow across one or several stacked meshes (or screens). The meshes are plane surfaces containing a uniform array of holes that either prevent or permit the downward passage of particles through the holes, depending on their relative size. Hence the screener is used as a type of filter to collect discharged particles of desired sizes for subsequent processing down the line (Fruchtbaum 1988a). For example, it may be required at some point in the process to remove finer particles from a mixture, or to capture coarser granules.

[•] Barley, dry beans, canola, coffee, corn, cotton, hay, hops, oats, peanuts, dry peas, potatoes, rice, sorghum, soybeans, sugarcane, sunflower, and wheat.

Industrial solids handling systems are often operated in a mode to continuously create and output a product, which requires reduction and/or elimination of costly downtime[∞] that is ultimately a detriment to profitability. The selection of proper and "efficient" screeners must take into consideration their effect on overall performance of the system of which they are integral components. It is particularly important to understand the unit operations that occur directly before and after the screener (Matthews 1981). Because of the very wide range of materials and processing conditions for which screeners are used, the selection task is further complicated by the oftentimes necessity of applying an empirical approach. Consider, for example, the complications which can occur when a screener in a processing loop cannot discharge material as fast as it is loaded. Often, market-driven demand to increase production results in the need to modify the system to handle a larger throughput of material. And this may be limited by the ability of the screeners to accommodate additional load (McCauley 1999).

Mixing and Segregation

The operation of mixing of powders has been a subject of great interest among the community involved with particle technology. Intimately involved with this is the subject of sampling in order to assess the quality of a blend. In common use in industry are various types of continuous and batch mixtures, such as V-blenders, paddle type and conical mixers, ribbon blenders, and drum mixers (Miyanami 1997). Working against the ability to mix dissimilar dry solids is the natural tendency of particles to "segregate", (Bates 1997) for which a principal factor is a discrepancy in size. It is noteworthy that the number of studies in the literature on segregation has seen rapid growth over approximately the last 20 years. Reported experiments have achieved moderate success in isolating and identifying specific mechanisms; however, the connection between them remains elusive. (Rosato and Blackmore 2000).

Pneumatic Conveying

Pneumatic conveying is the term used to describe any system where particulates are carried in a stream of a gas through a pipeline network. These systems are often used to convey fine particles[◊] (mean size smaller than 10 mm) over short distances of usually less than 1000 m, although there are exceptions (Klinzing *et al.* 1997). Systems are classified as either *dilute phase,* in which particles are transported in the gas stream via lift and drag forces under low pressure and high

[∞] Downtime may be caused by the need to clean the meshes or a screener and/or to repair or replace fragile components that deteriorate due to wear and adverse operating conditions.
[◊] See Table 1.1 in Klinzing *et al.* (1997) for an extensive list of materials.

velocities, or *dense phase*, in which particle masses are essentially pushed along the conduit under high pressures. As a note, while dilute phase systems are fairly well understood, there is little or no theory to aid in the design of dense phase systems. All pneumatic systems contain an essential component known as the *prime mover*, which supplies power to move the gas that carries the solids. Introduction of the particles into the system through a valve takes place in a feeding section or zone, followed by a horizontal conduit to allow the system to reach a steady-state flow velocity. The gas-solids mixture then enters the pipe network and is transported to a section in which particles are separated from the moving stream for further processing. The reader is referred to Klinzing *et al.* (1997) for a thorough treatment of pneumatic conveying systems. (See also Fan and Zhu 1998; and for examples of practical case studies, see Powder & Bulk Engineering 1999, Solt 1999, and Hartman 1996).

Mechanical "Horizontal" Conveyors and Elevators

There are many different types of mechanical conveying systems that are employed to transport materials in solids processing systems. Included in this category are screw, belt, and drag conveyors, vibrating feeders, vibratory conveyors (Johnstone 1998), bucket elevators, and package handlers. (See Fruchtbaum 1988b). In general, one can make a distinction between systems that mainly carry materials horizontally or up slight inclines, from those that convey mainly in the vertical direction (Kessler *et al.* 1998).

A belt conveyor can carry a wide variety of materials, such as coal, wood, ores, stones, chemicals, plastics, foods, *etc*. It consists of a closed loop (the belt) resting on supports (idlers), and wrapped around driving pulleys. One rather common example of this is the moving walkway found in large airports that transports people to and from the gates. The main criteria for selection of the belt conveyor are the carrying capacity, desired inclination and belt loading points. The screw conveyor is a metal helix wrapped around a shaft, all of which is enclosed in a trough. Rotation of the screw causes granular solids between the "threads" to move from the entrance to the exit. Bucket elevators are used to lift the material upward from the base to the top where discharge occurs. This is done with a series of "buckets" attached to a closed loop chain that rotates. Package handlers are conveyors that are either gravity-driven rollers, or a powered system using belts, rollers or chains. These systems can be used to transport containers that can be quite heavy (for example, luggage conveyors in the baggage claim areas of airports).

Characterization and Classification

The term "classification" is associated with the organization of particles by the size alone (Beddow 1980; Heiskanen 1993). The measurement of particle size and size distribution is fundamental and plays an extremely important role in the field of particle technology. For this purpose, there are a number of techniques and devices available (Allen 1990; Davies 1997), the selection of which depends on the application. There is an entire enterprise involved with the development of instruments capable of measuring particle properties, of which size is a fundamental constituent. The problem of making sensible comparisons between measurements obtained by various devices remains a subject of discussion (Lieberman 1999).

In the context of industrial solids handling, *classification* refers to the process (or unit operation) of separating solids from solids, while equipment designed for this purpose is placed into a category known as *classifiers*. These devices make use of the differences in behavior of fine and coarse particles in flows (particles and a liquid, or particles and a gas) in order to affect a size separation. A critical issue affiliated with classification concerns a manufacturer's ability to supply a customer with products having the required distribution of particle sizes. For the purpose of quality control and efficiency of processing, it is becoming increasingly important to have the ability to perform in-line characterization and analyses of materials moving through process loops.

The technology associated with the determination of particle properties (such as ability to absorb moisture, microstructure, porosity, reactivity, size, shape or morphology, and surface conditions) is known as *characterization*. These properties and their distributions ultimately dictate how a bulk solid will behave under various handling, processing, and storage situations. This being said, the ability to predict the flow behavior based on measured properties remains problematic, and is a topic of active research (Bell 1999). Further complexities come into play with the creation and use of extremely small particles for which interparticle forces become of even greater importance in the material's rheology.

SCIENTIFIC FOUNDATIONS OF PARTICLE TECHNOLOGY

A granular material or bulk solid is unique in that it can exhibit the properties of a solid or a fluid depending on the load and flow conditions. However, unlike fluids and solids, for which reliable analytical theories have been established for some time**, models capable of predicting the behavior of dry granular systems over the wide range of observed phenomena are generally unavailable. This statement by itself is quite startling given the plethora of applications and uses of particle-based products in our modern industrial society, and the fact that mankind has been involved with aspects of granular materials for thousands of

** For example, C. Navier derived the equations governing the behavior of pure fluids in 1823, while as early as 1738, the kinetic theory of gases had its roots in the ideas of D. Bernoulli (Chapman and Cowling 1970).

years. This leads to an interesting question as to how handling systems are contrived since, in order to design equipment, engineers require scale-up laws which dictate the needed parameters. The answer is that engineers are often compelled to design systems based on intuition and experience, the consequences of which often include system failures and breakdown at tremendous loss of productivity. This problem was highlighted in a 1985 report (Merrow 1985) which indicated that the performance of recently built solids processing plants was no better than those constructed in the 1960s. In a 1986 report by the International Energy Agency on Coal Research (Wood 1986), a number of urgent issues in solids transport were identified:

1) Identification of particle characteristics and their effects on flow mechanics
3) Influence of moisture content, pressure and temperature
4) Development of granular microstructure to incorporate into phenomenological theories
5) Effects of boundary conditions on the flow
6) Experimental methods to measure particle collision properties for use in computer simulations
7) Understanding the mechanics responsible for size sorting.

The recent article by Bell (1999) entitled "Industrial Needs in Solids Flow for the 21st Century" also emphasizes several critical issues related to the handling of powders. These include an understanding of the stresses and the role of ratholes in silo flows, the determination of powder flow properties in the context of particle property measurements, dense phase pneumatic conveying, and particle technology education.

Our present understanding of granular flows is very much indebted, in part, to the pioneering work of Bagnold (1954, 1956, 1966). Within the scope of solids handling is what he termed the *grain-inertia regime*, in which pressure generated in a flowing bulk solid is due primarily to the momentum transfer arising from particle-particle collisions. In contrast to this is the *quasi-static regime*, dominant at very slow deformations of the bulk solid, in which particles remain in contact and slide over one another, and stresses are produced via a changing network of contact forces. Between these limiting cases is the *transition regime*, where stresses are created by some combination of the latter mechanisms (Savage 1984). The development of continuum models based on the ideas of soil mechanics for the quasi-static regime, and of kinetic theory models for the grain-inertia regime, has been the subject of much research in an attempt to explain experimentally observed behavior. (See references cited in Savage 1984, 1993). In addition to theoretical models, there has been a parallel effort in which the equations of motion governing systems of dissipative[×], interacting particles are numerically

[×] The term "dissipative" is used to describe particles that lose energy in collisions. In contrast, systems composed of molecular particles observe energy conservation.

solved. The expression *discrete element* simulation (also known as *granular dynamics* and *distinct element* simulation) is commonly used to designate this method. This technique was independently developed in the early 1970s by Cundall (1974), although it can be actually regarded as an outgrowth of the molecular dynamics simulations employed in statistical physics studies (Alder and Wainwright 1956; Ashurst and Hoover 1973). The ability to assign properties to the particles, to apply realistic approximations of individual collisions, and to compute quantities which can be measured experimentally have made this method extremely attractive. These features, coupled with rapid advances in the capabilities of computers over the last 10 years, have resulted in a tremendous growth in the use and application of the method to investigate a wide variety of granular flow problems (Powder Technology 2000). Access to complete kinematic and dynamic data for each particle in the system, which is not generally available or easily obtainable in physical experiments, makes it possible to correlate the development of microstructure (*i.e.*, the physical arrangement of the particles, their velocities and contact forces) with calculated transport properties[#]. As an experiment carried out on the computer, the discrete element method holds great promise to connect developing theories with physical experiments performed in the laboratory.

PARTICLE TECHNOLOGY EDUCATION IN THE UNITED STATES

While flows of bulk solids are not sufficiently understood and extensive research to develop clear scientific principles is necessary before the design of handling systems becomes routine, the need to educate and train engineers in the field of particle technology is urgent and exigent. This is especially true in light of the fact that the probability of an engineer encountering problems related to solids handling and processing during his or her professional career is very high. In order for the US to remain competitive in the global market, universities must develop courses and programs that will train graduate and undergraduate engineers in particle technology. Industry recognition of this was emphasized in the cover story (Ennis *et al.* 1994) of the April 1994 issue of *Chemical Engineering Progress* entitled "Particle Technology: The Legacy of Neglect." (See also Bell 1999). Fourteen years earlier, a statement in the introduction to Beddow's book (1980) asserted:

> "But in one major industrial enterprise of this world, the United States of America, there is no single institution devoted to the study of matter in finely divided form, although there are numerous individual students scattered throughout the country."

[#] Transport quantities are parameters derived from flow statistics.

Table 10.4 Partial listing of institutions in the United States in which particle technology has a presence.

Institution	*Designation/Department*
California Institute of Technology (CA)	Granular Flow Group
Illinois Institute of Technology (IL)	Powder Materials Laboratory
City College of New York (NY)	Mechanical Engineering Department
Clark University (Worcester, MA)	Mechanical Engineering Department
Colorado School of Mines (CO)	Geomechanics Research Center
Massachusetts Institute of Technology (MA)	Computational Modeling of Materials
New Jersey Institute of Technology (NJ)	Mechanical Engineering Department
Northwestern University (IL)	Chaos & Mixing Group
Ohio State University (OH)	Chemical Engineering Department
Pennsylvania State University (PA)	Particulate Materials Center
Purdue University (IN)	Wassgren Research Group/ME Dept.
Purdue University (IN)	Chemical Engineering Department
Rutgers University (NJ)	Pharmaceutical Engineering Program
University of Chicago (IL)	Granular Physics Group
University of Colorado, Boulder (CO)	Chemical Engineering Department
University of Florida, Gainesville (FL)	Particle Science & Technology Center
University of Notre Dame (IN)	Granular Material Science
University of Pittsburgh (PA)	Granular Transport Group

Consider for example that there are four university particle research centers in the United Kingdom, and thirteen in Germany, while Switzerland, the Netherlands, Korea and Taiwan maintain university chairs in particulate research. In Japan, the particle technology research effort is impressive, as there are 28 centers, 16 government laboratories, a consortium of more than 300 companies under the Association of Powder Process Industry and Engineering, and the Council of Powder Technology.

During the last five or six years, particle technology education has begun to make an appearance in US universities with the development of centers, programs and course options. This can be seen from the listing in Table 10.4, which provides the names of several institutions where particle technology has a presence. The expansion and sustaining of funding for education and research in this important field, both from federal and industrial sources, will provide impetus for continued development and growth of programs concerned with the technology of particulates across the science and engineering disciplines.

REFERENCES

Agricultural Statistics (1999), pp. 1–50. US Department of Agriculture, Washington, DC: United States Government Printing Office

Alder, B. J. and Wainwright, T. E. (1956) Statistical Mechanical Theory of Transport Properties. Proceedings of the International Union of Pure and Applied Physics, Brussels, 1956

Allen, T. (1990) Particle Size Measurement, 4th edn. New York: Chapman & Hall

Ashurst, W. T. and Hoover, W. G. (1973) Argon Shear Viscosity via a Lennard-Jones Potential with Equilibrium and Nonequilibrium Molecular Dynamics. *Phys. Rev. Lett.*, **31**, 206–209

Bagnold, R. A. (1954) Experiments on a Gravity Free Dispersion of Large Solids Spheres in a Newtonian Fluid Under Shear. *Proc. R. Soc. London, Ser. A*, **225**, 49–63

Bagnold, R. A. (1956) The Flow of Cohesionless Grains in Fluids. *Phil. Trans. R. Soc. London, Ser. A*, **249**, 235–297

Bagnold, R. A. (1966) The Shearing and Dilation of Dry Sand and the 'Singing' Mechanism. *Proc. R. Soc. London, Ser. A*, **295**, 219–232

Bates, L. (1997) User Guide to Segregation, edited by G. Hayes. London: British Materials Handling Board

Beddow, J. K. (1980) Particulate Science and Technology. New York: Chemical Publishing Co., Inc

Brown, R. L. and Richards, J. C. (1970) Principles of Powder Mechanics, edited by P. V. Danckwerts. Oxford: Pergamon Press

Chapman, S. and Cowling, T. G. (1970) The Mathematical Theory of Non-Uniform Gases, 3rd edn. London and New York: Cambridge University Press

Charrier, J-M. (1990) Polymeric Materials and Processing. New York: Oxford University Press

Cundall, P. A. (1974) Computer Model for Rock-Mass Behavior Using Interactive Graphics for Input and Output of Geometrical Data. U.S. Army Corps. of Engineers, Technical Report. No. MRD, p. 2074

Davies, R. (1997) Size Measurement. In Powder Technology Handbook, 2nd edn., edited by K. Gotoh, H. Masuda and K. Higashitani, pp. 15–42. New York: Marcel Dekker

Ennis, B. J., Green, J. and Davies, R. (1994) The Legacy of Neglect in the U.S. Chemical Engineering Progress, April 1994, 31–43

Fan, L.-S. and Zhu, C. (1998) Principles of Gas-Solid Flows, edited by A. Varma. New York: Cambridge University Press

Food and Agriculture Organization of the United Nations 1999, http://apps.fao.org

Fruchtbaum, J. (1988a) Bulk Materials Handling Handbook, pp. 269–278. New York: Van Nostrand Reinhold

Fruchtbaum, J. (1988b) Bulk Materials Handling Handbook, Sections 2–9. New York: Van Nostrand Reinhold

Hartman, J. A. (1996) Engineering Study: Solving a Plugging Problem in Pneumatic Conveying Lines. *Powder and Bulk Engineering*, **10**(7), 19–22

Heiskanen, K. (1993) Particle Classification, edited by B. Scarlett. London: Chapman & Hall

Johnstone, L. (1998) Vibrating Conveyors. *Bulk Solids Handling*, **18**(2), 261–264

Kessler, F., Paelke, J. W. and Grabner, K. (1998) Vertical Conveyors for Bulk Materials – The Economic Solution. *Bulk Solids Handling*, **18**(3), 443–448

Klinzing, G. E., Marcus, R. D., Rizk, F. and Leung, L. S. (1997) Pneumatic Conveying. London: Chapman & Hall

Kuhar, M. S., Constantino, D., Dorn, C. A. and Wolf, T. (1999) 1999 State of the Industry Report. *Pit & Quarry*, **92**(6), 30–36

Lieberman, A. (1999) Particle Sizing Data: How to Correlate Results of Different Measuring Instruments. *Powder and Bulk Engineering*, **11**(2), 55–64

Matthews, C. W. (1981) Screening. In Solids Handling, edited by K. McNaughton, pp. 225–232. New York: McGraw-Hill

McCauley, N. (1999) Vibrating and Gyratory Screeners: Proper Installation Yields Top Performance. *Powder and Bulk Engineering*, **13**(12), 35–39

Merrow, E. W. (1985) Linking R&D to Problems Experienced in Solids Processing. *Chemical Engineering Progress*, **81**(5), 14–22

Mining Industry Council of Missouri (1999). http://www.momic.com/society.htm

Miyanami, K. (1997) Mixing. In Powder Technology Handbook, 2nd edn., edited by K. Gotoh, H. Masuda and K. Higashitani, pp. 609–625. New York: Marcel Dekker

Navier, C. L. M. H. (1823) Memoire sur les Lois de Mouvement des Fluides. *Mem. Acad. R. Sci., Paris*, **6**, 389–416

Nevada Mining Association (1999) http://www.nevadamining.org

Powder and Bulk Engineering (1999), Vol. 13, No. 3. Minnesota: CSC Publishing, Inc

Powder Technology (2000). Special issue on Numerical Simulation of Discrete Particle Systems, to appear. Ireland: Elsevier Science

Rice Situation and Outlook Yearbook, Sept. 1998. Economic Research Service, USA

Rosato, A. D. and Blackmore, D. (2000) Segregation in Granular Flows. Dordrecht: Kluwer Academic Publishers (To appear)

Rodriguez, F. (1996) Principles of Polymer Systems, 4th edn., pp. 2–4, Washington, D.C.: Taylor and Francis

Savage, S. B. (1984) The Mechanics of Rapid Granular Flows. In Advances in Applied Mechanics, Vol. 24, pp. 289–366. London: Academic Press

Savage, S. B. (1993) Mechanics of Granular Flows. In Continuum Mechanics in Environmental Sciences and Geophysics, edited by K. Hutter. pp. 467–522. International Center for Mechanical Sciences, Courses and Lectures No. 337. Vienna and New York: Springer-Verlag

Savage, S. B., Pfeffer, R. and Zhao, Z. M. (1996) Solids Transport, Separation and Classification. *Powder Technology*, **88**, 323–333

Solt, P. E. (1999) Pneumatic Points to Ponder. *Powder and Bulk Engineering*, **13**(7), 73–79.

Statistical Highlights of U.S. Agriculture 1998–99, National Agricultural Statistics Services, USDA, http://www.usda.gov/nass/pubs/stathigh/content.htm

Wood, P. A. (1986) Fundamentals of Bulk Solids Flow. Report No. ICTIS/TR31, IEA Coal Research, London

11. FLUIDIZATION AND FLUIDIZED BEDS

JOHN R. GRACE

Department of Chemical and Biological Engineering, University of British Columbia, 2216 Main Mall, Vancouver, Canada V6T 1Z4

INTRODUCTION

One of the most unusual types of equipment used by chemical engineers is the fluidized bed, a device in which solid particles, for example sand or catalyst particles, are made to behave like elements of a fluid, and hence are "fluidized". Once fluidized, the bed of particles not only looks like a liquid, but it has a number of fluid-like properties. For example:

- The surface of the bed becomes horizontal, and if disturbed returns quickly to the horizontal.
- The pressure increases with depth, as in a liquid.
- Objects of lower density float on the surface of the fluidized bed, while denser objects sink to the bottom.
- If a hole is opened on the wall of the containing vessel, particles pour out as a jet, just like water from a puncture in a water bed.
- Bubbles can rise through the bed and burst at the surface as in a liquid. Moreover, the bubble shapes and rise velocities are similar to those of bubbles in liquids.
- If one attempts to stir the bed, there is a resistance as in viscous liquids. The apparent viscosity for gas-fluidized beds is often approximately 0.2 to 1 Pa.s, the latter being similar to the viscosity of maple syrup.

The state of being fluidized, called "fluidization", turns out to be advantageous for a number of important chemical and physical processes designed and operated by chemical engineers. As we have seen above, fluidized beds involve particles, and we must turn to these before exploring how fluidized beds work and why they are useful in practice.

PARTICLES AND PARTICULATE MATTER

Solid particles occur widely in nature, for example as snow, hail or pollen in the atmosphere, sand on beaches, pebbles on the bottom of river beds, mineral ores exposed by ice, and volcanic ash. Many particles are also made and used by humans. Some of these, like milk powder, talcum powder, sugar, salt, toners for photocopiers, detergents, pills, and cosmetics, are themselves the end product sold to customers. Others, like industrial catalysts, pigments, glass beads sold as abrasives, metallurgical powders used as a source of minerals or to manufacture specialized products, coal particles, activated carbon particles employed to remove undesirable components from gases, and wood fibers extracted to make paper, are intermediates in industrial processes. All of these particles are beneficial in practical applications and important to chemical engineers. On the other hand, some particles are nuisances and are undesirable, for example, cancer-causing tobacco smoke grains or asbestos dust in the atmosphere, and these require pollution control and safety measures to prevent harm to humans and the environment.

Whether they occur naturally or are made by human processes, solid particles vary widely in their characteristics. Shapes can range from nearly spherical peas to rounded seeds or pebbles, irregular shapes found in crushed rock, flat flakes of graphite, and needle-shaped rice and wood fibers. For simplicity, the size of a particle is usually described by an "equivalent diameter," defined as the diameter of a true sphere whose volume is equal to that of the particle. In practice, particle equivalent diameters can vary over a very wide range, for example, from nanometers (10^{-9} m) to centimeters (10^{-2} m). When one considers that particle mass and volume are proportional to the cube of diameter, this means that particles can vary in volume over extraordinary ranges, the largest ones, for example, having 10^{21} times the volume of very small ones. The smallest "nanoparticles" are far too small to be seen with the naked eye, while pharmaceutical tablets or berries, for example, at the upper end of what chemical engineers consider to be particles, are large enough to weigh and measure individually. Some typical size ranges of naturally occurring and man-made particles are shown in Figure 11.1.

Other particle properties are also important. The particle density is nearly as important as its size and shape. Many particles, for example those used as industrial catalysts, contain interior pores, and the interior "porosity" or "voidage" (interior void volume divided by total particle volume) is then of direct interest, as is the distribution of pore diameters. Some catalysts are made to have hundreds of square meters of interior surface area per gram of catalyst. In particulate materials (often called powders), the distribution of particle sizes is also of considerable interest since, for example, a powder made of identical particles usually behaves quite differently from one having the same average particle size but a wide distribution of sizes from much smaller to much larger than the average. Hardness, ability to dissipate electrical charges, and surface characteristics are also important in some applications. In addition, interparticle forces are of great importance for fine particulate materials. These forces are usually described as van der Waals forces. An excellent discussion of their origin and importance in the context of fluidized beds has been provided by Visser (1989).

A widely used classification scheme describing certain properties of powders when fluidized by air at atmospheric temperature and pressure was introduced by Geldart (1973). Depending on their average diameter and density, particles are assigned to one of four particle groups:

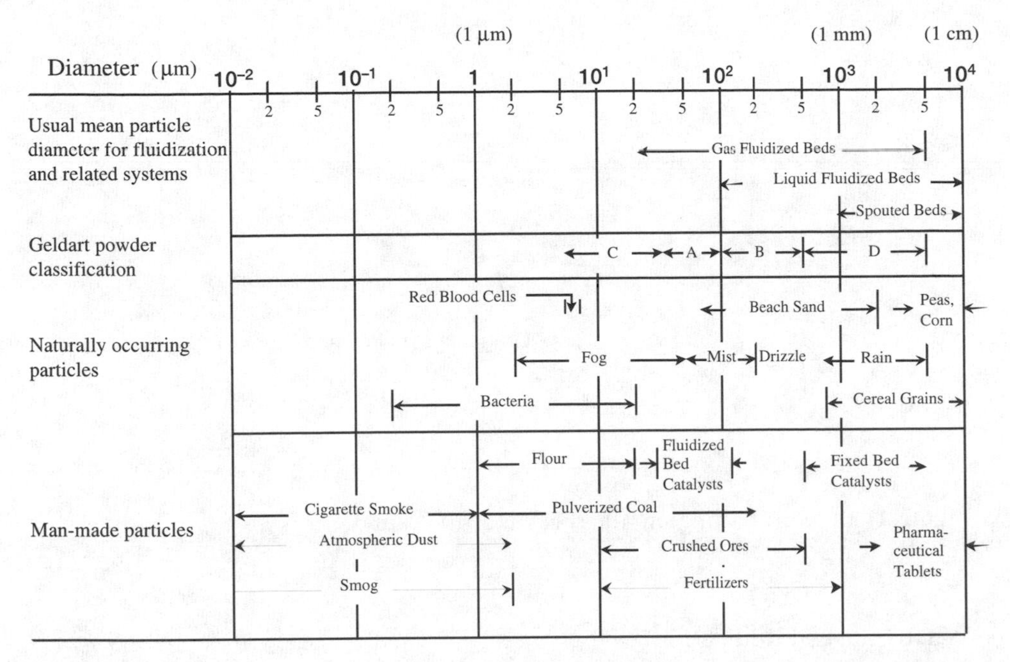

Figure 11.1. Particle size ranges of naturally occurring and man-made particles, and their size ranges for fluidized beds.

- *Group C*: These are the finest particles (usually smaller than about 30 μm). They behave in a cohesive manner, with interparticle forces sufficient to make it difficult to achieve fluidization at all. Flour is a typical example of a group C powder.
- *Group A*: Group A powders are the next smallest particles, typically having an average particle diameter between about 30 and 100 μm. They fluidize very smoothly, with relatively low apparent viscosity. Many catalysts are purposely prepared in such a way that they fall into this size range. The particles then fluidize smoothly, and they are small enough that they offer little resistance to the diffusion of gases into and out of the pores contained within the catalyst particles.
- *Group B*: Beach sand is a good example of a group B powder. The average particle size is between about 100 and 500 μm. Again, these particles fluidize well, but not with the same degree of fluidity and agitation as those in group A.
- *Group D*: Particles in group D are relatively coarse, usually at least 0.5 mm in diameter. When fluidized they tend to behave like more viscous fluids, and to show less mixing than group A and B powders. Wheat and corn are typical examples of group D particles.

The boundaries between these four groups are indicated in Figure 11.1 for a particle density of about 2000 kg/m^3.

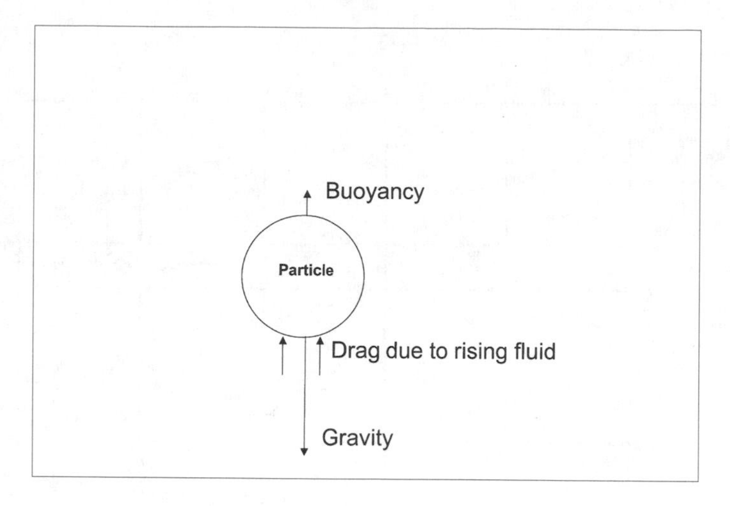

Figure 11.2. Forces acting on a particle in a rising fluid.

What Causes Fluidization?

Fluidization can occur when a gas or liquid is forced upwards through a bed of particulate material resting on (and supported by) a perforated or porous horizontal plate (called a "distributor"), for example a plate in which evenly spaced holes have been drilled to occupy typically 3 to 6% of the plate cross sectional area. Consider the particle shown schematically in Figure 11.2. We assume that the density of this particle is greater than that of the gas or liquid in which it is immersed, in which case the particle weight is greater than the buoyancy force. (It has been shown that particles can also be fluidized downwards by a liquid whose density is greater than the particle density.) Imagine that the upwards flow of the gas or liquid is gradually increased from zero. At first the gas or liquid simply percolates upwards through the spaces between the particles (usually referred to as "voids" or "interstices"). This percolation causes an upwards frictional force, known as "drag" on the particles. Buoyancy also acts upwards, while the weight of the particles (i.e., the force caused by gravity) acts downwnards. As the flow of the gas or liquid is increased, the drag also increases, while the weight and buoyancy remain unchanged. Hence a point is reached where the drag is just sufficient to counterbalance the weight minus the buoyancy. At this point, the interstitial flow of gas or liquid is sufficient to suspend the particles, and they become fluidized. The volumetric flow of gas or liquid divided by the cross-sectional area at this point is called the "minimum fluidization velocity", usually designated by the symbol U_{mf}. For reference purposes, U_{mf} is often compared with the terminal settling velocity, designated v_t, of the particle, which is the final constant velocity achieved by a particle in free fall through the same gas or liquid.

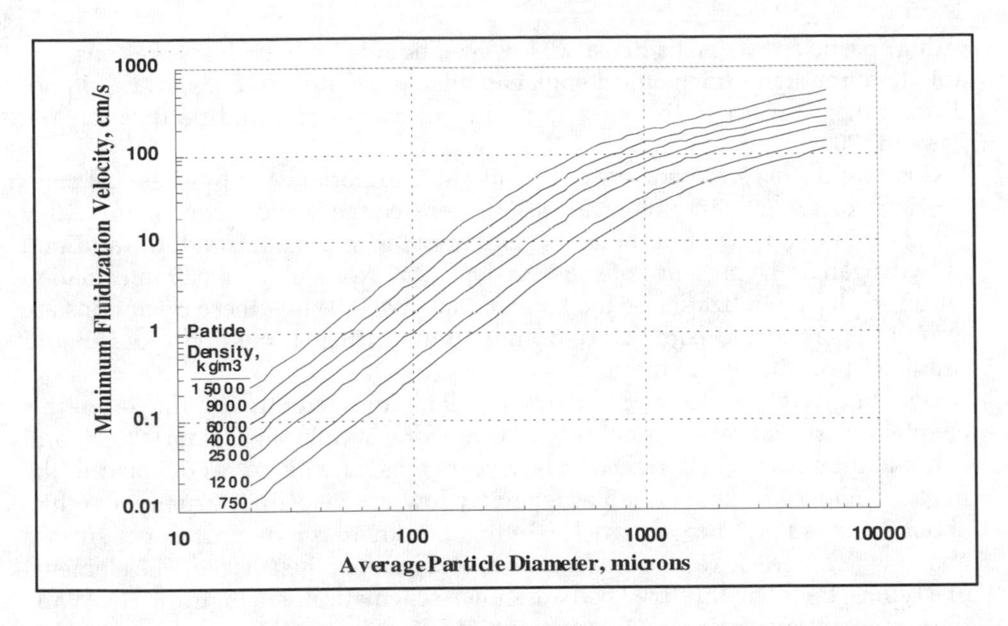

Figure 11.3. Minimum fluidization velocities for particles of different diameters and densities in air at atmospheric temperature and pressure.

Typically U_{mf} is about $v_t/70$ for small particles (e.g., particles smaller than about 60 μm) and $v_t/7$ for large particles (e.g., particles larger than about 2 mm). Typical minimum fluidization velocities for particles in air and water at 20°C and 1 atmosphere pressure are shown in Figure 11.3. It is seen that U_{mf} varies from about 1 mm/s to 1 m/s as the particle diameter increases from about 50 μm to 1 mm, the range of particle sizes of greatest interest for practical applications of gas-fluidized beds.

TYPES OF FLUIDIZATION

Fluidized beds are first categorized according to the type of fluid which is causing the fluidization. Hence it is possible to talk of:

- *Gas-fluidized beds*: The supporting fluid is air or some other gas;
- *Liquid-fluidized beds*: The supporting fluid is water or some other liquid;
- *Gas-liquid fluidized beds* (often called "three-phase fluidized beds"): In this case, both a gas and liquid are flowing, usually, but not always, in the same upwards direction.

Liquid-fluidization is of some importance in natural phenomena, as in "sinking sand," and there are some industrial applications, e.g., in "classifying" (i.e. separating) particles by size or density. However, liquid-fluidized beds are considerably less important to chemical engineers than gas-fluidized beds, so we do not discuss them further in this chapter. For a full review of liquid-fluidization, see Epstein (2001).

Gas-liquid fluidized beds are of considerable importance in biochemical engineering, for example in biological wastewater treatment and fermentation. They are also very significant in the upgrading of hydrocarbons through the addition of hydrogen in the presence of solid catalysts. An excellent source of information on three-phase fluidization is the book of Fan (1989). While these operations are important, space and complexity do not allow further discussion of gas-liquid fluidized beds in this chapter.

Gas-fluidized beds, to which attention will be confined through the rest of this chapter, are the subject of a number of texts, for example see Geldart (1986) and Kunii and Levenspiel (1991). For many years most attention was devoted to the *Bubbling Fluidized Bed* operated at relatively low gas velocities, usually less than about 0.3 m/s for group A particles. In this case there is a distinct upper surface above which particles are splashed as bubbles burst at the surface. The elements of a typical bubbling fluidized bed are shown schematically in Figure 11.4a. While there is some carryover (or "entrainment") of particles, these particles are readily captured and returned to the bottom of the bed.

In recent years the *Circulating Fluidized Bed* has also become popular. A schematic diagram is shown in Figure 11.4b. In this case, the gas upwards velocity

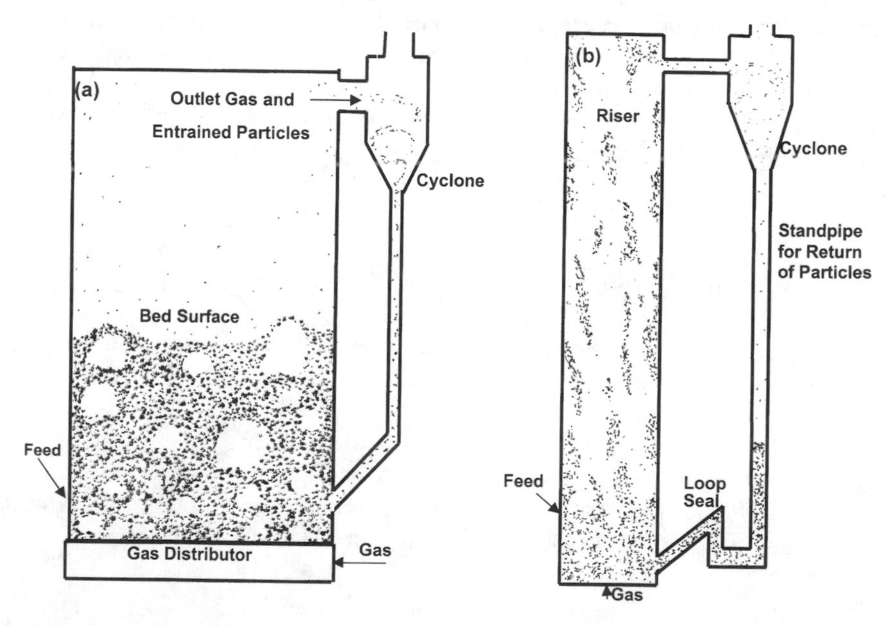

Figure 11.4. Typical geometry and major components of: (a) Bubbling fluidized bed system; (b) Circulating fluidized bed system.

is sufficiently high (typically 4 to 12 m/s) that the solid particles are completely suspended by the gas, so that instead of having a distinct top bed surface, the particles are in continuous circulation in a loop: they are first carried upwards through a "riser," next removed from the gas by a separation device (usually a centrifugal device called a "cyclone"), and then returned to the bottom of the riser through a vertical or angled pipe known as a "standpipe" or "downcomer". Typically the solid particles occupy 5 to 20% of the total volume of the riser. Circulating fluidized beds have become especially important for environmentally clean combustion of coal and for catalytic cracking of hydrocarbons to produce gasoline and a range of other petroleum products. For extensive coverage of circulating fluidized beds see Grace, Avidan and Knowlton (1997).

FLOW REGIMES OF GAS-FLUIDIZATION

Once a bed of particles has been fluidized by a gas, it can behave in one of several quite different characteristic manners, each of which is called a "flow regime". These flow regimes are analogous to the flow regimes which are seen when gas is passed through a true liquid at different flow rates (Bi and Grace 1996). They are shown schematically in Figure 11.5.

1. Bubble-Free Expansion (Figure 11.5a):
   This type of behaviour is found only with particles belonging to group A in the Geldart classification. For a restricted interval of gas velocity when the gas flow is increased above that needed for minimum fluidization, the bed expands relatively homogeneously, with the particles simply moving a little further apart to accommodate the additional gas flow. This flow regime is of little practical interest.
2. Bubbling Fluidized Beds (Figure 11.5b):
   One of the most fascinating aspects of gas-fluidized beds is the spontaneous appearance of bubbles once the gas flow is increased above that required for minimum fluidization (for groups B and D particles) or the so-called "minimum bubbling velocity" for group A powders. The bubbles arise because the bed is no longer able to accommodate all the gas flow while maintaining a homogeneous structure. Bubbles form at the distributor or at the end of gas jets attached to the distributor. They have the shape of spherical caps (rounded top and flat or indented bases) and rise with a velocity of about 0.5 to 1 m/s, depending on the bubble size, growing as they rise due to coalescence with other bubbles. Splitting may also occur, especially when the particles are small enough to be in group A of the Geldart classification. In groups B and D, the bubbles can become very large, sometimes as large as half a meter or more in diameter, causing considerable surging and splashing at the bed surface and large stresses on any tubes or other fixed surfaces that may be located inside the bed. The bubbles also transport particles in their wakes, and this transport is

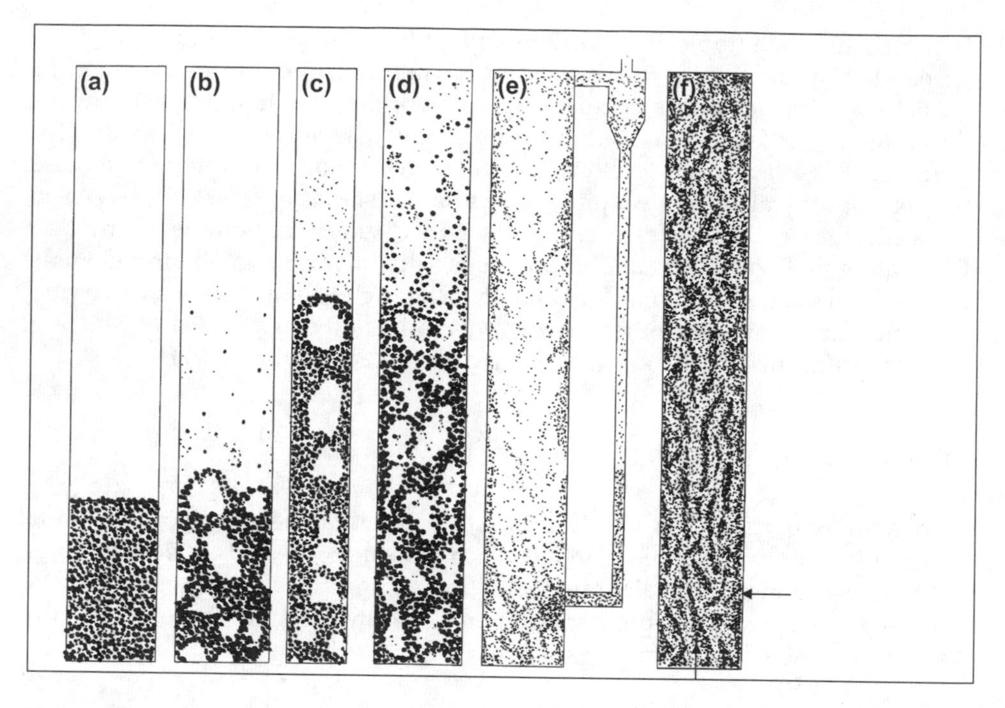

Figure 11.5. Flow regimes for gas-solid fluidized beds: (a) Bubble-free (homogeneous) expansion; (b) Bubbling; (c) Slug flow; (d) Turbulent fluidization; (e) Fast fluidization; (f) Dense suspension upflow. Gas velocity and upwards flux of particles both increase in moving from left to right in the diagram.

responsible for rapid particle mixing, temperature uniformity and excellent bed-to-surface heat transfer, some of the key advantages of fluidized beds in many applications.

3. Slug Flow (Figure 11.5c):
   When the bubbles grow large enough that they occupy most of the cross-sectional area of the column, they are called "slugs" and the bed is said to be encountering "slug flow". This flow regime is analogous to slug flow first discovered in liquid-vapour flows generated when rapid boiling occurs in vertical channels. It is usually not advantageous in fluidized bed processes, but is often encountered in laboratory scale equipment.
4. Turbulent Fluidization (Figure 11.5d):
   Turbulent fluidization is a transitional flow regime which occurs when gas bubbles are no longer stable. Instead small voids continuously break up and reform, causing a chaotic churning motion in which interspersed regions of high voidage dart obliquely upwards, while particles are vigourously agitated. The gas and particles are in excellent contact in this flow regime, and, in addition, maximum bed-to-surface heat transfer usually corresponds to

turbulent fluidization conditions. Hence it is quite common for applications of fluidized beds to exploit the characteristics of turbulent fluidization.

5. Fast Fluidization (Figure 11.5e):
   Fast fluidization is found when particles are continuously fed to the bottom of a tall vessel in which gas travels upwards at relatively high gas velocities (typically 4 to 12 m/s) leading to an overall particle flux of typically 20 to 100 kg/m^2s. This usually occurs within a circulating fluidized bed (CFB), especially in CFB combustors. Fast fluidization results in a central rather dilute core, where gas rises rapidly, conveying well separated particles, surrounded by a relatively dense annular region on the outside of the vessel, where the particles mostly travel downwards in strands and clusters at velocities of about 1 m/s. Horizontal mixing is quite rapid, and heat transfer at the wall is quite favourable, though not as favourable as in the bubbling and turbulent fluidization flow regimes.
6. Dense Suspension Upflow (Figure 11.5f):
   The final flow regime requires both high gas velocities (at least 5 m/s) and high carrying rates of particles (typically 300 to 1000 kg/m^2s.) (Try to imagine a tonne of particles being conveyed through one square meter of area each second!) Now there is no downflow at the wall and the suspension works its way upwards with fine-scale wisps and strands of particles observable at the outer wall of a transparent vessel. These conditions are advantageous for catalytic reactions in which catalyst particles, in contact with the gaseous reactants, rapidly lose their catalytic activity (called "deactivation"). To counter the deactivation, the particles are transported quickly through the reactor vessel, separated from the gas and then sent to a second reactor where they are reactivated. Modern processes that produce gasoline and other hydrocarbon products make use of this flow regime. Even though there are major advantages and practical applications, chemical engineers have only recently begun to understand the nature of this flow regime.

ADVANTAGES AND DISADVANTAGES OF GAS-FLUIDIZED BEDS

As we have seen, fluidized beds are subject to varied and often complex behaviour. Their complexity is sometimes viewed as a *disadvantage*, because it is sometimes impossible to predict exactly what will happen when a new unit is built. Another disadvantage is that fluidized beds can cause particles to break up (i.e. undergo "attrition") as they collide with each other and with the walls of the equipment. Sometimes one is less concerned about the particles breaking up than about wear, usually called "erosion" (similar to sand blasting), on internal surfaces of the equipment. Fine particulates are difficult to separate from gases. Hence when they are entrained from fluidized beds, they represent potential sources of atmospheric pollution, and therefore require special clean-up equipment such as filters or electrostatic precipitators. For reaction applications, fluidized beds can have other disadvantages. First, in the bubbling flow regime there is a tendency for the gas to pass through the bed with limited contact with the solid particles. Second, the particles and, to a lesser extent, elements of the gas do not all spend equal times in the bed, and this decreases the overall effectiveness of the fluidized bed as a chemical reactor.

The above paragraph covers undesirable features of gas-fluidized beds. The fluidized bed also has significant *advantages*, some of the key ones being as follows:

- A fluidized bed has a very uniform temperature. Hence it is able to avoid "hot spots" which might lead to explosions or harmful reactions.
- A gas-fluidized bed has excellent heat transfer characteristics. Heat exchanger tubes therefore require less surface area than in other common types of equipment involving gas flows.
- The particles can be treated almost like liquids in that they can be added and removed continuously. For example, as we have already mentioned, some reactions require continuous treatment of the catalyst to restore its activity, and the fluidized bed provides a ready means of accomplishing this.
- Pressure losses though fluidized beds are generally reasonable. Essentially, one must simply provide adequate pressure to balance the weight minus buoyancy, and increases in gas flow above that required for minimum fluidization generally have little influence on the overall pressure loss of the gas passing through the bed.
- Fluidized beds can be scaled up to very large size. For example, there are many industrial units in operation with cross-sectional areas in excess of 100 square meters.
- Fluidized beds are usually very tolerant of variations in particle properties and significant changes in operating conditions. For example, the particles commonly cover a spectrum of equivalent diameters of at least 25:1, and there is usually a broad range over which the gas velocity may be turned down to respond to changes in demand.

SOME APPLICATIONS OF GAS-FLUIDIZATION

From the preceding section, it will be clear that fluidized beds cannot be used in all applications. However, fluidized beds are now employed throughout the world for a large number of practical applications. More details are given by Hetsroni (1982) and Kunii and Levenspiel (1991). Here are some examples:

A. Purely Physical Applications:

Drying: Wet particles, for example wet coal or wet fertilizer pellets, can be readily dried in fluidized beds using dry air as the fluidizing medium. There are many units in operation. When the material to be dried is sticky, it is common to vibrate the fluidized bed.

Granulation: In spray granulation processes, a liquid is sprayed downwards onto the surface of a vigourously fluidized bed and the particles then grow as the liquid evaporates, creating particles of a desirable size. These units are used, for example, in the pharmaceutical industry.

Coating: If a hot object is dipped into a fluidized bed of plastic particles, the particles can be made to melt onto the surface of the object, forming a uniform coating. This technique has been used, for example, to apply coatings to toothpaste tubes.

Blending and Classification: When a fluidized bed is operated just above the point of minimum fluidization, particles of higher density or larger size tend to migrate to the bottom, while those of lower density or smaller diameter move to the top. Hence the fluidized bed can be used to classify the particles, i.e., to separate them by density or size. On the other hand, when the bed is vigourously fluidized, the mixing processes caused by bubbles and/or turbulence ensure that particles of different characteristics are well blended. Hence fluidized beds are also used to blend particles.

Heat Transfer Baths: The favourable heat transfer offered by fluidized beds means that they can be used for rapid cooling and heat treatment of hot objects. They can also be used to provide uniform temperature environments in the laboratory over a wide range of temperatures.

Medical Beds: An unusual application of fluidized beds is in treating burn victims in hospitals. The patient lies on a sheet stretched across the top of the fluidized bed. Like a water bed, the fluidized bed provides uniform support to the body, but in addition the gentle air flow through the sheet and gentle bubbling are found to be soothing.

B. Catalytic Gas-Phase Reactions:

Fluid Catalytic Cracking: During World War II, the Allied troops required large volumes of aviation fuels, petroleum and diesel fuel for aircraft, tanks and transportation equipment. The existing technology was incapable of supplying the required quantities of these fuels. Development of a new fluidized bed process, called fluid catalytic cracking (FCC), allowed large hydrocarbon molecules to be broken down into simpler molecules, removing an important bottleneck in the war effort. Since then, FCC units have been built around the world and are now dominant reactors in the petroleum industry.

Other Catalytic Fluid Bed Processes: Several other fluidized bed catalytic processes are important in petroleum refining. For example, fluidized bed reformers are used to upgrade octane to higher octane number isomers. The Fischer-Tropsch process has been carried out for a number of years in fluidized beds as a means of synthesizing hydrocarbons from carbon monoxide and hydrogen, thereby producing gasoline in areas of the world where oil is scarce.

A number of other large-scale processes based on fluidized bed reactors are also in operation worldwide. Such products as acrylonitrile, aniline and ethylene oxide are routinely produced in fluidized beds.

Polyethylene and Polypropylene: One of the techniques for producing key polymers for making plastic products involves polymerization reactors where the polymer grows around tiny catalyst grains in fluidized bed reactors.

C. Gas-Solid Reactions:

Combustion and Incineration: Among the reactions where the solid particles are converted, rather than simply acting as catalysts, is combustion. Fluidized bed combustors and incinerators have been built worldwide, ranging from small industrial boilers right up to huge power stations. There are major environmental benefits, in particular the ability to capture sulphur *in situ* and major reductions in NO_x emissions (hence reducing acid rain in the atmosphere). Fuels include coal, wood wastes, agricultural biomass, shredded tires and municipal waste.

Roasting of Ores: In extracting metals like zinc and copper, ores containing sulphur are heated with air in fluidized bed furnaces to drive off sulphur dioxide (which can then be used to make sulphuric acid). The solid products, usually metal oxides, are then further treated to recover the pure metals.

Calcination of Limestone and Phosphate Rocks: In fluidized bed calciners, carbon dioxide and/or water vapour are driven off from carbonates, rocks that contain carbon impurities, or hydroxides.

Cokers: Fluidized bed cokers are large reactors used to thermally crack, and thereby upgrade, hydrocarbon feedstocks. A heavy hydocarbon deposit, referred to as coke, is laid down on the particles causing them to grow in size.

SOME VARIATIONS ON THE FLUIDIZATION THEME

Some variants of fluidized beds have also been investigated by chemical engineers. Occasionally attempts have been made to augment the gravity force by spinning the bed on its axis in a so-called "centrifugal fluidized bed", with the gas then forced to travel radially inwards through the particles. While this can result in compact equipment, the idea has not yet proven to be commercially viable. Research has also been done on beds where the particles are magnetic and are subjected to magnetic fields which then modify the fluidization behaviour. Electrical fields, ultrasonics, external vibrations, pulsing the gas, and internal baffles of many different geometries are other devices that have been explored by chemical engineers in an effort to enhance the performance of fluidized beds, often with some success.

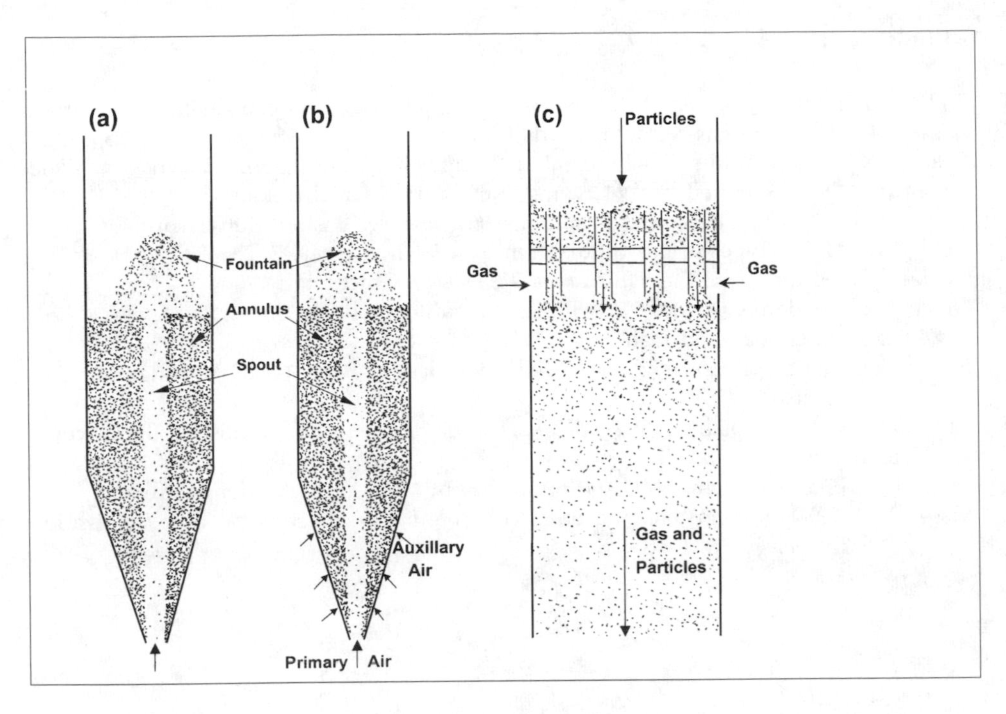

Figure 11.6. Schematics of particulate systems which are closely related to fluidized beds: (a) Spouted bed; (b) Spout-fluid bed; (c) Downer reactor.

Three other cases deserve attention, none of which is strictly speaking a fluidized bed, but all of which share some properties of fluidized beds and are commonly included when fluidized beds are discussed. The first is the *Spouted Bed* (see Mathur and Epstein 1974), which differs from the normal fluidized bed in that the gas all enters through a single hole at the base, causing a central jet or "spout". Usually the bottom section of the containing vessel is a diverging cone. Typically only about two-thirds to three-quarters of the weight of particles is supported by the gas. Spouted beds were first used for drying wheat and other agricultural products, and they are sometimes used in place of fluidized beds when the average particle diameter is bigger than about 1 mm. A schematic diagram of a spouted bed is shown in Figure 11.6a. *Spout-fluid Beds* (see Figure 11.6b) are hybrids of fluidized beds and spouted beds in which a central gas jet (as in a spouted bed) is supplemented by gas entering along the outer cone (distributed evenly as in a fluidized bed). *Downer Reactors* (Figure 11.6c) distribute particles across the top of a vertical column so that they fall vertically downward, with gas simultaneously travelling downwards. The particles then tend to fall evenly spaced and to spend nearly equal times in the reactor, advantages in achieving uniform reaction or drying of the particles.

REFERENCES

Bi, H. T. and Grace, J. R. (1996) Regime transitions: analogy between gas-liquid cocurrent upward flow and gas-solids upwards transport. *Intern. J. Multiph. Flow*, **22S**, 1–19

Epstein, N. (2001) Liquid-solids fluidization. Chap. 9 in *Handbook for Fluidization and Fluid-Particle Systems*, edited by W-C. Yang. New York: Marcel Dekker

Fan, L. S. (1989) *Gas-Liquid-Solid Fluidization Engineering*. Boston: Butterworths

Geldart, D. (1973) Types of gas fluidization. *Powder Technology*, **7**, 285–292

Geldart, D., editor (1986) *Gas Fluidization Technology*. Chichester: Wiley

Grace, J. R., Avidan A. A. and Knowlton T. M., editors, (1997), *Circulating Fluidized Beds*. London, Chapman and Hall

Hetsroni, G., editor (1982) *Handbook of Multiphase Systems*, Chapter 8. Washington: Hemisphere Publishing

Kunii, D. and Levenspiel, O. (1991) *Fluidization Engineering*, 2nd ed. Boston: Butterworth-Heinemann

Mathur, K. B. and Epstein, N. (1974) *Spouted Beds*. New York: Academic Press

Visser, J. (1989) Van der Waals and other cohesive forces affecting powder fluidization. *Powder Technology*, **58**, 1–10

12. GENETIC ENGINEERING

SHINJI IIJIMA

Department of Biotechnology, Nagoya University, Nagoya, 464-8603, Japan

INTRODUCTION

Before the 1970s, DNA was the most difficult molecule to analyze, since DNA molecules are composed of only four kinds of nucleotides that seemed chemically featureless. However, development of so-called recombinant DNA technology changed the situation entirely. Now, DNA has become the easiest of the biological macromolecules to analyze. For instance, DNA sequence determination is much easier than direct amino acid sequence determination.

Recombinant DNA technology includes several principal methods which gave technical breakthroughs and had a dramatic impact on biology and biotechnology. These are DNA analyses (gel electrophoresis, hybridization, DNA sequence determination), DNA cloning, DNA amplification (PCR) and currently progressing DNA microarray and bioinformatics. Among them, PCR, microarray and bioinformatics are bringing about a revolution in genetic engineering and its application in fields such as diagnosis.

In this chapter, we discuss the basic techniques and recent progress of recombinant DNA technology.

RESTRICTION ENDONUCLEASE

Restriction endonucleases were found in the late 1960s to be an important system of bacterial self defense: restriction endonuclease cuts incoming foreign DNA (e.g., bacteriophage) but is inactive against the chromosomal DNA of the host bacterium itself, since host chromosomal DNA is specifically modified so as to avoid digestion.

These enzymes recognize specific 4- to 8-base pair sequences and cleave both DNA strands at this recognition site. For example, a restriction enzyme *Eco*RI recognizes GAATTC and makes staggered cuts in both DNA strands, generating DNA fragments that have a single stranded tail at both 5′ends (Figure 12.1). Since the single strand tails of the DNA fragments are complimentary to those on all DNA fragments generated by the same restriction endonuclease, these single strand regions produce intra- or inter-molecular base pairing. This base paring of sticky ends allows different DNA species cut with the same restriction endonuclease to be ligated by an enzyme, DNA ligase.

Although almost all restriction endonucleases recognize a specific sequence of 4~6 nucleotides, some special enzymes have longer recognition sequences. For

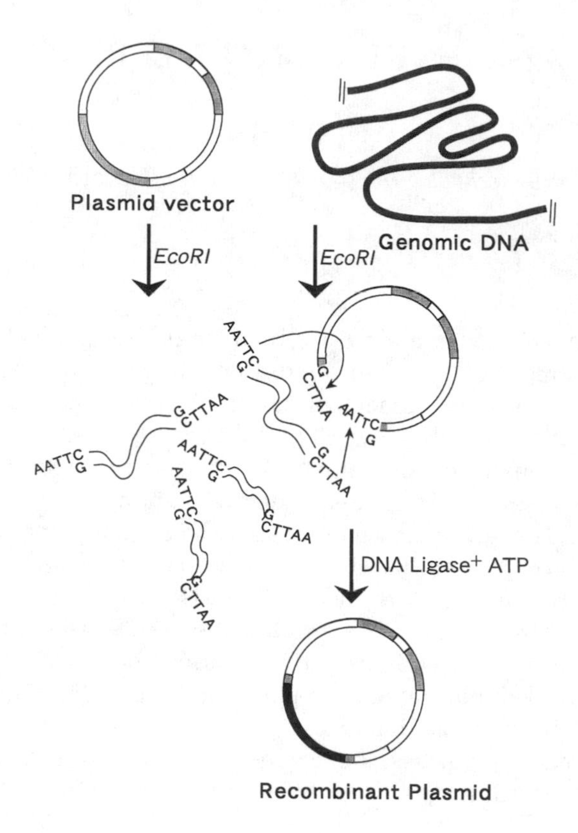

Figure 12.1 Construction of recombinant plasmids by restriction endonuclease digestion and ligation.

instance, *Not*1 recognizes a sequence of 8 nucleotides GCGGCCGC and *sfi*1 recognizes a sequence of 13 nucleotides ($GGCCN_5GGCC$). The frequency of the recognition sequence of *Eco*RI (6 nucleotide recognition) in chromosomes is expected to be once every 4096 base pairs and that of *Not*1 is once every 65, 536 base pairs. Since the size of a gene is 0.5~20 kilo-base pairs, the size of DNA fragments generated by restriction endonucleases is suitable for analyses, cloning and other applications.

DNA fragments generated by restriction endonuclease digestion or other techniques are usually separated by gel electrophoresis on the basis of their length and the molecular weight of the DNA fragment can easily be estimated. Gel electrophoresis can resolve lengths of up to 20 kilo-base pairs. The DNA bands can easily be visualized with a fluorescent dye and the minimum detection level is ca. 1 ng. Agarose or polyacrylamide gels are used for this purpose. Since it takes just 1h for electrophoresis, gel electrophoresis combined with restriction endonuclease digestion to generate DNA fragments is the most basic and principal technique in recombinant DNA technology.

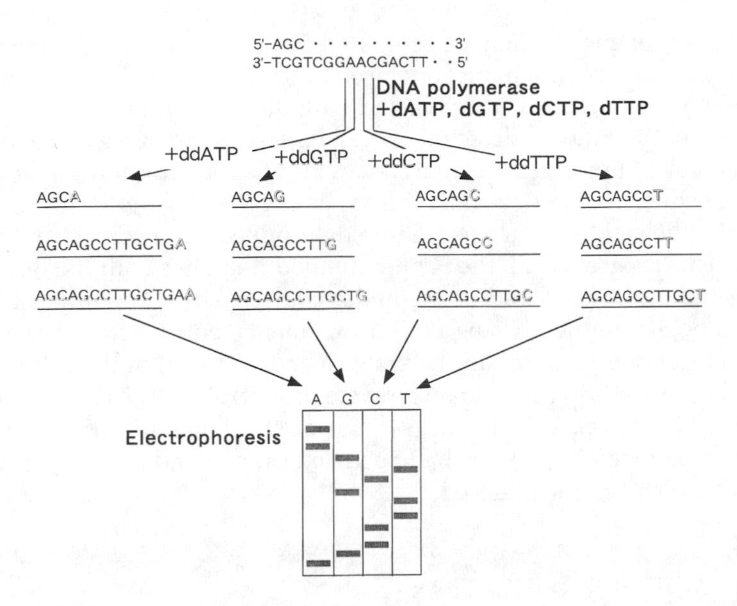

Figure 12.2 Determination of a DNA sequence using the Sanger method. Dideoxynucleotides (ddATP, ddGTP, ddCTP, ddTTP) are dye labeled.

DNA SEQUENCE DETERMINATION

In the late 1970s, a chemical method was developed by A. M. Maxam and W. Gilbert that allowed the nucleotide sequence of any DNA fragment to be determined. A few years later, a method which used DNA polymerase to make partial copies of the fragment to be sequenced was also developed by F. Sanger.

The Sanger method, which is also called the dideoxy-termination method, is now widely used for DNA sequence determination and automated or semi-automated instruments were developed more than ten years ago. In the Sanger method, fluorescent dye labeled primers or terminators (dideoxynucleotides) are used. The terminators are analogs of the normal deoxyribonucleoside triphosphates that lack 3′-hydroxyl group. In a polymerase reaction, template single strand DNA, excess amounts of primers and four deoxyribonucleoside triphosphates (dATP, dGTP, dCTP and dTTP), and fixed amounts of dideoxy-terminators are included. Due to the enzyme reaction, nucleotides are incorporated into the growing strand one by one to complement the template. Chain elongation by DNA polymerase occasionally stops at each step when a dideoxyribonucleotide is incorporated, instead of a corresponding normal deoxyribonucleotide, since the absence of the 3'-hydroxyl group in dideoxynucleotides prevents addition of the next nucleotide. If the mixing ratio of dideoxy-terminators and normal deoxyribonucleotides is suitable, a set of DNAs of different lengths complementary to the template and terminating at each of the different nucleotides of the sequence is produced by the reaction (Figure 12.2).

The products of this reaction are separated by polyacrylamide gel electrophoresis after heat denaturation. In recently used systems, dideoxy-terminators of four normal deoxyribonucleotides are labeled with different fluorescent dyes and the fluorescence of the bands is detected using a laser scanner. In regular experiments, 20~30 reaction samples can be analyzed with one electrophoresis and 200~700 nucleotide sequences are determined from one reaction sample. Therefore, more than 10,000 nucleotide sequences can be determined by one electrophoresis experiment. This progress with the Sanger method, together with the development of automated DNA-preparation methods, enables huge amounts of DNA sequences to be determined. Using these instruments, entire genome sequences of a wide varieties of organisms, including human beings are now undertaken with world wide cooperation (the human genome project) . By 1997, the complete DNA sequences of *Escherichia coli* (4.6 mega-base pairs), a lower eucaryote *Saccharomyces cerevisiae* (13 mega-base pairs) had been determined and a large part of human genome had also been sequenced.

NUCLEIC ACID HYBRYDIZATION

DNA is composed of four deoxyribonucleotides (A, G, C and T) and forms a double helix structure with two strands. This structure is stabilized through the formation of Watson-Crick base pairs: adenine on one strand forms two hydrogen bonds with thymine in a corresponding position on a complementary strand and guanine forms three hydrogen bonds with a corresponding cytosine on a different strand (Figure 12.3). This base paring ensures the correct transfer of genetic information to offspring in addition to the stabilization of the double helix structure.

These hydrogen bonds formed by Watson-Crick base pairs can be broken by heating DNA or subjecting it to an alkaline pH. The temperature needed to separate DNA strands depends on the G, C content of the DNA, since a G-C base pair gives three, and an A-T pair gives two, hydrogen bonds. After separation of the strands, if the process is slowly reversed (that is, slowly reduce the temperature), complementary strands readily re-form double helices. This process is called renaturation. If two different DNA species that have partly complementary sequences are mixed, denatured and then renatured, hybrid DNA molecules that have double helix structures partly in complementary regions, as well as the original DNA double helix molecules, can be formed as shown in Figure 12.4.

This phenomenon is the molecular basis of hybridization. In regular DNA-DNA hybridization, DNA fragments are separated by gel electrophoresis and DNA is then transferred to a membrane by laying nitrocellulose or niron membranes over the gel. Then, the membrane is dipped in an alkaline solution to denature the DNA on the membrane. An isotope or fluorescent dye labeled DNA probe is denatured by heating and the membrane is submerged in the solution containing the labeled probe. After hybridization for a suitable time, unbound probe

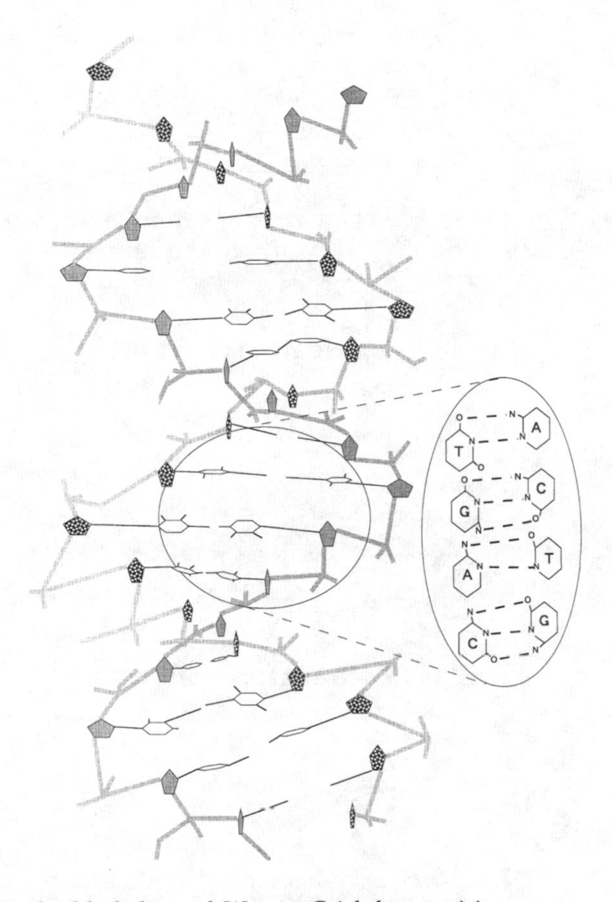

Figure 12.3 DNA double helix and Watson-Crick base pairing.

molecules are washed away and bands that hybridized to the labeled probe are visualized by autoradiography. RNA is composed of ribonucleotides (A, G, C and U instead of T), and DNA and RNA that have complementary sequences can also hybridize. In this case, U forms a base pair with A.

Hybridization has lead to many important discoveries in biology after its development in 1975. For instance, cellular oncogenes, that are the cellular counterparts of viral oncogenes, but can not cause neoplastic transformation, were found by this technique. In one experiment, an oncogene *src* derived from the Rouse sarcoma virus was isotope-labeled as a probe and hybridized with the cellular DNA of normal chicken cells. The existence of a complementary sequence to viral *src* in normal chicken chromosomes indicated that the oncogene *src* of the sarcoma virus originated from a gene in normal cells. By similar hybridization experiments, it was shown that most oncogenes originated from normal cells. This finding was a great breakthrough in the study of carcinogenesis.

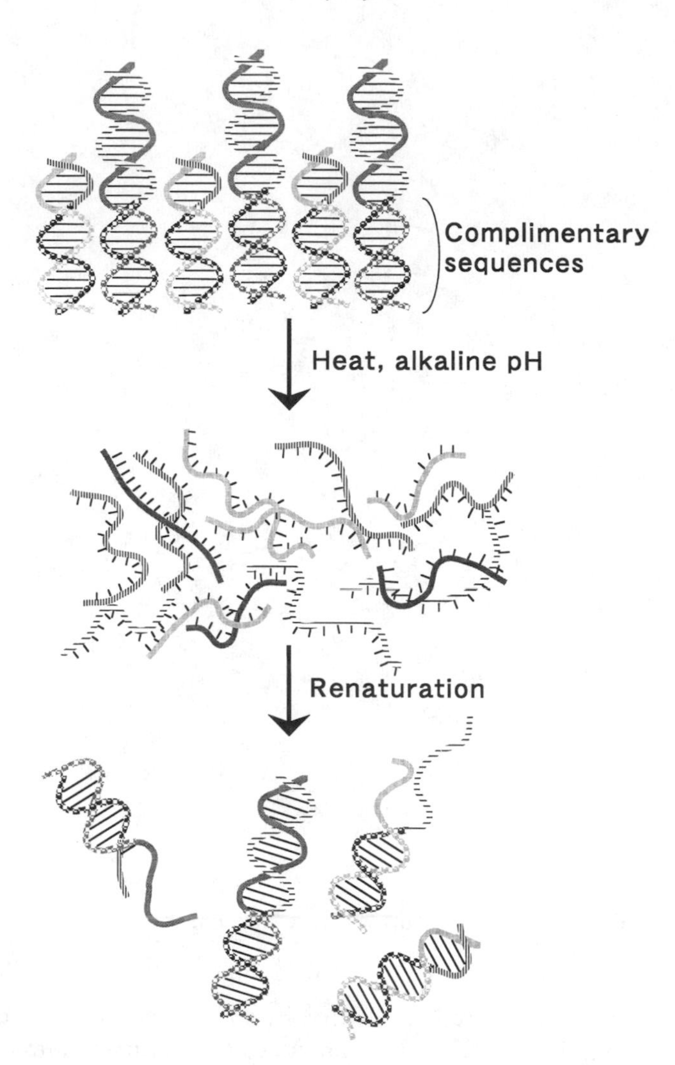

Figure 12.4 Melting and renaturation of the DNA double helix is the molecular basis for hybridization.

As shown in the example of oncogenes, nucleic acid hybridization is a very useful technique to search for homologous sequences to be referred as probes. This technique is also used for the analyses of expression levels of target genes, as in Nothern blotting, by detecting homologous messenger RNA. To locate a specific sequence in a chromosome, *in situ* hybridization is also performed. Since the chemical synthesis of oligonucleotides with a size of 20~30 nucleotides is now easily performed, hybridization using these oligonucleotides is widely used in molecular and cellular biology.

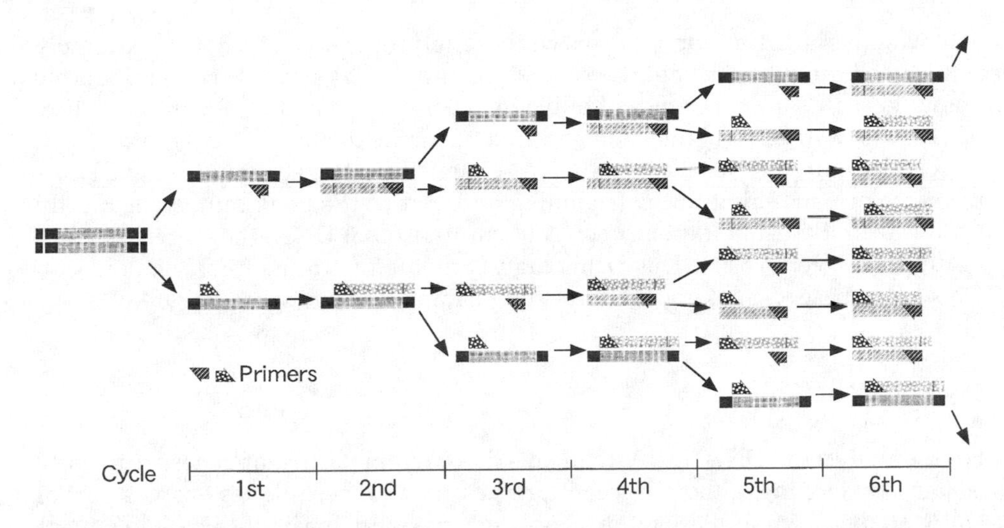

Figure 12.5 PCR can amplify a DNA sequence from only one molecule via *in vitro* DNA synthesis.

AMPLIFICATION OF DNA USING THE PCR TECHNIQUE

The polymerase chain reaction (PCR) was developed almost 10 years ago and is widely used now in genetic engineering, since it provides a quicker and less expensive alternative for cloning. With the PCR technique, selective DNA fragments can be amplified to sufficient amounts even for gel electrophoresis from a trace amount of DNA (DNA mixture). PCR is based on repeated cycles of *in vitro* DNA polymerization from the primer. As shown in Figure 12.5, double-stranded DNA is separated into single strands by heat-treatment as the first step, then samples are gradually allowed to cool down. During cooling, two synthetic oligonucleotide primers anneal with each strand. Each primer is designed to complement sequences at opposite ends of the region to be amplified. Then, DNA polymerase synthesizes the new strand *in vitro*. After polymerization is completed, the next cycle begins with heating. To continue this cycle 25 to 80 times, all the components such as enzymes, double stranded DNA, excess amounts of two primers and four deoxyribonucleoside triphosphates are included in the reaction mixture. To ensure repeating cycles of the reaction, DNA polymerase has to be heat-stable, since the heating temperature of each cycle is above 90°C. The use of heat-stable DNA polymerase from thermophilic bacteria enables PCR. After several cycles of reaction, newly synthesized DNA fragments serve as templates and the predominant DNA is identical to the DNA fragment which has a primer sequence at each end. Since a single cycle of reaction takes 3 to 5 min, whole PCR procedure takes several hours.

Nowadays, PCR is applied to a wide variety of fields, since it is extremely sensitive: it can detect a single copy of DNA, and it works with very small sample amounts such as a trace amount of blood. Therefore, it is used for the detection of very early stage viral infection and in forensic medicine. In the research field, PCR is also extremely useful. If the DNA sequence of a target gene is known, DNA primers corresponding to the beginning and end of the gene can be chemically synthesized. By using these primers and chromosomal DNA, the target gene can be amplified very easily. This technology, combined with the establishment of a DNA data base, is making fundamental changes in cloning strategy.

DNA CLONING

For the analyses of DNA, DNA fragments that contain particular genes are usually isolated by cloning, although recent progress in PCR techniques enables partial DNA analysis without isolation. Since genetics and the molecular biology of *Escherichia coli* have been studied extensively, this bacterium is widely used for host organisms and plasmids (extra-chromosomal-circular DNA) and phages are used for cloning vectors.

Plasmids are circular, double stranded-DNAs that are separated from chromosomes and autonomously replicated. Thus, cloning vectors enable autonomous replication of introduced DNA as carriers. Plasmids occur naturally in bacteria. Artificially constructed, drug-resistant factor derivatives are used as plasmid vectors in *E. coli*. To clone large size DNA from eucaryotic cells, phage vectors are used preferentially. For subcloning, cloning of short sequence ca. below 5-kilo bases or amplification of DNA, plasmid vectors are usually used.

For cloning of a particular gene, chromosomal DNA is cut by a suitable restriction enzyme and a mixture of DNA fragments is ligated to the vector DNA cut with the same restriction enzyme by a DNA ligase. Then, *E. coli* is transformed by this ligated DNA mixture (Figure 12.1). This gene library, constructed by *in vitro* DNA dissection, is screened to identify particular DNA fragments which contain the gene of interest. For the screening, hybridization using synthetic oligodeoxyribonucleotide probes designed and synthesized from the amino acid sequence of a particular gene product is used. In this case, bacterial colonies or phage plaques are directly transferred to a nitrocellulose membrane by overlaying membranes on agar plates on which bacterial cells or phages grow. Then denaturation and hybridization are performed essentially in the same way as DNA-DNA hybridization.

In bacteria, proteins are coded by an uninterrupted stretch of DNA. In contrast, eucaryotic genes are often disrupted by non-coding sequences (intron), and freshly produced messenger RNA (primary transcript) is modified through three main processing steps including 5′-capping, poly A tail addition and splicing before it exits the nucleus. By splicing, intron sequences are removed and coding sequences (exon) joined together. After splicing, eukaryotic messenger RNA is

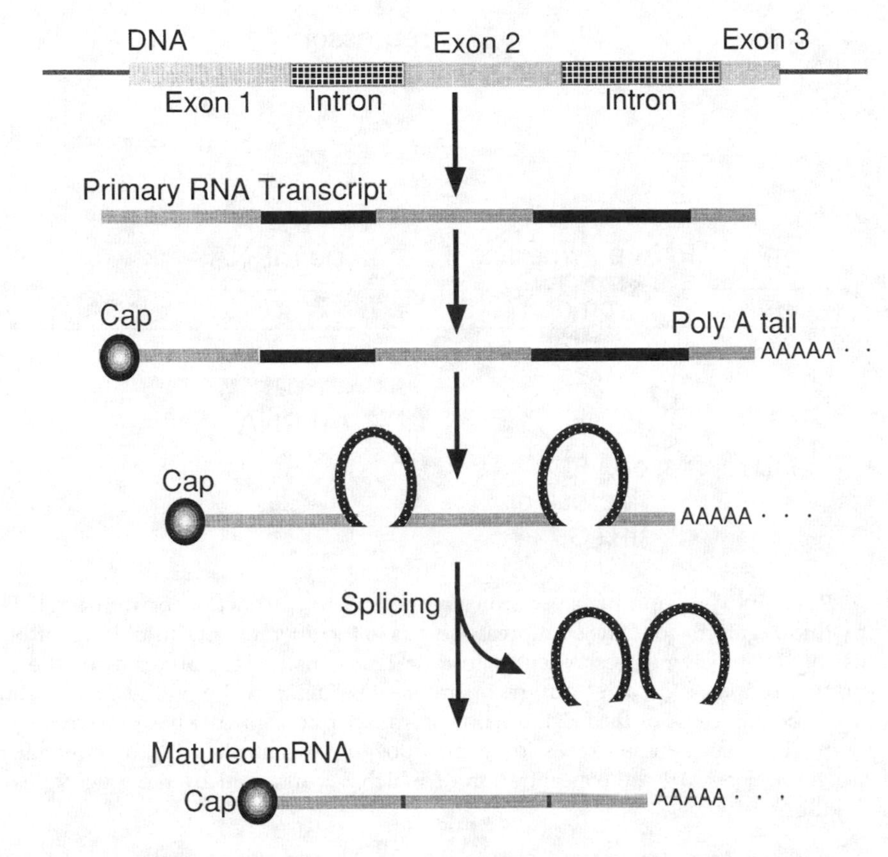

Figure 12.6 Eucaryotic genes are disrupted by non-coding intron sequences. Primary RNA transcripts are matured through three different modifications: 5′ capping, 3′ poly A tail addition and splicing, by which coding sequences (exons) are joined.

functional (Figure 12.6). Therefore, to produce an eucaryotic gene product in *E. coli*, complementary DNA to the matured messenger RNA after splicing is enzymatically synthesized by reverse-transcriptase of RNA tumor viruses. Then, a DNA product named complementary DNA (cDNA) is ligated to a suitable vector and *E. coli* is transformed. To produce abundant amounts of specific proteins in *E. coli* or other particular host organisms, many expression systems have been designed. Since promoters of eucaryotic origin do not work in *E. coli,* the regulatory sequences have to be changed to that of *E. coli*. Now, strong and inducible promoters such as the *E. coli* lactose operon (*lac*) and tryptophan operon (*trp*) are used (Figure 12.7). By using these expression systems, cloned gene products can be produced with up to 30% of *E. coli* total cellular protein under suitable conditions. As shown in Figure 12.7, gene expression from *lac* and *trp* promoters can be induced by simply adding inducing chemicals (inducer) to the culture medium.

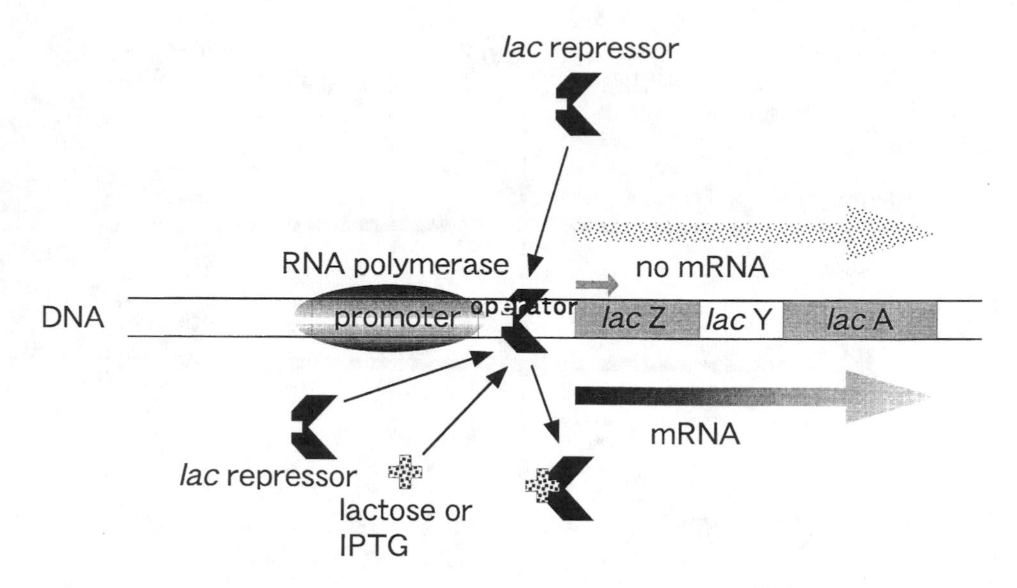

Figure 12.7 Gene expression of *lac* operon. In the absence of lactose or inducer IPTG (isopropylthio-β-galactoside), the *lac* repressor binds to the operator site, inhibiting mRNA synthesis by RNA polymerase. When lactose or IPTG makes complexes with the *lac* repressor, the complexes detach from the operator site. Therefore, by adding IPTG, transcription can be induced. For the full induction of transcription, glucose has to be removed from the medium, since gene expression is also positively controlled by catabolite-activated protein and c-AMP, the concentration of which is controlled by the glucose level in the medium.

DNA MICROARRAY

Hybridization is a very effective method in recombinant DNA technology for identifying homologous DNA or RNA sequences. DNA microarray is based on hybridization, but its striking difference is the miniaturization of its size. Several thousand to several hundred thousand different DNA or synthetic oligonucleotides are immobilized on a slide glass or small silicon wafer chips and this enables simultaneous analyses of huge numbers of genes by hybridization. Note that hundreds of DNAs can be immobilized on a 10 cm × 10 cm membrane by regular dot hybridization. Therefore, the microarray has had profound effects on medical and basic molecular biology, and biotechnology.

Up to now, two types of microarray have been developed. One is a micro array of cDNA. This type of microarray is prepared by spotting DNA solution onto a slide glass or niron membrane. Due to the technical improvements in the spotting machine, the DNA solution can be spotted reproducibly with a spot diameter of 200 μm and the distance of each spot is 300 μm. After hybridization, hybridized spots can be detected with a laser scanner if hybridized messenger RNA is labeled with fluorescent dye. By using this type of microarray, the expression level of genes can be assessed very easily. For instance, gene expression profiles in cancer progression are now under investigation for about 9 typical cancers which may give important information on the mechanism of carcinogenesis and cancer diagnosis. The human genome contains hundreds of thousands of genes. Therefore, once a complete collection of human cDNA is established, a wide variety of applications will be possible.

The second type of microarray is composed of huge numbers of synthetic oligonucleotides (20–25 mer in length) that have different sequences. In this array (chip), hundreds of thousands of oligonucleotides are directly synthesized onto the surface of the substrate within an area of 1–2 cm^2. Simultaneous microsynthesis of oligonucleotides can be performed using nucleotide derivatives with a photo-reactive protection group and photolithographic techniques which are used in semiconductor production. This microchip is suitable for production in large numbers and for high density immobilization of huge numbers of oligonucleotides. This microchip enables the DNA sequence determination and detection of single nucleotide polymorphism via the genetic diversity that causes genetic diseases as well as measurements of messenger RNA expression levels.

BIOINFORMATICS

The human genome project will be completed by the year 2002 and the project is collecting huge numbers of human DNA sequence data. Furthermore, biological and biochemical data such as cDNA sequences and their expression levels under various conditions, including particular diseases and linkage of mutation to genetic diseases, are currently being compiled. For the practical application of this vast amount of data, sophisticated common data management is required. Bioinformatics is defined as a computer-assisted method that transforms huge numbers of DNA, and protein sequences and accompanying data to a ubiquitous assay system for medicine and biotechnology. Now, many researchers are trying to establish various data bases that can be accessed from anywhere in the world at any time.

With DNA sequences, various computer-assisted techniques are being developed for the identification of genes from genomic sequences derived from the human gene project. The use of this sequence data will change the status of molecular biology.

CONCLUDING REMARKS

The discovery of restriction enzymes in late 1960s led to the development of recombinant DNA technology. In late 1970s, a cloning technique enabled the mass production of human gene products which exist in very low amounts in the human body, such as human growth factor and insulin in *E. coli*. After the 1980s, the development of PCR and a high-throughput-type automated DNA sequence determination method facilitated analyses based on DNA sequences. This methodology has had a great impact on medical research and biotechnology. Progress in microarrays and bioinformatics over the last five years has led to the extensive use of sequence data in diagnosis and medication.

13. BIOREACTORS

SHINTARO FURUSAKI

Department of Chemical Systems and Engineering, Kyushu University, Fukuoka 812-8581, Japan

WHAT IS A BIOREACTOR?

"Bio-" means something related with functions endowed with living features. So, chemical reactors using such functions may be called as bioreactors. However, traditional fermentation reactors such as wine, soy sauce and sake fermentors, are not generally considered as bioreactors. They existed hundreds of years ago. The term "bioreactor" is generally applied to recently-developed reactors or reactor systems. Various types of reactors using immobilized biocatalyst systems, membrane bioreactors, reactors for modern environmental treatment systems, and reactors associated with recombinant DNA or cell fusion techniques are examples of bioreactors.

DEVELOPMENT OF BIOREACTORS

Bioproducts coming from bioreactors are very diverse. They are sugars, lipids, proteins or peptides, hormones, vitamins, acids including amino acids, antibiotics and cytokines. Modern and large scale bioprocesses started with penicillin production, whose large scale production was initiated in the early 1940's. Streptomycin production followed penicillin as a cultivation process using large-scale fermentors. The next major large-scale bioprocess was for the production of sodium L-glutamate. These large fermentors became possible by the development of biochemical engineering, which deals with transport phenomena in bioprocesses. In 1969, Tanabe Seiyaku, Co. in Japan started the industrial production of L-amino acids with immobilized enzyme, *i.e.* immobilized aminoacylase. L-acetyl amino acids were hydrolyzed to form L-amino acids, but D-types were not hydrolyzed. Thus, the mixture of L- and D-acetyl amino acids could be separated since only L-types were converted to amino acids. The immobilization technology became important in the development of new bioprocesses. In recent biotechnology, genetic manipulation has become important for the creation of the biocatalysts with desirable properties.

Industrial production of useful substances such as lymphokines and cytokines by animal cells and by recombinant microorganisms started in the 1980's in the United States, and have since been industrialized in many countries. Plant cell culture was industrialized in Japan in 1984 for the production of a pigment, shikonin, and in 1988 for the large-scale production of ginsenoside by a carrot cell, *Panax ginseng*.

Table 13.1 Examples of products from bioreactors.

	Biocatalysts	*Products*
Enzymes	glucose isomerase	fructose
	rennin	cheese
	aminoacylase	L-amino acids
	glucoamylase	glucose
	thermolysin	Aspartame
	aspartase	L-aspartic acid
	lipase	fatty acids, lipids
	nitrogenase	acryloamide
	penicillin acylase	6-amino penicillic acid
Microorganisms	*Escherichia coli* (recombinant cells)	cytokines, insulin, etc.
	Bacillus subtilis	proteases, amylases
	Saccharomyces sp. (yeast)	ethanol
	Aspergillus sp.	citric acid
	Lactobacillus sp.	lactic acid
	Acetobactor sp.	acetic acid
	Streptomyces sp., *Penicillium* sp.	antibiotics
	Clostridium sp.	acetone, butanol
Animal cells	mammalian cells	cytokines, vaccins, antibodies, prostaglandin
	insect cells	antiseptic virus, proteins
Plant cells	*Catharanthus roseus*	ajmalicin
	Digitalis lanata	digoxin
	Panax ginseng	saponin
	Lithospermum erythrorhizon	shikonin (pigment)
	Taxus sp.	paclitaxel (anti-tumor drug)

Several types of bioreactors have been developed, such as stirred vessels, bubble columns and reactors using immobilized biocatalysts. Control systems have been developed significantly through the introduction of computer control, automatic sampling systems, and on-line monitoring systems. Recently, artificial intelligence was introduced into the control of bioprocesses. The construction of expert systems and the application of fuzzy control algorithms are eagerly studied for various fermentors or bioreactors. In fact, analytical treatment of cellular metabolism and growth/reaction behavior is successfully applied these days, and is called as metabolic engineering. This is just one of the contributions of chemical engineering to biotechnology.

PRODUCTS FROM BIOREACTORS

As written above, bioreactors use various kinds of biocatalysts, *e.g.* enzymes, microorganisms, animal cells and plant cells. Examples of the products from bioreactors are shown in Table 13.1.

Reactions using biological functions are catalyzed by enzymes. Enzymes are used *in vitro* in reactors as catalysts to produce various biological products. The enzymes may be solubilized in aqueous phase or often in organic phase. In the latter case, enzymes are lipophilized by attaching amphipathic groups such as polyethylene glycol chains using chemical bonds. In contrast, surfactants are used to solubilize enzymes in organic solvents. The surfactant-bound enzymes are dissolved in organic solvents and convert hydrophobic substrates to useful products. The activity sometimes increases dramatically compared with native enzymes. Different reaction characteristics also appear, which cannot be seen in aqueous media. Also, enzymes may be entrapped in amphiphilic gels like polyethylene glycol or polyurethane to be used in an organic phase. These are several examples of enzyme immobilization, which will be described in the following sections.

When many enzymes are involved in production or the enzymes are unstable in use, whole cells are used in the production system. This is called fermentation or cell culture. Microorganisms or animal/plant cells may be used. Microorganisms are used in various bioprocesses to produce organic acids, saccharides, proteins, etc. Ethanol fermentation is a well-known example of fermentation. Because cofactors such as ATP and NADH are consumed and regenerated in the metabolic pathway, the whole yeast or bacteria cell must be used to produce ethanol. Moreover, genetically modified microorganisms can produce various proteins effectively. Mammalian cells and insect cells can produce proteins with sugar chains, which are effective for human therapy. Plant cells can produce those products which cannot be produced by the other cells. Products include anthocyanins, menthols and various kinds of alkaloids. Several types of bioreactors using free biocatalysts, *i.e.* not immobilized, are shown in Figure 13.1.

IMMOBILIZED BIOCATALYSTS

Biocatalysts such as enzymes and cells can be used as they are dissolved or suspended in fermentation media. However, these biocatalysts are small in size and it is not easy to separate them from the fermentation broths. Instead, they may be attached on solid supports by several methods. By this way, it becomes easier to separate them from the reaction media for reuse. Methods of immobilization are briefly described in Table 13.2. For the case of enzyme immobilization, chemical or ionic bonding is used in many processes, while for the case of cell immobilization, entrapment in a gel is most popular. Several types of bioreactors using immobilized biocatalysts are shown in Figure 13.2.

Membranes are also used to immobilize biocatalysts. The principles are shown in Figure 13.3. In the diffusion type, substrate diffuses into the catalyst side through the membrane, and reaction proceeds. The product diffuses back through the membrane to the substrate side. Thus, the substrate is gradually converted to the product by this process. In the case of a permeation design, the reaction medium is transferred through the membrane into the sweep side. The biocatalyst can be dissolved or suspended in the reactant-side cell, or immobilized in the membrane. In the latter case, the bioreaction occurs inside the membrane. The reactant is converted while it permeates through the membrane.

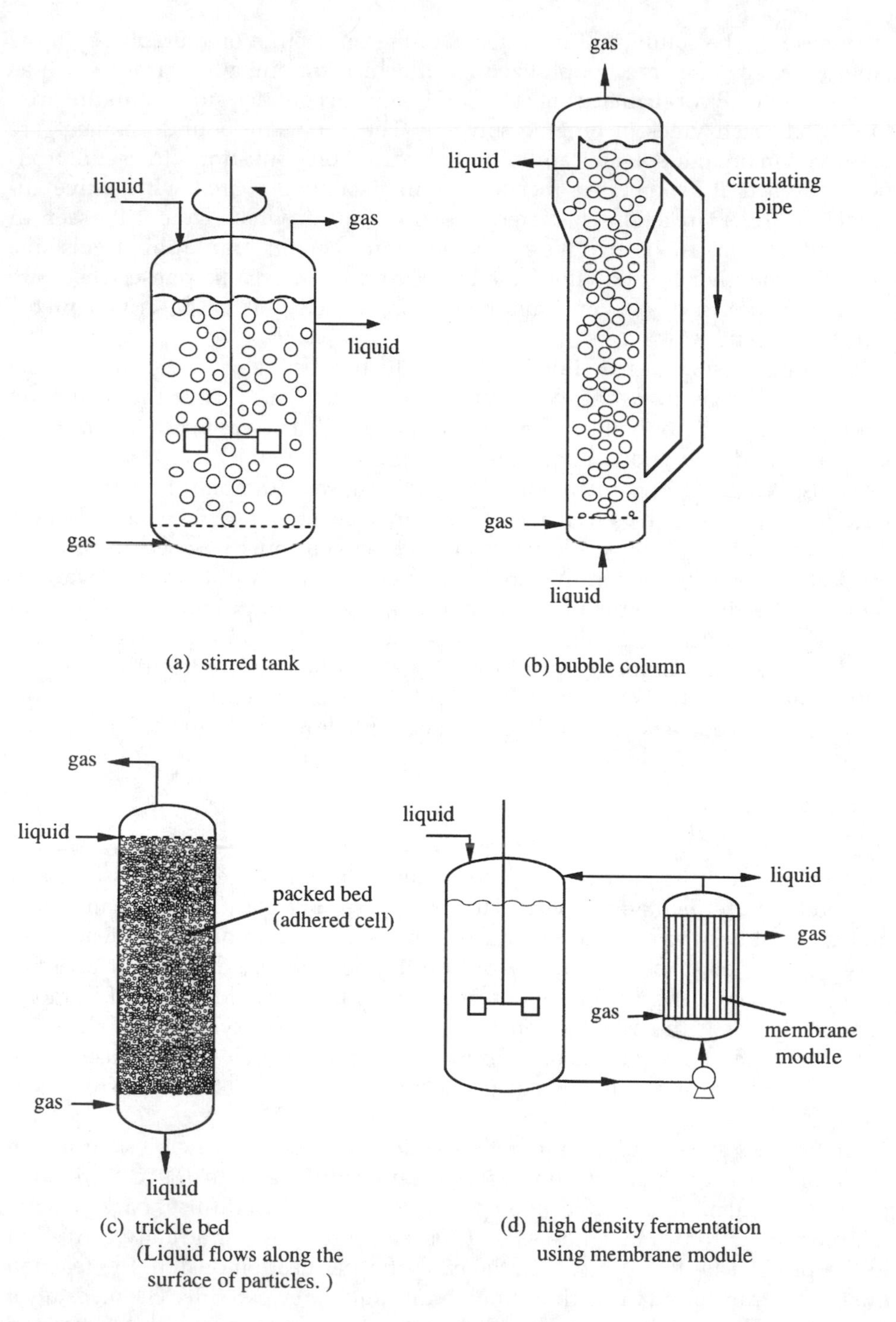

Figure 13.1 Examples of bioreactors using free biocatalysts.

Table 13.2 Methods of biocatalyst immobilization.

Method	*Immobilizing material*
Physical adsorption	silica gel, alumina gel, hydroxyapatite, ceramics
Ionic bonding	ion exchange resins, charged solids
Bridge formation	bridging agent (glutaraldehide, isocyanates, etc.)
Covalent bonding	polymer particles with functional chemical groups
Microencapsulation	cellulose derivatives, collagen, phospholipids, etc.
Entrapment	agarose, poly acrylamide, Ca alginate, κ-carrageenan etc.

BEHAVIOR OF IMMOBILIZED BIOCATALYST

In this section, kinetic behavior of immobilized biocatalyst particles will be described. In order for reactions to occur, substrate must reach the biocatalyst, *e.g.* an enzyme. Since the biocatalyst is immobilized inside the support particle, the transfer process by diffusion is an important step that possibly controls the

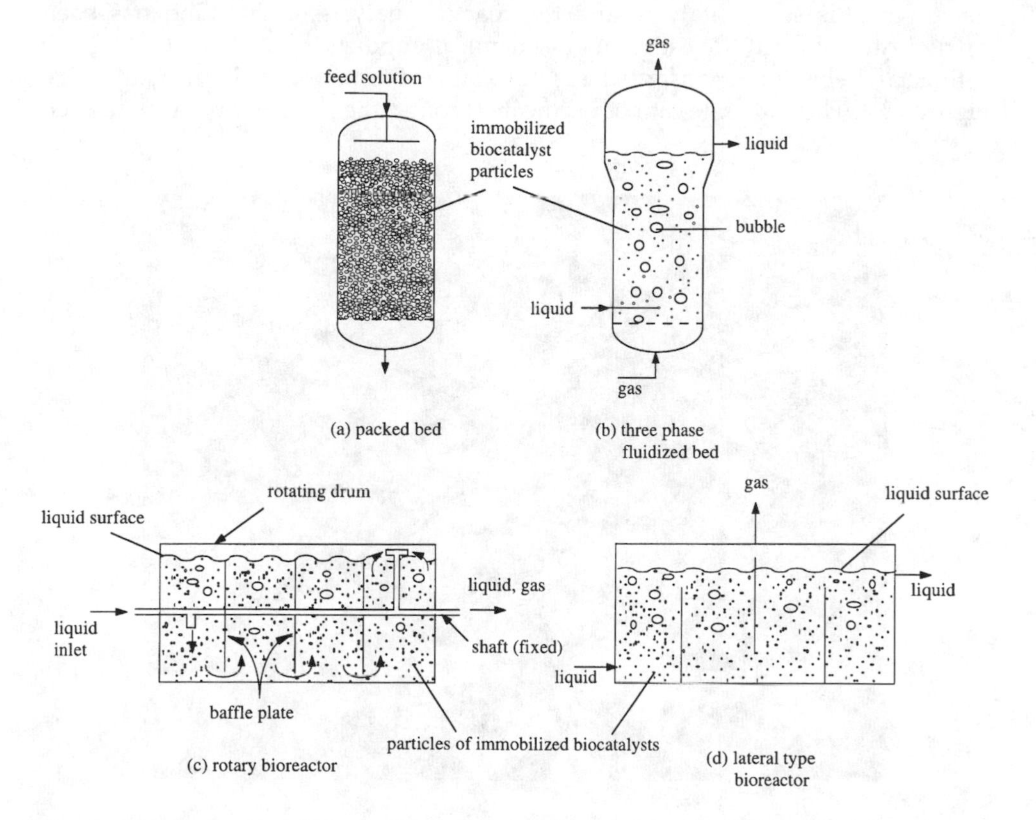

Figure 13.2 Examples of bioreactors using immobilized biocatalysts.

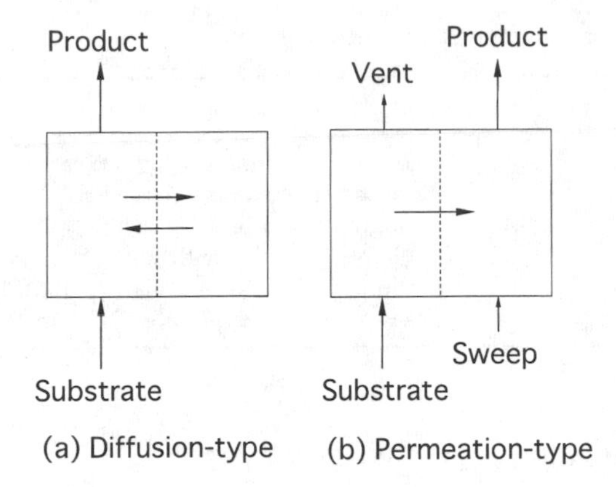

Figure 13.3 Basic style of membrane bioreactors.

reaction. Thus, simultaneous diffusion and reaction should be treated in the analysis of the immobilized biocatalyst. The analysis is similar to that applied for heterogeneous solid catalysis, and the reactor analysis of this kind has been carried out by many investigators of chemical engineering.

Special behavior is observed for the case of immobilized growing cells, Figure 13.4. The substrate cannot permeate through the cell body because the cell

Figure 13.4 Picture of immobilized yeast.

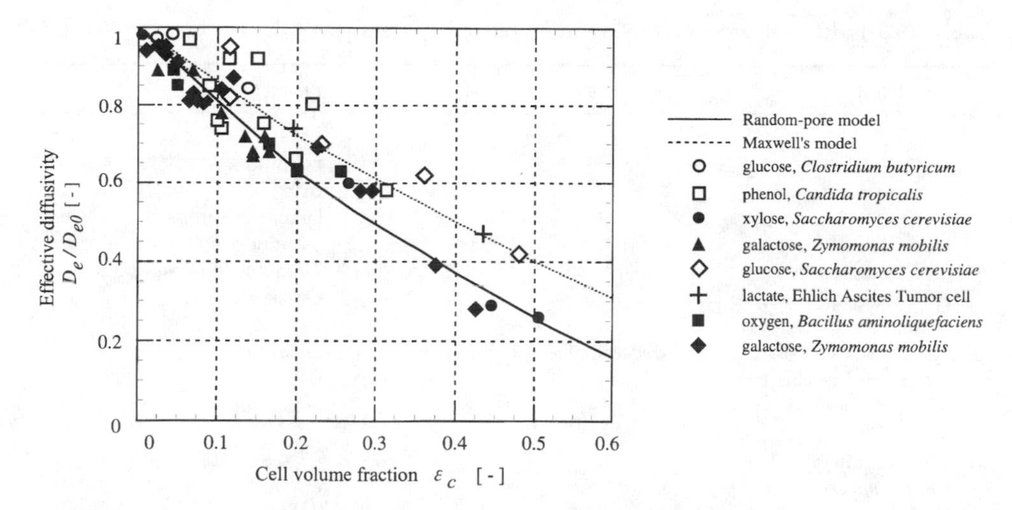

Figure 13.5 Effective diffusivity of substrates in immobilized cells.

membrane regulates their permeation. This phenomenon has been analyzed with Maxwell's model (Maxwell, 1904) or the so-called Randam Pore Model by J.M. Smith and N. Wakao (1962). This was formulated in the author's articles (Furusaki and Seki, 1985; Seki and Furusaki, 1985). The diffusion coefficient in the gel holding living cells is expressed by Eq. 1 following the Random Pore Model.

$$D_e = D_{e0}\ (1 - \varepsilon_c)^2 \qquad (1)$$

where D_e is the effective diffusivity in the cell-containing gel beads, D_{e0} is the diffusivity in the gel without cells, and ε_c is the fraction of volume occupied by cells. From the equation, we know that the diffusivity decreases with an increase in ε_c, or with cell growth. A similar result can be seen with Maxwell's model, as shown in Figure 13.5. The Maxwell model, which was originally derived for electro-conduction in particle-dispersed systems, ignores the influence of the nearby cells on diffusion of materials. Therefore, the Random Pore Model describes the case of beads with densely-grown cells, which is generally encountered in microbial immobilization. Normally, cells are dense in the peripheral area of the particles, because the oxygen and nutrient concentrations are high in this region.

According to the above transport behavior, the overgrowth of the immobilized living cells reduces reactivity. In some cases, oxygen concentration in the central part of the particle reaches below a critical value so that the cells cannot survive. The cells in the center are inactive or even dead. The product distribution from the cells will also change. So, the growth of the immobilized cells should be controlled to stay in a clear optimal regime.

Table 13.3 Industrial processes using immobilized biocatalysts.

Biocatalyst	*Product*
1. Enzyme	
Aminoacylase	L-amino acids
Glucose isomerase	fructose
Lactase, β-galactosidase	lactose-free milk
Penicillin amidase	6-aminopenicllinic acid
7-ACA amidase	7-aminocephalosporanic acid
Aspartase	L-aspartic acid
Fumarase	L-malic acid
Aspartase + L-aspartic acid 4-decarboxylase	L-alanine
α-Glucosyltransferase	palatinose
Nitrile hydratase	acrylamide
2. Microorganism	
Saccharomyces cerevisiae	ethanol
Penicillium sp.	penicillin
3. Animal cell	
Human cell	interferon, cytokines

WHY IS IMMOBILIZATION USEFUL?

In the previous sections, immobilization procedures are described. Industrial processes using immobilized biocatalysts are summarized in Table 13.3. In this section, the merit of immobilization will be described.

1. Biocatalysts can be used again and again.

Consider enzymes dissolved in the reaction media. In order to be used again, the enzymes must be removed from the media by membrane filtration or precipitation by adding salt or non-solvents. It is easy to separate biocatalysts from the reaction media if they are immobilized in particles or with a membrane. This enables repeated use of enzymes many times, unless the enzyme activity decreases significantly. This is true for immobilized cells as well.

2. Continuous operation becomes easy.

Since the separation of biocatalysts and reaction media is easy, the feed of substrates and the removal of products can occur continuously. This provides a steady state operation. When the volume of the feed solution becomes large, continuous operation reduces the cost of operation significantly.

3. The density of biocatalysts in reactors can be made large.

Microorganisms or animal/plant cells can grow in gel supports. The density of cells often exceeds that in suspended culture systems. The microbial density in gels may reach over 10^{10} cells/cm^3 which is more than 100 times higher than that in suspension culture. In this case, the number of cells in the whole reactor will be higher as well. As understood easily, higher cell density results in higher conversion and hence a significant reduction in production cost.

4. Separation of biocatalysts is easy.

After reaction, bicatalysts must be removed from the broth. Since they are immobilized in relatively large supports, it is easy to separate the immobilized biocatalysts from the broth. If the reactor is a stirred tank, filtration with a mesh filter can be applied. Often, packed beds are used as bioreactors with immobilized biocatalysts. Here, the biocatalysts stay in the packed bed reactor while substrate and product flow through. Thus, the biocatalysts can be used repeatedly.

CELL CULTURE TECHNOLOGY

Since the 1980's, bioprocesses using cell culture have been developed. Mammalian cells are used to produce various antibodies and cytokines. There are two types of cells, ordinary cells and cancer cells. The former must grow attached to solid walls. So some sort of carrier is needed in industrial use. Often, positively charged matrices are used as carriers for this kind of cells. The cells in the latter category are immortal. They can grow in suspension or when immobilized. Thus, they can be treated similarly to microorganisms. Insect cell, a sort of animal cell, is also investigated by many researchers. They grow fast and have a strong viability. Production of bacculovirus is one of the applications of insect cells. Bacculovirus is thought to be harmless to human and mammals, but kills pests. It can be used as a pesticide or antiseptic. Also it can be used as a DNA vector for insect cells, *e.g. Spodoptera frugiperda* (Sf-9), to make vaccines, proteins and other useful substances. One of the important topics of the application of insect cells is the production of glycoproteins for medical use. The application of animal cells to produce various kinds of pharmaceuticals is thus increasing.

Plant cells are becoming more important for human health care, environment preservation, and the production of foodstuff. Plant cells grow slowly, so the sterilization must be done very carefully. If properly treated, plant cells produce chemicals without the influence of seasons or climate. Production by cultured cells is shorter than by intact cells; for example, we can obtain products by cultured cells in several weeks but by intact cells in the order of year.

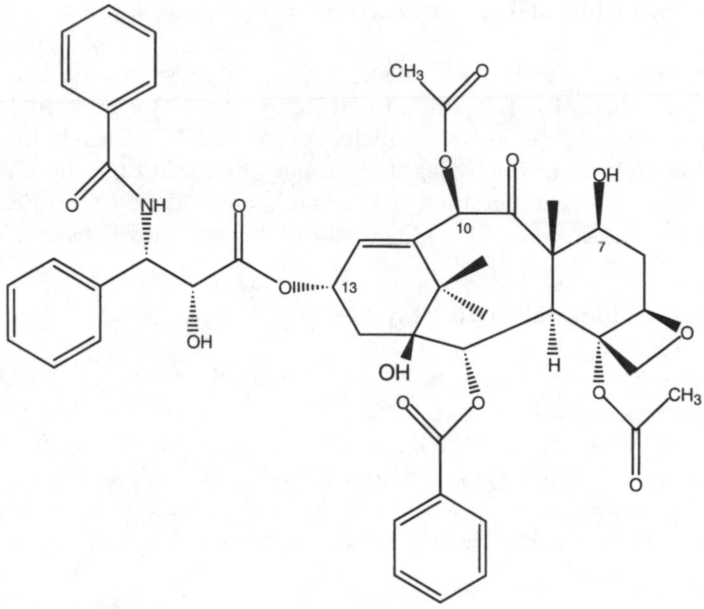

Figure 13.6 Molecular structure of paclitaxel.

An example of the products produced by plant cells is paclitaxel. It is used as an anti-cancer drug. Its molecular structure is shown in Figure 13.6. Since there are many enantiomers, the chemical synthesis is not feasible for production. It has been extracted from leaves of yew trees, but recently the cell culture of *Taxus* sp. (yew) has been developed (Fett-Neto *et al.*, 1992). Since the amount of paclitaxel in the leaves is very small, the consumption of leaves is large. However, by the cell culture process paclitaxel can be produced without fear of extermination of the yew species.

Since paclitaxel produced by yew is excreted from the cells, perfusion culture is used, where the medium with substrate is fed continuously and product is taken out from the cultivating reactor. Plant cells are adversely affected by fluid stress, so immobilization is effective in this culture system. Also, production of paclitaxel is inhibited by the product itself, called feed-back inhibition. The product should therefore be removed simultaneously with production. Thus, an immobilized reactor and a separator of paclitaxel are combined to form an effective production system. This proposal is shown in Figure 13.7.

Regarding bioreactors for plant cell culture, devices to remove the adverse effects of fluid dynamic stress are necessary. One example is immobilization with gel matrices; another is membrane reactors. To achieve gentle mixing, lateral reactors may be used as well. For cultivation of somatic embryos which are expected to be used for artificial seeds, an important tool for production of

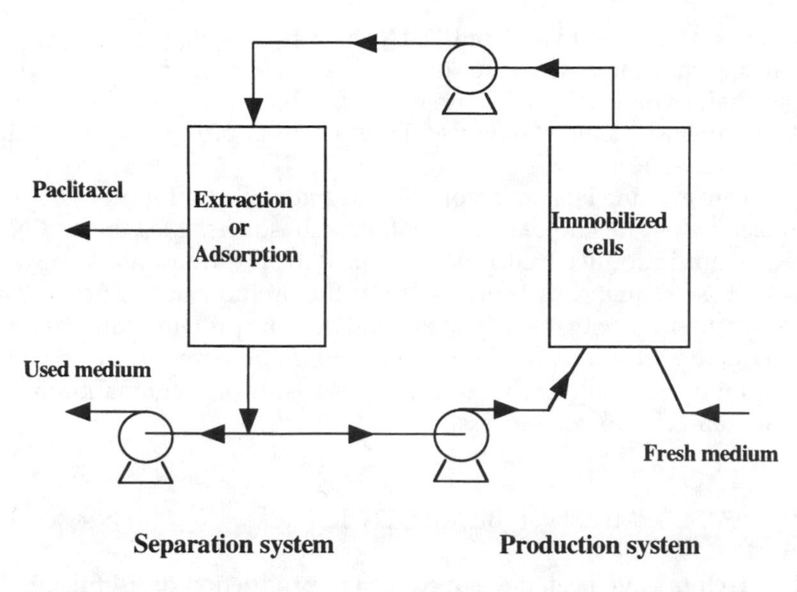

Figure 13.7 A production scheme for paclitaxel equipped with a separation system.

foodstuff, the effect of stress is more serious because of their irregular shapes. Reactor design for somatic embryos is definitely needed. Artificial seeds are considered to be important for food production and afforestation of desert areas.

Another type of culturing plant cells is the hairy root culture. Hairy root is formed when cells are infected by *Agrobacterium rhizogenes*. Ri plasmid from this bacterium is introduced to the chromosome of the plant cell, from which hairy roots come out. The root grows very fast and produces secondary metabolites. However, the growth is observed in a quite unorganized way. Thus, control for the uniform distribution of hairy root in the bioreactor is a salient problem for stable operation. Several new types of the reactor for hairy roots are now being investigated.

RECENT DEVELOPMENT OF BIOCATALYSTS

As described above, several kinds of biocatalysts such as microorganisms, organelles, animal cells and plant cells are used in bioreactors. Genetic manupilation is now commonly applied to modify biocatalysts, particularly microorganisms and animal cells. By changing the amino acid sequence, biocatalyst reactivity and stability can be improved. In some cases, stable viability is observed under extraordinary conditions such as high temperature (above 100°C), or in strong acidic or basic environments. This genelic change is done by cutting the DNA chain and inserting a new base into it. Once the desired chain is prepared, polymerase chain reaction (PCR) is a useful technique to multiply the DNA so it can be introduced into host cells. Here, the hybridized DNA chain is annealed to form two single chains, then a base oligomer of 10–20 DNA called a primer is hybridized with the single chain. Thereafter, the complete DNA chain is formed by the function of DNA polymerase. Thus, the desired DNA chain is proliferated to be applied for gene manupilation.

One of recent exciting bioreactor topics is the microreactor for protein synthesis. In this reactor system, the reaction contains ribosomes. Messenger RNA and amino acids are introduced into the reactor. With the ribosome located in the reactor, the base sequence is translated into the amino acid sequence, and the protein is synthesized with the help of tRNA. Thus, the protein synthesis is carried out in the microreactor *in vitro*. The microreactor is prepared from silicone chips by a lithography technique. This technique is a sort of combination of reactor design and semi-conductor processing.

FUTURE DEVELOPMENT OF BIOREACTORS

So far, bioreactors have been developed for the production of useful substances. They use enzymes, micororganisms and animal/plant cells. Future application will be for environmental remediation. The target may be garbage, soil and/or ash. These amounts are enormous and the bioreactors treating them must be different from conventional bioreactors. Large scale but energy efficient reactor types are desired. Detailed study of the fluid dynamics and transport in such bioreactors must be done for development.

Another important use of bioreactors is for the production of food. Some time of the 21st century, the global population will reach 10 billion. Then, the supply of food is expected to become seriously short. Biochemical engineering must contribute to supply foodstuff in order to overcome this crisis.

REFERENCES

Fett-Neto, A. G., DiCosmo, F., Reynolds, W. F. and Sakata, F. (1992) Cell Culture of *Taxus* as a Source of the Antineoplastic Drug Taxol and Related Taxanes. *Bio/Technology*, **10**, 1572–1575

Furusaki, S. and Seki, M. (1985) Effect of Intraparticle Mass Transfer Resistance on Reactivity of Immobilized Yeast Cells. *J. Chem. Eng. Japan*, **18**, 389–393

Maxwell, J. C. (1904) A Treatise on Electricity and Magnetism, 3rd edn, vol. 1, pp. 435–441. Oxford: Clarendon Press

Seki, M. and Furusaki, S. (1985) Effect of Intraparticle Diffusion on Reaction by Immobilized Growing Yeast. *J. Chem. Eng. Japan*, **18**, 461–463

Seki, M. and Furusaki, S. (1996) An Immobilized Cell System for Taxol Production. *CHEMTECH*, **26**(3), 41–45

Wakao, N. and Smith, J. M. (1962) Diffusion in Catalyst Pellets. *Chem. Eng. Sci.*, **17**, 825–834

14. SEPARATION AND PURIFICATION IN BIOTECHNOLOGY

SHIGEO KATOH

Professor, Department of Chemical Science and Engineering, Kobe University, Nada-ku Rokkodai, Kobe 657-8501, Japan

THE ROLE OF CHEMICAL ENGINEERING IN BIOTECHNOLOGY

Many useful materials have complicated structures which make their chemical synthesis difficult but they have long been produced by utilizing microorganisms in processes such as fermentation and cell culture. In recent years, the production of many high-value pharmaceuticals, such as human interferons, interleukins and peptide hormones, has become possible by the use of bacteria that have been engineered using gene technology and also by cells with desired characteristics obtained by gene technology and/or cell fusion. The biochemical industry is expanding rapidly as a means of producing valuable chemicals. To take one example, recombinant human interferon, which is now widely used for the treatment of cancer and hepatitis, is produced by the use of recombinant bacteria as shown in Figure 14.1 (Staehelin *et al.* 1981). Recombinant bacteria producing interferon were first obtained by introducing plasmids containing the interferon gene. Among them, strains with high productivity were screened and culture conditions suited to their growth were determined. Production processes and equipment were then designed, operated and optimized to produce interferon effectively. Formation of by-products, however, was unavoidable and so it was necessary to separate and purify interferon to the degree of purity to satisfy the pharmacopoeia activity and safety requirements.

Chemical engineering plays an important role in conventional chemical and biochemical industries. In both of these fields production processes consist of physical, chemical and biological conversions together with mass and heat transfer processes: the role of chemical engineers is to design and operate the processes needed for effective and efficient production. In both the conventional and biochemical industries new processes constantly need to be devised. Most of the products of the biochemical industry are either foodstuffs or pharmaceuticals which directly affect our lives and high purity and safety are essential. To satisfy these requirements, separation processes have become increasingly important and account for a large part of the production costs. Management of the downstream processes is, therefore, a major task in the biochemical industries. We will look at some of the requirements for separations in this field and see how they are developed and managed by chemical engineers.

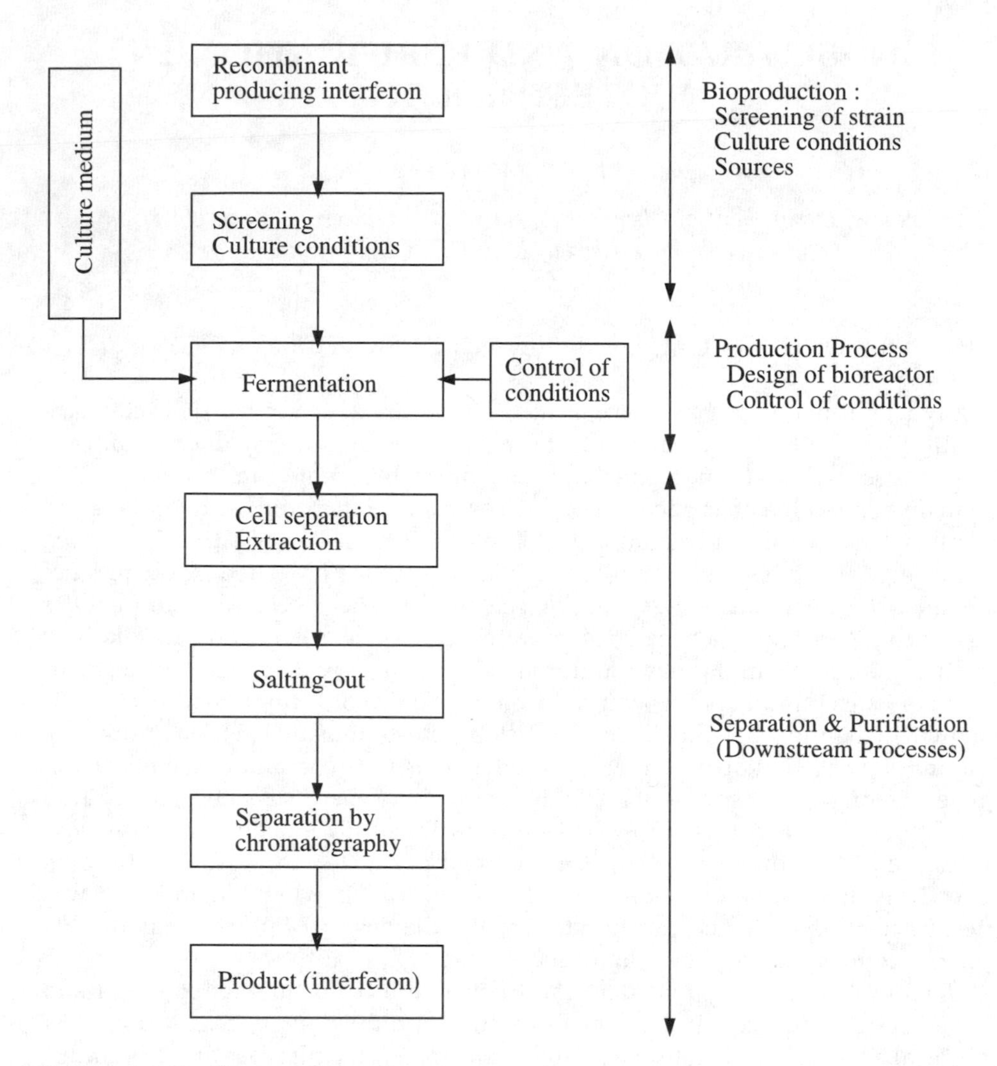

Figure 14.1 Production of interferon by recombinant microorganisms.

REQUIREMENTS IN THE SEPARATION AND PURIFICATION OF BIOCHEMICAL PRODUCTS

Proteins and polypeptides are the most important of the various biochemical products. A scheme for the synthesis of proteins *in vivo* is shown in Figure 14.2. A gene encoding a protein in DNA is transcribed to mRNA by RNA polymerase. Then, according to the information about the amino acid sequence (the primary structure) written in mRNA the protein is synthesized. The characteristics of the amino acid sequence determine partial conformations such as α-helices and

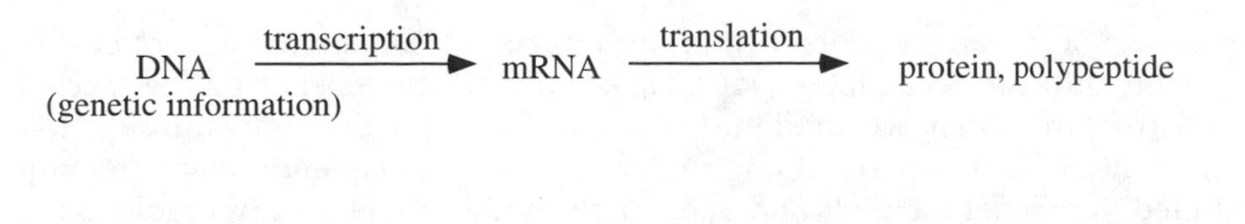

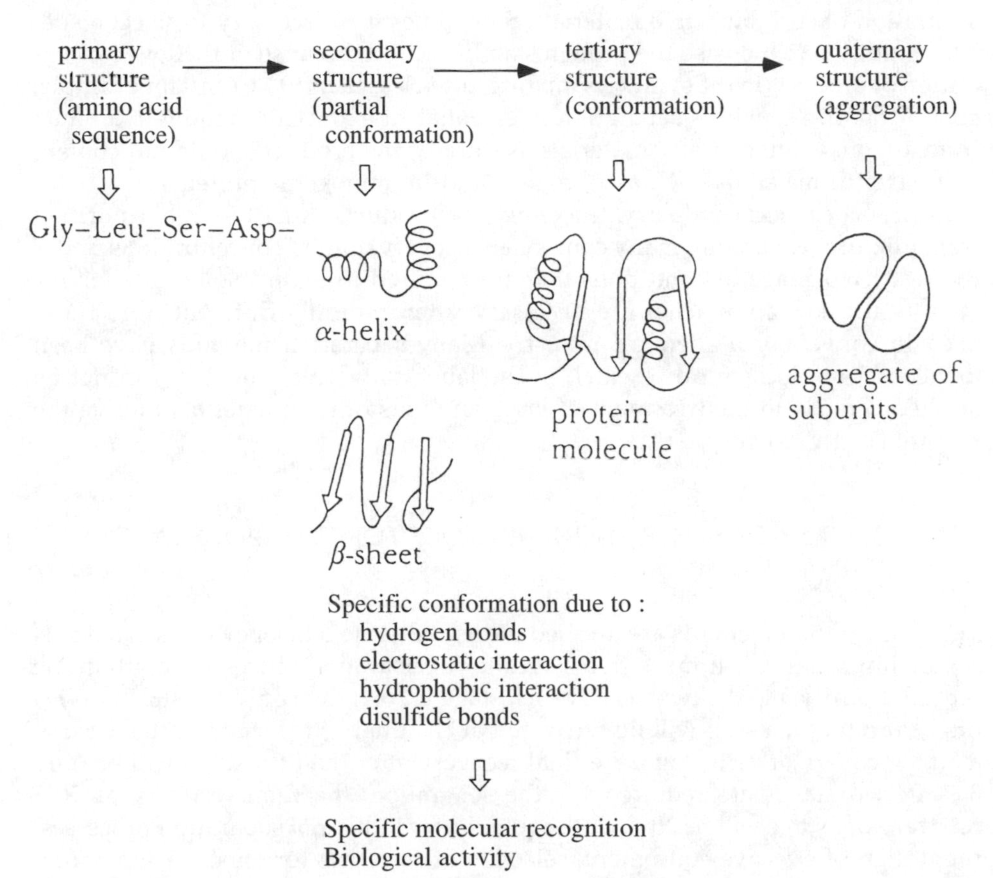

Figure 14.2 Specific conformation of proteins.

β-sheets (the secondary structure) and a specific steric structure of the protein (the tertiary structure). Sometimes several subunits aggregate to form an oligomeric protein (the quaternary structure). The specific biological activity of the protein depends on the steric structure. Since tens or even hundreds of the twenty kinds of amino acids make peptide bonds to construct protein molecules, an enormous variety of proteins with different characteristics can exist.

In the separation and purification of proteins there are many potential difficulties. First, because of the small differences in the chemical and physical characteristics originating from the similarity of their constitutions as polypeptides, separation of the target protein from the mixture of contaminating proteins is extremely difficult. Second, the biological activity, which depends on the specific steric structures as shown in Figure 14.2, may be lost if the temperature is too high or the pH or ionic conditions are inadequate. Consequently there are only a limited number of methods and operating conditions that are applicable to the separation of such bioactive materials. Sometimes it is necessary to select a specific condition to increase the protein stability. Third, because of the low concentration of products in the synthesis mixture, often less than 10^{-6} g/cm^3 for example, concentration as well as separation is essential to extract the required product from a large volume of raw materials. Naturally, the product should not contain any harmful materials such as pyrogens and immunogenic proteins.

In the biochemical industry, therefore, the products must be highly purified from mixtures containing many components at very diluted concentrations under the restricted range of conditions that are required to retain biological activity. As a result new approaches are necessary, often radically different from those used in conventional chemical industry. Many separation methods have been developed in biochemical research in the laboratory scale, but they can not be applied directly to the treatment of large amounts of raw materials to give the required high recovery rates.

DEVELOPMENT OF SEPARATION METHODS FOR THE BIOCHEMICAL INDUSTRIES

When separation methods are applied industrially the efficiency of each stage is of key importance. If it takes five stages for a separation processes to attain the required purity and the recovery at each stage is 70%, then only 17% ($0.7^5 = 0.17$) of the product material will be recovered at the final (fifth) stage. In the case of a 50% recovery at each stage, the final recovery after the fifth stage will be only 3%. If two stages are required for the separation, the final recovery at 50% recovery per stage will be 25%. Consequently, for industrial separation processes, the design of effective equipment, selection of materials for effective separation and determination of optimum operating conditions are crucial if we are to achieve separations with high purity and recovery.

One of the approaches used to achieve the separations required in the biochemical industries is to combine different separation principles which depend on differences of chemical and physical properties such as solubility, molecular size, electric charge and hydrophobicity, between the target material and the contaminants. For example, by applying several separation methods based on different principles in series or by using a separation method based on two or more principles simultaneously, higher degrees of purification can be attained. Some examples are shown later in Figure 14.4 and Figure 14.6.

Another approach is to utilize molecular recognition based on bioaffinity. As shown in Figure 14.2 the steric structure of a protein is determined by the amino

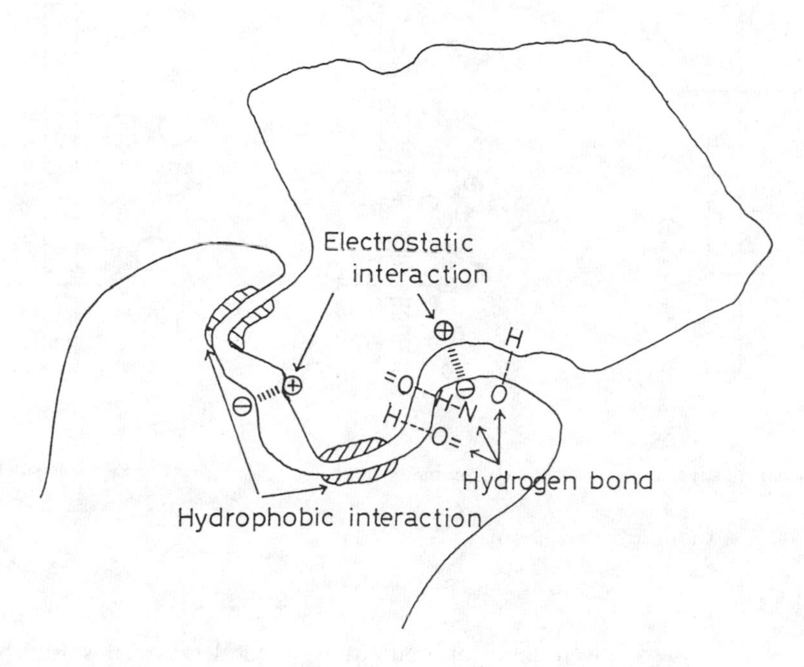

Figure 14.3 Factors leading to bioaffinity.

acid sequence, this steric structure allowing specific recognition between a pair of molecules and various kinds of biological function. This is known as bioaffinity. As shown schematically in Figure 14.3 such affinity results from hydrogen bonds and electrostatic and hydrophobic interactions between a pair of molecules with complimentary steric structures. If one of the two structures differs slightly, the molecules fail to recognize each other and so bioaffinity has a high specificity between complimentary structures. By using separation methods based on bioaffinity, it is possible to separate bioactive materials from mixtures with high selectivity and purity and thus these techniques should have great potential in the biochemical industries. Chemical engineering plays the important role of linking mechanistic studies of the specificity of enzyme reactions, selective transport in biomembranes, signal transmission by hormones and the recognition between the cells, and the development of the new separation methods that utilize these processes.

PRINCIPLES OF SEPARATION METHODS

One important separation method is based on the recognition of difference in molecular size by using a membrane containing pores of a specific size.

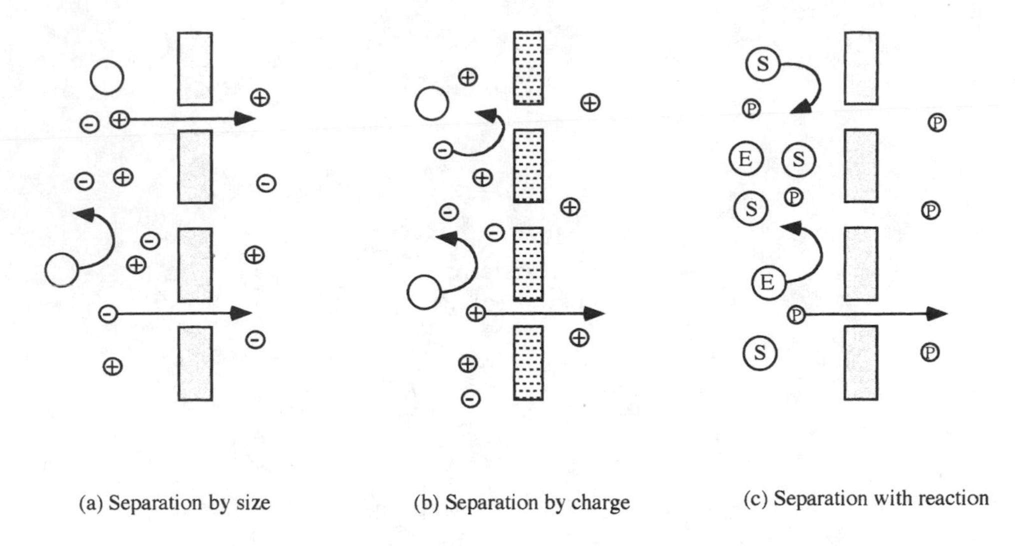

Figure 14.4 Separation by membranes.

Figure 14.4(a) shows how a solution containing several types of solute having different molecular sizes is forced to move through the membrane by the application of pressure. A solute molecule larger than the pores cannot pass through the membrane, unlike the smaller molecules which pass through the membrane to the right hand side. According to the size of molecules, membrane separations are categorized into micro-filtration (about 10 μm–0.1 μm), ultrafiltration (about 0.1 μm–0.002 μm), and reverse osmosis (about 0.002 μm–). Micro-filtration is used for separation of microorganisms from fermentation broth and particulate materials. Ultrafiltration is often used for separation and concentration of proteins, while reverse osmosis is useful for desalination of sea water. In the case shown in Figure 14.4(a), any difference in charge between the smaller molecules is not recognized. If, however, the membrane is negatively charged (Figure 14.4(b)) the negatively charged smaller molecules cannot pass through the membrane and so only the positively charged smaller molecules are separated to the right hand side. Thus we can give the membrane a more highly specific function for separation and attain a higher level of purification. Further, as shown in Figure 14.4(c), if an enzyme (E) decomposes a large substrate (S) into a small product P on one side of the membrane, only the product can pass through the membrane.

By using not only membranes but also columns packed with particles of a suitable cross-linked gel or charged polymers, it is possible to separate solutes as a result of the differences in the partition coefficients between liquid and particle phases; this is known as liquid column chromatography. For example, gel chromatography is a separation method based on the size of molecules, which determines the extent of diffusion of the molecules into the pores of cross-linked

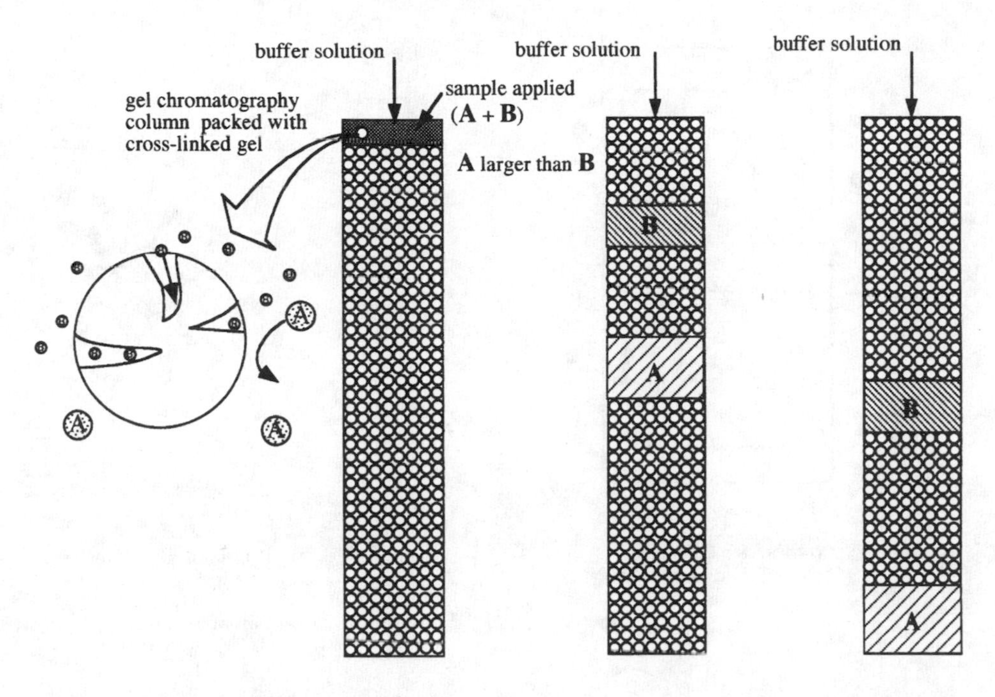

Figure 14.5 Schematic diagram showing chromatographic separation of solutes with different molecular size.

gel packed in a column. If a small volume of a sample solution containing two types of solutes having different molecular sizes is supplied to the top of the column and a buffer solution is continuously supplied to the column, the large molecules (A) are excluded from most of the pores of packed gel, whereas the small molecules (B) can diffuse further into the pores. Thus, as schematically shown in Figure 14.5, the latter molecules take longer time to move down the column and retarded in appearance from the column. Proteins are also amphoteric, changing the sign and amount of their charge depending on the pH. Thus, after supplying a multi-component protein solution to a column packed with particles of anion exchange polymer, to which negatively charged materials are electrostatically adsorbed, a solution with high ionic strength or low pH which weakens the electrostatic forces of the ion-exchange particles is supplied to the column. Each protein in the feed is then eluted from the column according to its electrical character. This chromatographic separation method, based on the electrostatic interaction, is called ion-exchange chromatography. There are many other chromatographic methods utilizing other kinds of interactions, and they are widely used in biochemical industries. Charged solutes with different sizes and charges can also be separated by electrophoresis which utilizes their difference in transfer rate in an electrical field.

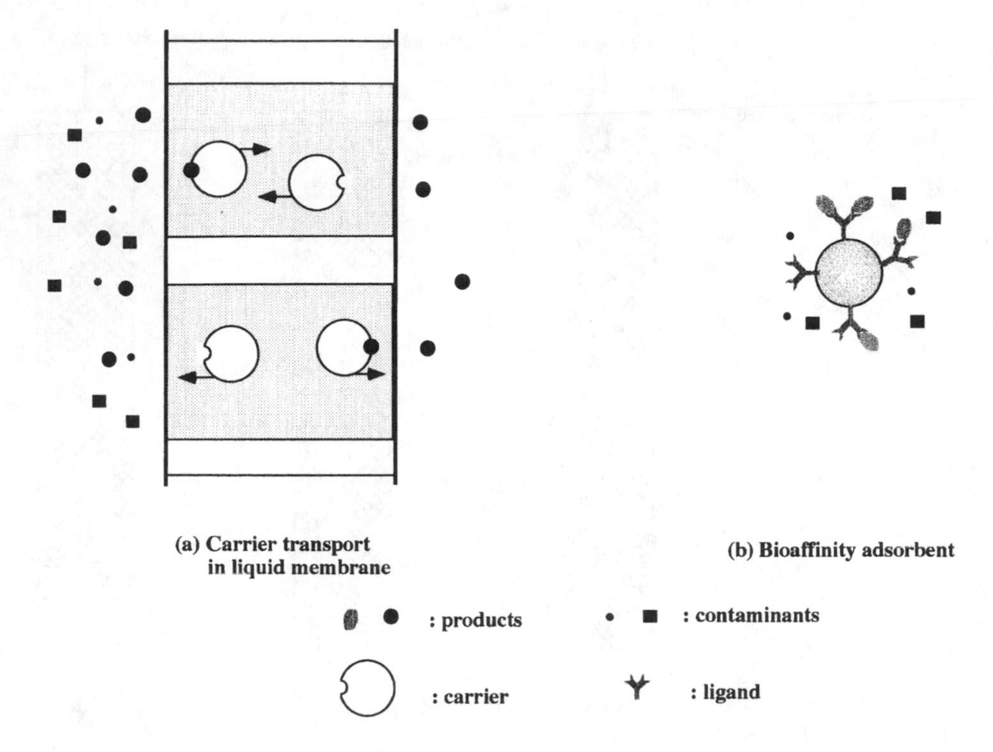

Figure 14.6 Bioaffinity separation.

SEPARATION METHODS UTILIZING BIOAFFINITY

Complemental structures of biological materials, especially those of proteins, cause often specific recognition and various kinds of biological functions, as illustrated in Figure 14.3. There are many pairs of materials, for example enzyme-inhibitor, enzyme-substrate (analogue), enzyme-coenzyme, hormone-receptor and antigen-antibody. Thus, bioaffinity is useful for separation of specific biological materials.

Figure 14.6(a) illustrates how a hydrophobic solvent containing a carrier (○) which is also hydrophobic and shows a specific interaction with a hydrophilic solute A (●) is soaked in a hydrophobic porous membrane to form a liquid membrane (Cussler 1971). Hydrophilic solutes (•, ■) other than A have very low solubility in the solvent and cannot pass through the membrane. Only A, which can combine selectively with the carrier in the solvent, can permeate through the membrane. By using many kinds of carriers in biomembranes and artificial carriers, selective separation is realized by liquid membrane systems.

By immobilizing one of the interacting components (the ligand), which shows a specific affinity to a solute B (●), a selective adsorbent for B is obtained as shown

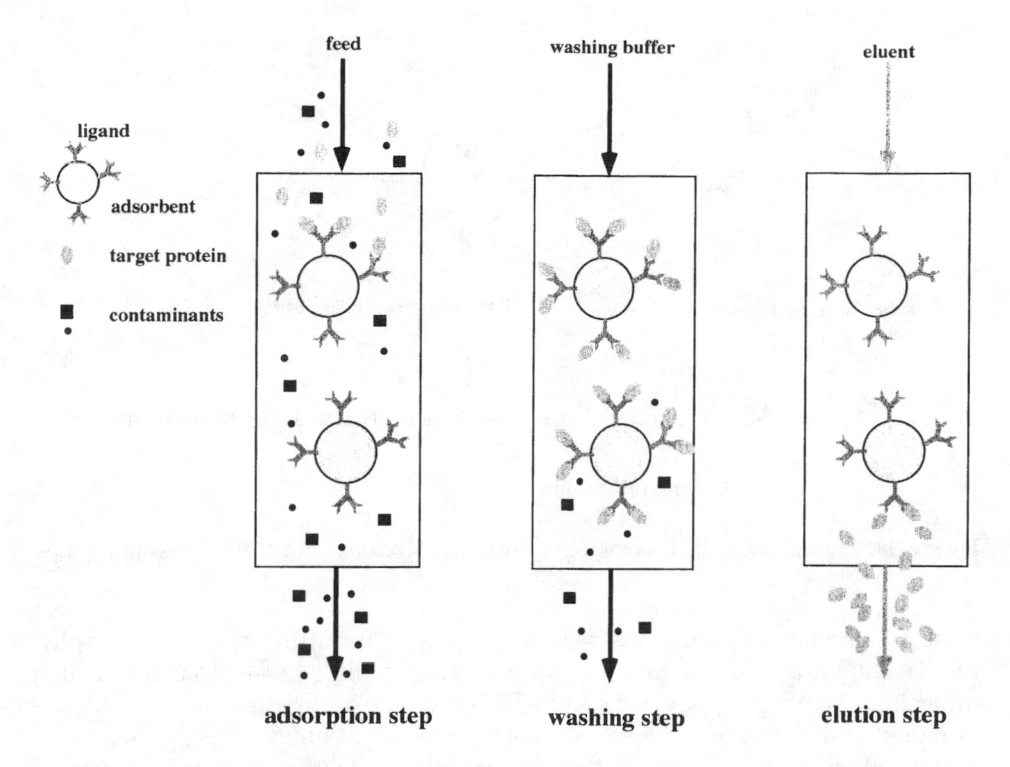

Figure 14.7 Schematic diagram of affinity chromatography.

in Figure 14.6(b) (Katoh *et al.* 1978). This can then adsorb B selectively from mixtures containing many contaminants. The adsorbent is packed into a column (Figure 14.7) and the crude feed solution containing B fed to the column. Only B is adsorbed while the contaminants (•, ■) flow out of the column (adsorption step). After washing the column with a buffer solution (washing step) the desired product B is desorbed (elution step) by changing the conditions, altering the pH or ionic strength for example. This separation process is known as affinity chromatography and, being based on specific bioaffinity, is a particularly useful method to separate bioactive materials with high selectivity. For industrial applications of affinity chromatography selection of ligands which have a high affinity and selectivity for the adsorption step and which lose affinity under mild conditions for the elution step is most important.

By utilizing the immunosystem, which is one of the vertebrate's self-defense systems, ligands for affinity purification of most bioproducts can be obtained. Immuno-response to the invasion of heteroproteins, bacteria or viruses, known as antigens, produces proteins called antibodies which reject them by combining or by activating specific physiological functions. Since the antibody has a high affinity and specificity against the antigen, the antibody produced by the target

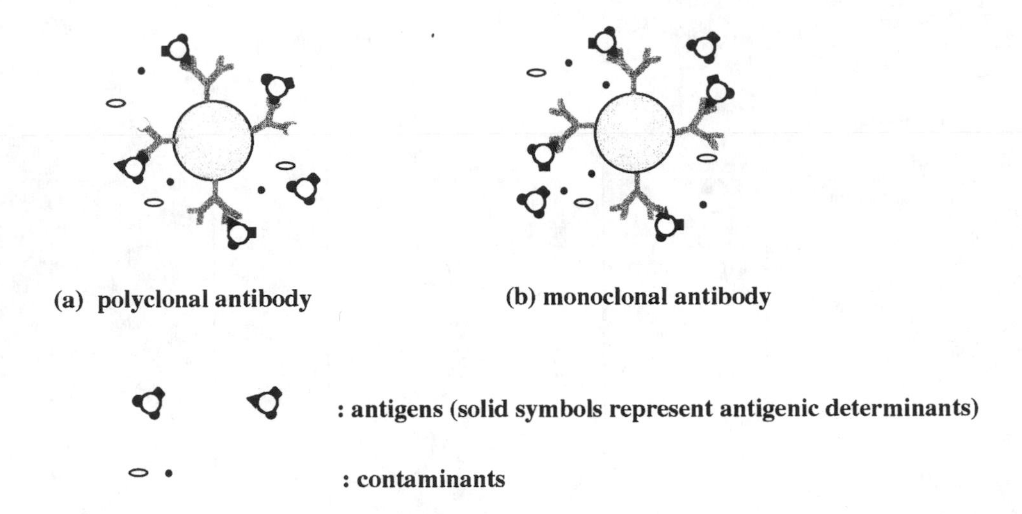

Figure 14.8 Immunoaffinity chromatography with polyclonal and monoclonal antibodies.

material for separation may be used as the ligand for affinity chromatography; this is known as immunoaffinity chromatography. Biological products of high molecular weight have several parts which produce antibodies (antigenic determinants) and can produce heterogeneous antibodies known as polyclonal antibodies which react with one of the antigenic determinants. By using polyclonal antibody as a ligand, the target material (), that is the antigen, can be adsorbed and separated as shown in Figure 14.8(a). Another protein (), however, similar to the target material, which has the same antigenic determinants as the target material, is also adsorbed. Using an antibody which can react with a specific antigenic determinant (a monoclonal antibody) enables antigens with similar constructions to be separated as shown in Figure 14.8(b). Monoclonal antibodies are of great use for the separation of bioactive materials whose activity differs to a great extent as a result of only a slight difference in structure.

Affinity chromatography is likely to be widely used as a powerful separation method in the biochemical industries because it can attain high purity and recovery with only a few separation stages.

EXAMPLES OF BIOSEPARATION AND FUTURE DEVELOPMENTS

Many separation techniques have become available and the choice of a particular technique for a specific application depends on the nature of the product. As an example, the purification factor and recovery of the separation process for interferon shown in Figure 14.1 (Staehelin *et al.* 1981) are shown in Table 14.1. It was purified 1500-fold from the crude extract by several separation steps, and affinity

Table 14.1 Purification of recombinant interferon.

Purification step	*Total protein mg*	*Specific activity units/ng*	*Purification factor*	*Recovery %*
Salting-out by ammonium sulfate	37,100	2.0×10^5	1	100
Immunoaffinity chromatography	30	2.3×10^8	1,150	95
Ion-exchange chromatography	20	3.0×10^8	1,500	81

chromatography step was especially effective for purification.

In the future it will be important to consider and design effective separation processes as one of the very first stages in the development of a new process. The selection of strains and culture conditions producing high concentrations of biological product and low levels of contaminants is of course important. The process of cell disruption and extraction which is often used in the first stage of the process to release the required products from the cells would become unnecessary when the process uses recombinant microorganisms that secrete products from the cells. Combination of production by secretion and affinity separation is shown to be very effective to obtain bioactive products with high purity (Katoh and Terashima 1994).

It is not always easy to find a suitable affinity ligand to a target protein to be purified. In such a case, if a fused protein of the target protein with a tag protein showing a bioaffinity to a specific ligand is produced by the gene technology, the affinity separation of this fusion protein would be possible by use of the predetermined specific ligand (Moks *et al.* 1987).

With the development of biotechnology many kinds of useful biological products will be produced on an industrial scale and used in our everyday lives. This will require an increasing role for chemical and biochemical engineers if the enormous potential of the field is to be achieved.

REFERENCES

Cussler, E. L. (1971) Membranes Which Pumps. *AIChE J.*, **17**, 1300–1303

Katoh, S., Kambayashi, T., Deguchi, R. and Yoshida, F. (1978) Performance of Affinity Chromatography Columns. *Biotech. & Bioeng.*, **20**, 267–280

Katoh, S. and Terashima, M. (1994) Purification of Secreted a-Amylase by Immunoaffinity Chromatography with Cross-Reactive Antibody. *Appl. Microbiol. Biotechnol.*, **42**, 36–39

Moks, T., Abrahmsen, L., Österlöf, B., Josephson, S., Östling, M., Enfors, S.-O. *et al.* (1987) Large-Scale Affinity Purification of Human Insulin-Like Growth Factor I from Culture Medium of *Escherichia coli*. *Bio/Technology*, **5**, 379–382

Staehelin, T., Hobbs, D. S., Kung, H., Lai, C-Y. and Pestka, S. (1981) Purification and Characterization of Recombinant Human Leukocyte Interferon (IFLrA) with Monoclonal Antibodies. *J. Biol. Chem.*, **256**, 9750–9754

15. MEMBRANE TECHNOLOGY

TAKESHI MATSUURA

Industrial Membrane Research Institute, Department of Chemical Engineering, University of Ottawa, P.O. Box 450, Stn. A, Ottawa, Ont. K1N 6N5, Canada

MEMBRANE SEPARATION

In every membrane separation process there is a membrane that is placed between two phases. One phase is called feed and the other is called permeate (Figure 15.1). A flow of mass is induced from feed to permeate by applying various driving forces. When the feed consists of two or more components, and some of those components flow faster than others through the membrane, separation of the feed mixture takes place. Different driving forces can be conceived, but the difference in pressure, concentration and electrical potential between the feed and permeate side are by far the most popular in industrial applications.

REVERSE OSMOSIS, NANOFILTRATION, ULTRAFILTRATION, MICROFILTRATION

Reverse osmosis is a membrane separation process by which a solution is separated into its components. A solute, which is either a small organic molecule or

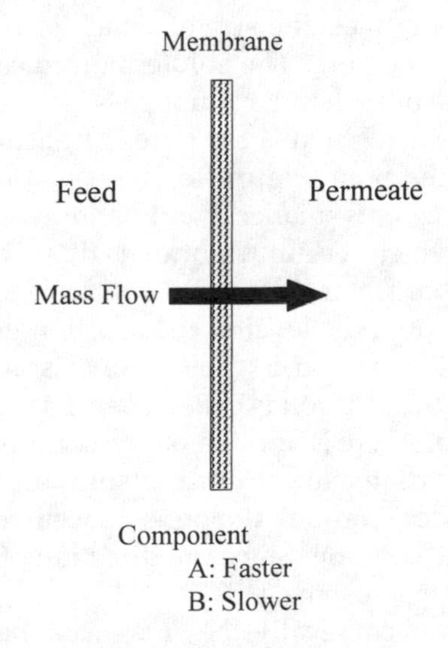

Figure 15.1 Schematic representation of membrane separation.

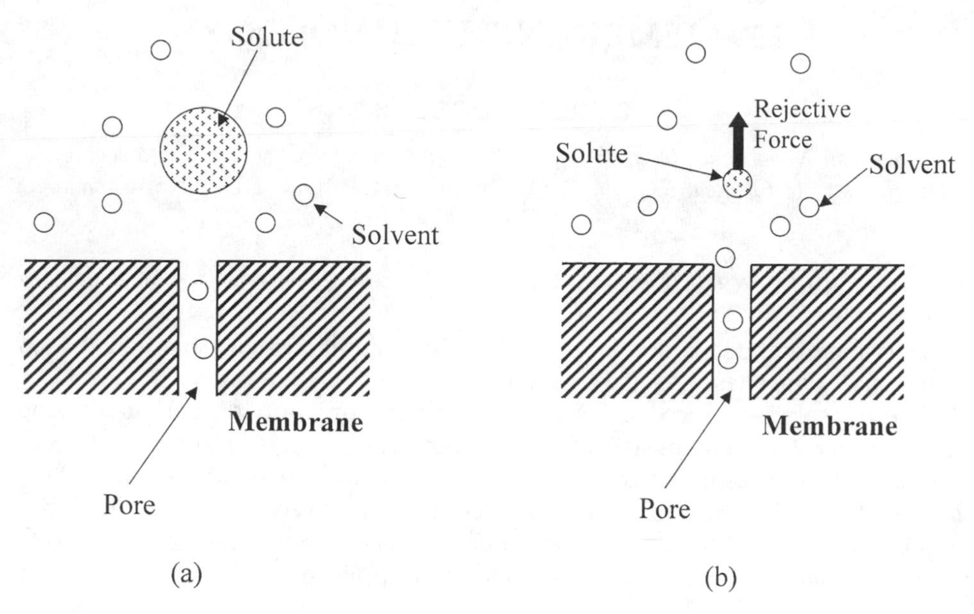

Figure 15.2 Mechanism of membrane separation.

an inorganic electrolyte, or a mixture of solutes, is separated from solvent, which is in most cases water. Pore sizes as small as less than one nanometer (10^{-9} m) are required to enable the separation of such small substances. When a high pressure is applied on a feed solution that is brought into contact with one side of a membrane, water molecules, the size of which is as small as a fraction of 1 nanometer, pass through the pore. The solute molecules or electrolytes that are dissolved in water, on the other hand, cannot pass, since they are either too large to enter the pore or they are rejected at the pore entrance by some forces working against their entry into the pore (Figure 15.2).

When a membrane with pores smaller than 1 nanometer is placed between pure water and a solution in which sodium chloride is dissolved, *e.g.* sea water, water starts to flow from the pure water side to the solution side (Figure 15.3). As more water flows into the solution, the level of the solution goes up until the flow of water will stop. This phenomenon is called osmosis and the difference in the levels of pure water and the solution is called osmotic pressure. When a pressure higher than the osmotic pressure is applied on the solution side, the flow of water is reversed; *i.e.* water starts to flow from the solution to the pure water side. Sodium chloride cannot flow through the pore; hence pure water can be obtained from the sodium chloride solution. Since the direction of water flow is reversed, this process is called reverse osmosis.

Figure 15.4 illustrates schematically the cross-sectional structure of a reverse osmosis membrane. A dense skin layer with a pore size less than 1 nanometer

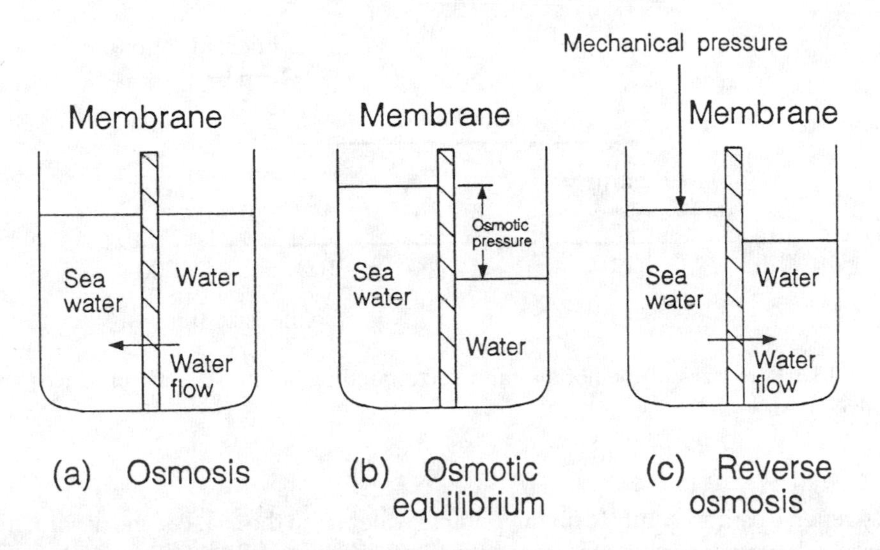

Figure 15.3 Principle of reverse osmosis.

is supported by a porous sublayer that provides the membrane with mechanical strength. The entire thickness of the membrane is about 0.1 mm, whereas the thickness of the dense layer is only 30 to 100 nanometers (Matsuura 1994). A membrane with the aforementioned structure is called an asymmetric membrane. The top dense layer and the porous sublayer of an asymmetric membrane may be prepared from the same material, or they may be prepared separately from different materials.

The most successful application of the reverse osmosis process is in the production of potable water from sea water. Fishing boats, ocean liners and submarines

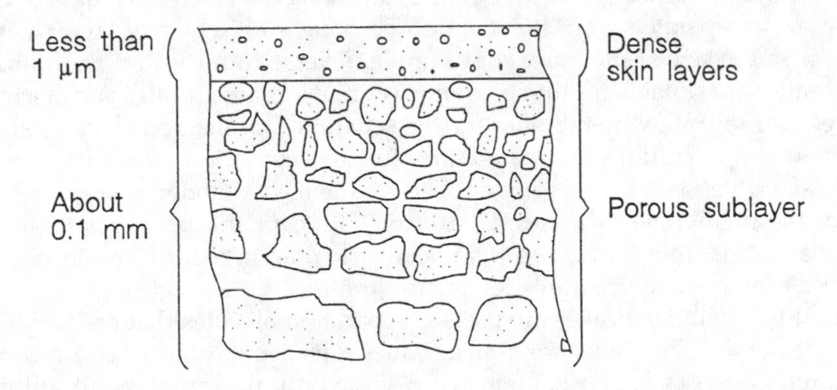

Figure 15.4 Schematic representation of the cross-section of an asymmetric membrane.

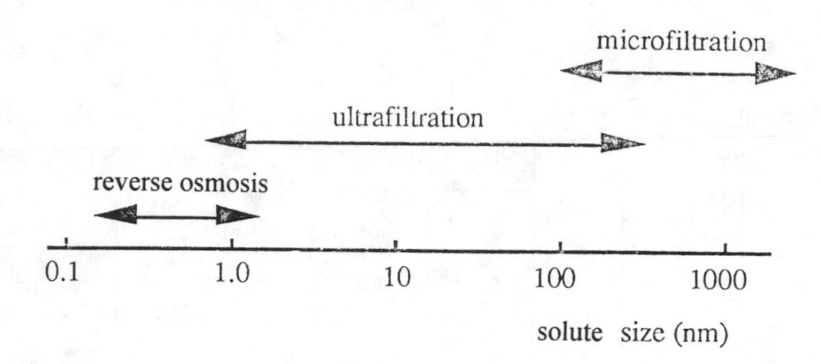

Figure 15.5 Pore sizes of membranes and corresponding solute sizes (Reproduced from Mulder 1991).

carry reverse osmosis units to obtain potable water from the sea. For the desalination of brackish water, in which the concentration of sodium chloride is much less than in sea water, a lower pressure is required since osmotic pressure is much lower than that of sea water. The reverse osmosis process is also being used to produce ultrapure water for semiconductor manufacturers. Market analysis by Business Communications Company reports that the present market for reverse osmosis modules and equipment, currently estimated at $914 million, will grow by 8% a year to $1.3 billion by the year 2003 (Membrane News 1999). The announcement of the asymmetric cellulose acetate membrane by Loeb and Sourirajan (1962) in 1960 made the membrane desalination process industrially practical and opened up the avenue for membrane applications to various separation processes.

As pore sizes of membranes become progressively larger than 1 nm, the membranes are called nanofiltration, ultrafiltration and microfiltration membranes. Figure 15.5 shows the range of the pore sizes of membranes belonging to each group, along with the sizes of molecules and particles. As the figure indicates, there is no clear boundary between the groups. It is obvious that the size of the solute to be separated from the solution becomes progressively larger with an increase in the size of the pore. For example, reverse osmosis and nanofiltration membranes are suitable for the separation of small organic solutes and electrolyte solutes, ultrafiltration membranes for the separation of macromolecules such as proteins, and microfiltration membranes for the separation of particles such as bacteria and yeasts. The pressure applied on the feed side decreases as the pore size of the membrane increases; *i.e.* 10 to 60 atmospheric pressures for reverse osmosis and nanofiltration, 1 to 10 atmospheric pressures for ultrafiltration and below 2 atmospheric pressures for microfiltration.

Nanofiltration membranes are used to separate molecules that are larger than those treated by reverse osmosis membranes. The separations of dye molecules and amino acids and their oligomers are some typical examples. Ultrafiltration membranes are used for milk concentration, recovery of proteins from cheese whey, recovery of starch, concentration of egg products, and clarification of fruit juice and beer. Ultrafiltration membranes are also used in pharmaceutical, chemical and pulp and paper industries. Microfiltration membranes are used to remove particles whose sizes are 0.1 micrometer or larger. Sterilization in food processing and pharmaceutical industries is one of the most important applications of microfiltration. Microfiltration is also used to remove particles in the production of ultrapure water that is necessary to manufacture integrated circuits.

For practical purposes, membranes should be packed as compactly as possible in a container. Such devices are called modules. There are four types of modules; *i.e.* plate and frame, spiral-wound, tubular and hollow fiber modules. The plate and frame module appeared in the earliest stage of industrial membrane applications. Its structure is simple and the membrane replacement is easy. As illustrated in Figure 15.6a, spacer-membrane-support plates are stacked alternately and pressed from both ends by oil pressure. The feed flows in the module, inwards and outwards, alternately, enabling the entire membrane surface to be covered by the feed stream. The membrane permeate is collected from each support plate. The module diameter is 20–30 cm. The total membrane area is up to 19 m^2, depending on the height of the module. In a spiral-wound module a permeate spacer is sandwiched between two membranes, the porous support side of the membrane being in contact with the permeate spacer (Figure 15.6b). Three edges of the membranes are sealed with glue to form a membrane envelope, the open end being connected to a central tube with holes. The membrane envelope is then wound spirally around the central tube together with a feed spacer. The feed flows through the feed spacer parallel to the central tube, whereas the permeate flows through the permeate spacer spirally, perpendicular to the feed flow direction, and is collected into the central tube. The diameter of the spiral wound module is 10–20 cm.

In a tubular module, membrane is coated on the inner wall of a support tube (Figure 15.6c). A number of such tubes are encased in a container of cylindrical shape. The feed liquid flows inside the tube, and the permeate flows from inside to outside of the membrane tube and is collected at the permeate outlet. There are also tubular modules, in which the feed is supplied to the outside of the membrane tube. Hollow fiber membranes are fibers of 0.1 to 1.5 mm with a hollow space inside. The feed is supplied to either the inside or outside of the fiber and the permeate passes through the fiber wall to the other side of the fiber. The fiber wall has the structure of an asymmetric membrane, the active dense layer being in contact with the feed liquid. A bundle of hollow fibers are mounted in a pressure vessel and the open ends of U-shaped fibers are potted into a head space. In a typical example of the DuPont permeator, the feed solution is distributed from the central distributor and flows radially through the hollow fiber bundle (Figure 15.6d). The permeate, on the other hand, flows inside the hollow fiber parallel to the distributor and collected at the open end of the fiber. The hollow fiber module features a very large (membrane surface area/module space) ratio.

a)

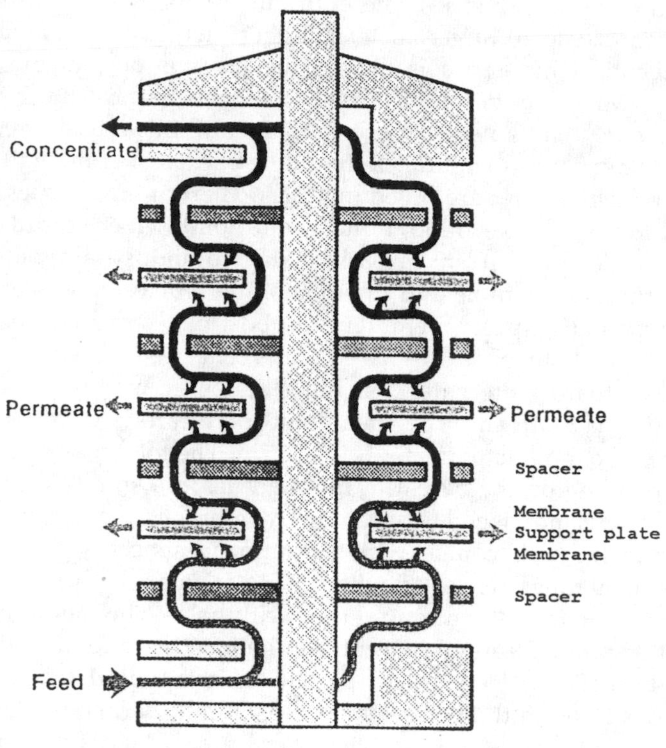

b)

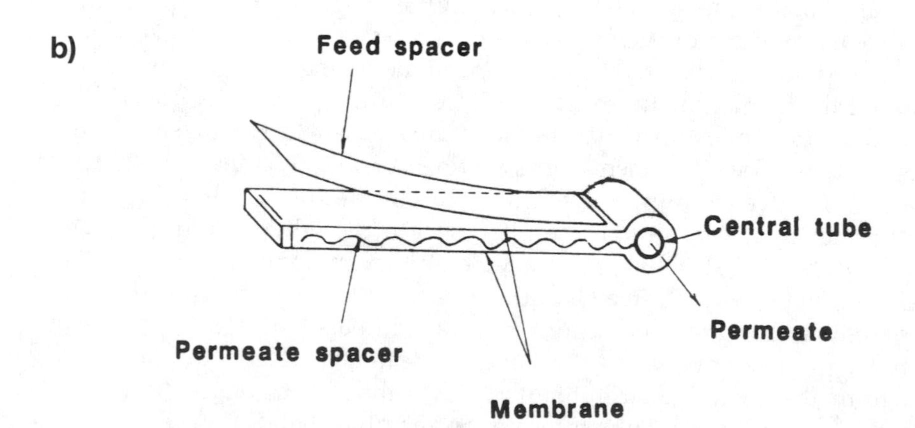

Figure 15.6 Various membrane modules.
a. plate and frame, b. spiral-wound.

c)

d)

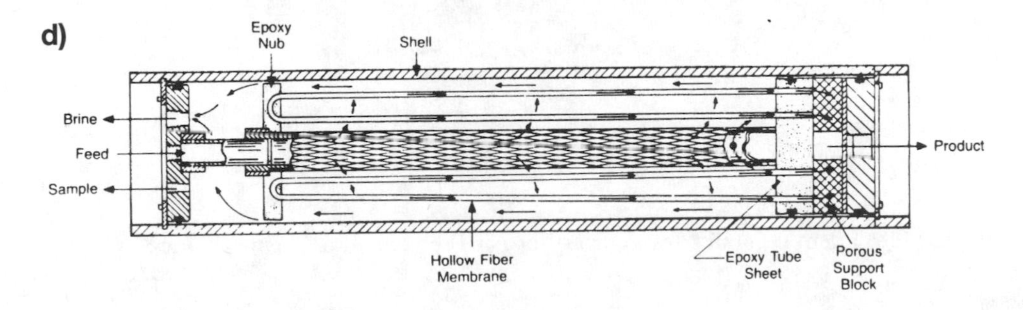

Figure 15.6 Various membrane modules.
c. tubular (Reproduction from Technical Bulletin of Nitto Electric Industrial Co. Ltd.),
d. hollow fiber (Reproduction from Technical Bulletin of E.I. du Pont de Nemours Inc.).

MEMBRANE GAS SEPARATION

It has been known for more than 100 years that gas mixtures can be separated using membranes, but the industrial use of membrane gas separation occurred only about 20 to 25 years ago. In Figure 15.1, both feed and permeate are in gas phases. The driving force is the pressure difference from feed to permeate. Gas flows through the membrane from the feed side to the permeate side. Separation of a gas mixture takes place because the flow rates of gas components through the membrane are different. Gas separation membranes are broadly classified as

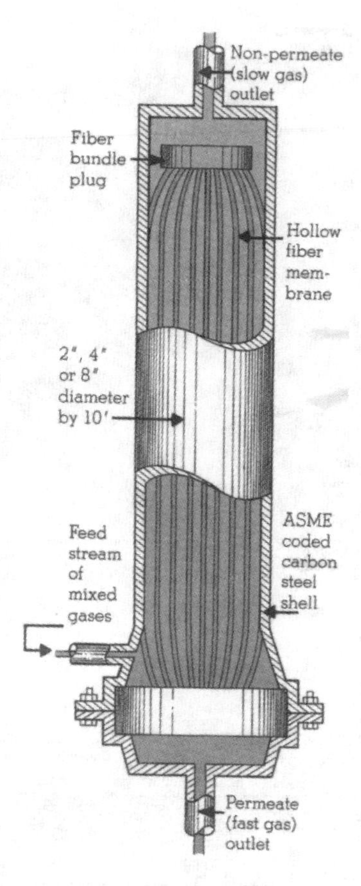

Figure 15.7 Prism separator (by courtesy of Air Products and Chemicals Inc.).

cellulosic and noncellulosic. Whereas cellulosic membranes have either spiral-wound or hollow fiber configurations, most noncellulosic membranes have hollow fiber configurations. The first attempt of membrane gas separation using a dry asymmetric reverse osmosis membrane was by Sourirajan (1963). The first accomplishment in the development of industrial gas separation membrane from cellulosic material was done by Schell (1979). The next accomplishment was in the development of composite hollow fiber membranes, using noncellulosic materials. The prism separator developed in the middle of the 1970s (Henis and Tripodi 1980) is based on the asymmetric hollow fiber of polysulfone material (Figure 15.7). The thin outer layer (0.1 to 1% of the entire thickness) is responsible for the selectivity of the membrane. However, it is practically impossible to produce a thin outer layer without pinholes that render the membrane useless because of its low selectivity. Therefore, these pinholes are filled by coating a layer of materials such as silicone rubber of high permeability, but low selectivity. There

Figure 15.8 A hydrogen recovery plant (by courtesy of Air Products and Chemicals Inc.).

are other industrial gas separation membranes based on noncellulosic materials such as polyimide.

Major applications of membrane gas separation are hydrogen recovery, carbon dioxide removal, separation of air, and helium recovery from natural gas.

Hydrogen recovery by membrane technology is important during the synthesis of natural gas from naphtha (Laverty and O'Hair 1986), in the adjustment of hydrogen to carbon monoxide ratio to 1.0 which is required for oxo synthesis, and in hydrogen recovery from purge gas from an ammonium synthesis reactor or hydrogenation purge streams. Hydrogen recovery is also important in other petroleum and petrochemical industries (Figure 15.8).

Natural gas contains CO_2, the concentration of which should be reduced to enhance the fuel value of natural gas. There is also a promising future for the use of membrane technology in the offshore removal of CO_2 from natural gas, since in the presence of water, CO_2 can be very corrosive and attack pipelines. The removal of CO_2 also reduces the quantity of gas to be transported on shore. The compact size of the membrane module, as compared with the adsorption process, also favors the membrane technology (Laverty and O'Hair 1986). The separation of CO_2 from the associated hydrocarbon gases from oil fields that are exploited in enhanced oil recovery (EOR) is another important area of CO_2/hydrocarbon separation. Purification of hydrocarbons from carbon dioxide is also required for such applications as upgrading biogas produced by the fermentation of biogas. Landfills generate methane that is contaminated with carbon dioxide (approximately 50%). Several landfills currently use membranes to provide a high quality methane. Recently, attention has been focused on the global warming effect (greenhouse effect). Among several greenhouse gases (such as CO_2, N_2O, CH_4 and CFCs), carbon dioxide is said to be responsible for half of the greenhouse effect (Conserving the Global Environment 1991). Membrane gas separation technology can be used for the removal of carbon dioxide from the CO_2 emission source such as power stations, steel works and chemical plants.

One of the applications of oxygen/nitrogen separation is to produce oxygen-enriched gas that is used in the medical field as a heart-lung oxygenator. In this application, relatively small-size gas separation modules are needed. Oxygen enriched gas is also used in a furnace to improve combustion. Injection of oxygen-enriched air, containing 25 to 35% oxygen, into a furnace leads to a higher flame temperature. Energy saving of 24 to 26% can be expected. Usually, membranes are preferentially permeable to oxygen gas, and nitrogen gas is concentrated in the feed stream. Nitrogen gas also finds applications in various sectors of the chemical process industries. For example, nitrogen is used as an inert purge gas in many processes throughout the chemical industry and is used also in the synthetic natural gas industry. Nitrogen can also be used for the purpose of enhanced oil recovery, instead of CO_2. A nitrogen blanket is an effective method of preserving fruits and vegetables, in which nitrogen-enriched air can be used. Figure 15.9 shows a compact nitrogen generator.

Helium is abundant in natural gas resources. According to estimates by Alberta's Energy Resources Conservation Board, helium present in proven reserves of Alberta natural gas amounts to 900 million m^3. The concentration of helium in natural gas in Alberta ranges from 0.01 to 1.5%, averaging about 0.05%. Using membrane gas separation technology, helium can be concentrated to 99.997%.

PERVAPORATION

According to Figure 15.1, feed is a liquid mixture, usually under atmospheric pressure, whereas vacuum is applied on the permeate side in pervaporation process. While the liquid mixture flows though the membrane from the high pressure to the low pressure side, phase transition from liquid to vapor takes place. Separation of the liquid mixture is enabled because the flow rates of the liquid components through the membrane are different. Carrier gas flow can also be supplied to the permeate side instead of applying vacuum. The first record of this phenomenon is by Kober in 1917, but the first major effort for industrial applications of this process was done in late fifties by Binnig and others of the American Oil Company (Amoco) (Binnig *et al.* 1961). Most of their work was for applications in petrochemical processing.

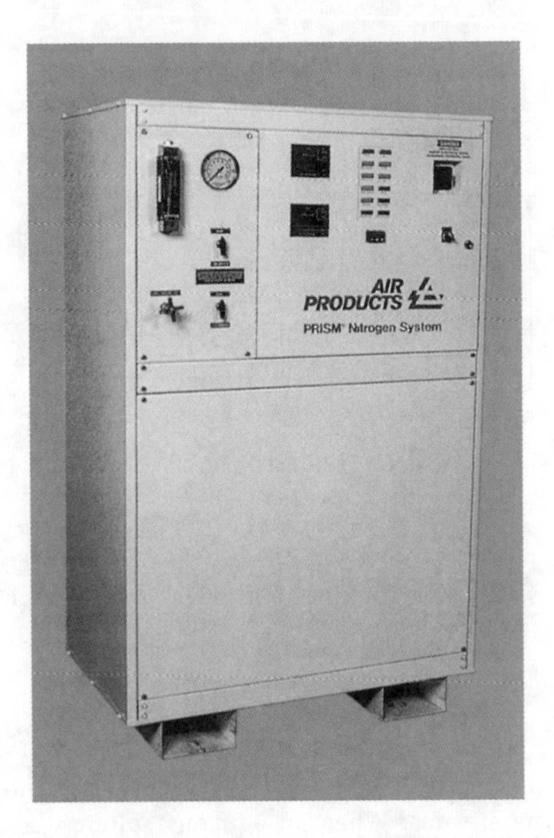

Figure 15.9 A compact nitrogen generator (by courtesy of Air Products and Chemicals Inc.).

The first attempt of commercial applications of the pervaporation process was in dehydration of ethyl alcohol produced by fermentation process. The fermentation broth contains 5 to 10% ethyl alcohol. When ethyl alcohol is concentrated by distillation, an azeotropic solution is formed at ethyl alcohol concentration of 95.6%, and further removal of water becomes impossible. Several attempts were made to concentrate ethyl alcohol by reverse osmosis. But there was a limit in the highest ethyl alcohol concentration achievable by reverse osmosis, since the osmotic pressure of ethyl alcohol solution increases rapidly when ethyl alcohol concentration becomes more than 10%. The pervaporation process was considered as an alternative membrane separation process. The distillation process is more economical for concentrating ethyl alcohol from 10 to 85%, whereas the pervaporation process is more economical for concentrating ethyl alcohol from 85 to 99%. Therefore, the combination of distillation and pervaporation in a hybrid system is the most economical way of producing ethyl alcohol from fermentation

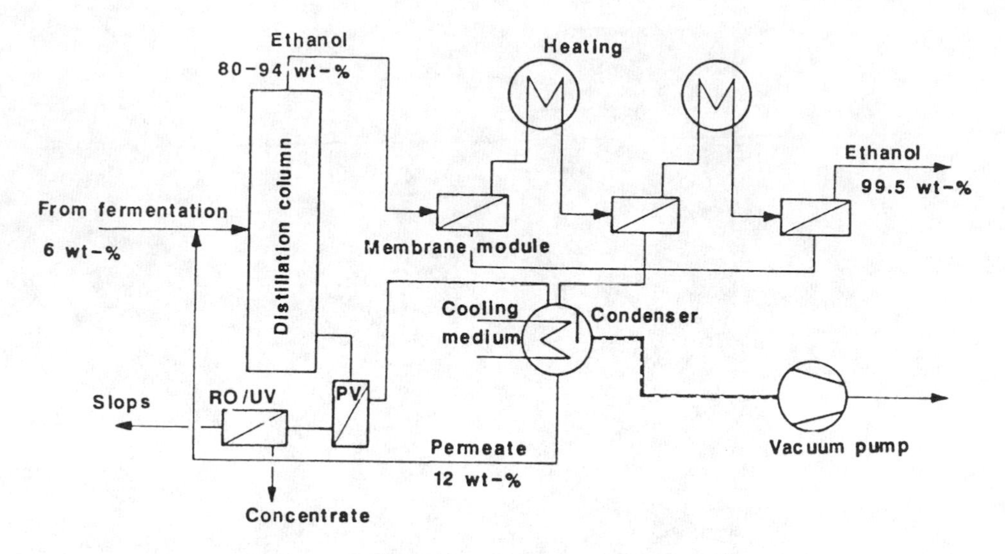

Figure 15.10 Process diagram of combined distillation/pervaporation (Reproduced from Tidball and Tusel 1986).

broth. The process diagram of such a hybrid process is shown in Figure 15.10. Figure 15.11 shows the picture of a commercial scale pervaporation plant.

Another application of pervaporation is removal of volatile organic compounds (VOCs) from waste streams. When an aqueous solution containing VOCs is brought into contact with a hydrophobic membrane such as silicone membrane, VOCs pass through the membrane faster than water because of its high affinity to the membrane and high volatility. Thus, it is possible to remove VOCs from waste water streams. For example, when water contaminated by benzene is supplied as a feed stream, benzene can be removed from the feed by using a silicone based membrane and concentrated in permeate 100 to 500 times. Flavour components of foods can also be concentrated in permeate by pervaporation.

The most interesting applications of pervaporation would be the separation of azeotropic mixtures. Distillation is useless for separating an azeotropic mixture, since its compositions in the vapor phase and the liquid phase are the same. The separation is possible by pervaporation due to the difference in affinity of the components to the membrane. There is a potential for this application to become a powerful tool in chemical process engineering, if membranes that can tolerate many organic compounds can be developed.

RECOVERY OF VAPOR FROM AIR

The strong affinity between volatile VOCs and silicone rubber membrane enables also the separation of VOCs from air. This technology is increasing its importance

Figure 15.11 A commercial scale pervaporation plant (by courtesy of Sulzer Canada Inc.).

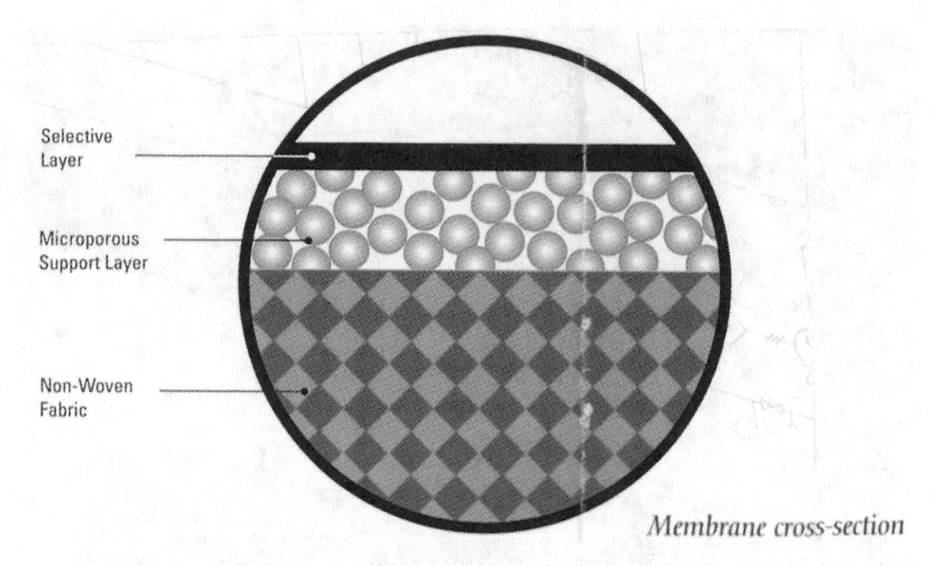

Figure 15.12 Membrane cross-section (by courtesy of Membrane Technology and Research Inc.).

in various process industries as the environmental regulations become more stringent. A composite membrane based on a selective silicone rubber layer coated on top of a porous polyetherimide support layer is illustrated schematically in Figure 15.12.

One of the modules is of a spiral-wound design, as illustrated in Figure 15.13. The feed air is circulated laterally through the module. The organic vapor passes through the membrane preferentially, enters the permeate channel of the membrane envelope, and spirals inward to the central permeate collection pipe. The feed gas stream is compressed to 1 to 2 atmospheric pressures and passed through the membrane modules. The treated clean air is discharged to the atmosphere or recycled to the process. Permeate vapor enriched in VOCs is compressed and supplied to a condenser. The condensed solvent is transferred to a solvent holding tank. A single stage system is generally able to remove 80 to 90% of organic solvent from feed air and produce permeate with a concentration five to ten times higher than in the feed.

Porous membranes can be used for drying air. When a moist air is supplied as a feed stream and brought into contact with a membrane, water is condensed in the membrane pore, preventing the flow of air through the pore, since the pore is full of water. Water, on the other hand, flows through the pore by the pressure difference as a driving force. Some dehydration systems are operated at 2069 kPa (300 psig) and produce dry air of 0°C dew point from the feed air of 40°C dew point.

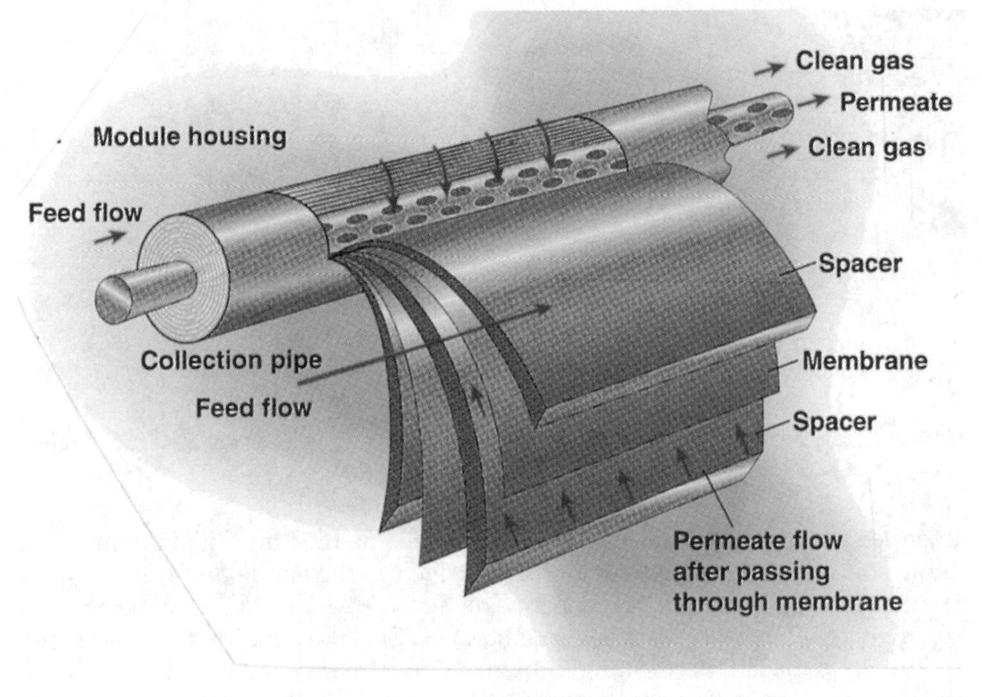

Figure 15.13 Membrane module (by courtesy of Membrane Technology and Research Inc.).

ELECTRODIALYSIS

There are membranes that are either negatively or positively charged. Typical examples are membranes made of styrene-divinyl benzene copolymers with sulfonic acid group or with quaternary ammonium group, as illustrated in Figures 15.14a and 15.14b. When a sodium chloride solution is brought into contact with a negatively charged membrane (Figure 15.14a), sodium ions with a positive charge will enter the membrane and flow through the membrane under an electrical potential difference across the membranae. On the other hand, chloride ions with a negative charge are repelled by the membrane due to the coulombic repulsive force. Thus, negatively charged membranes are permeable only to cations (positively charged ions like sodium ions), and therefore are called cation exchange membranes. Similarly, when the membrane is positively charged (Figure 15.14b), it is permeable to anions, and therefore is called an anion exchange membrane. Suppose anion and cation exchange membranes are placed alternately, as illustrated in Figure 15.15, and an electrical potential is applied by a cathode and an anode stationed at both ends of the membrane assembly. When

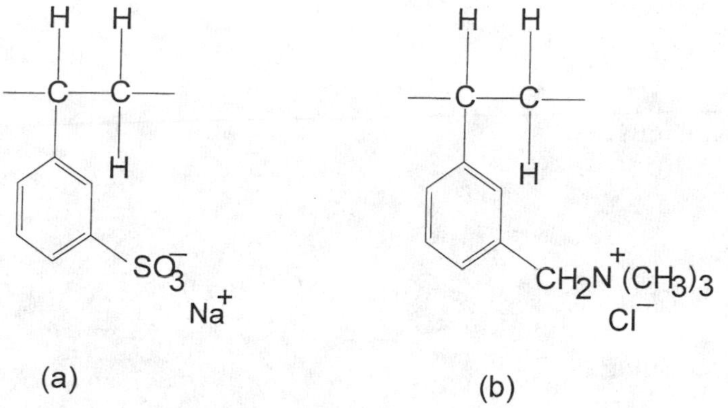

Figure 15.14 Chemical structures of negatively (a) and positively (b) charged membranes.

an aqueous sodium chloride solution is supplied as feed into the bottom of the dilute solution chamber, sodium ions are driven to the anode and start to move to the right. The sodium ions can permeate through cation exchange membranes but are rejected by anion exchange membranes. Chloride ions, on the other hand, are driven to the left, permeating through anion exchange membranes and being rejected at cation exchange membranes. As a result, both sodium and chloride ion will leave the dilute solution chamber and will be concentrated in the neighboring concentration chambers. Thus, an electrodialysis system consists of dilution chambers, where sodium chloride solution is diluted, and concentration

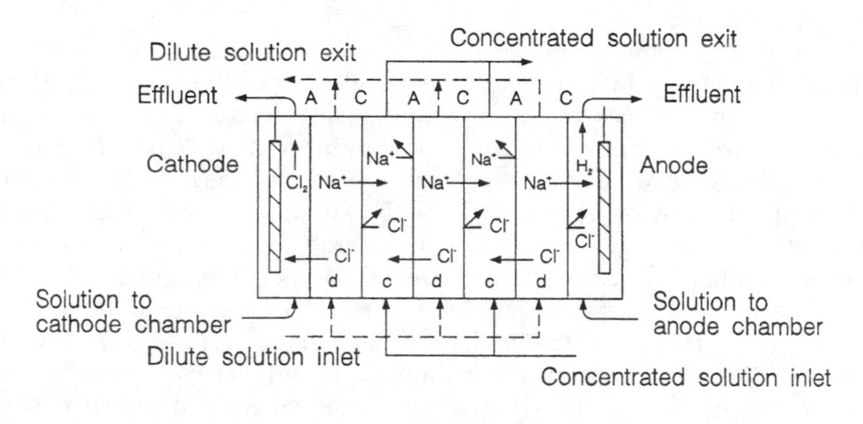

Figure 15.15 Principle of electrodialysis.

chambers, where sodium chloride solution is concentrated, each placed alternately. Therefore, electrodialysis can be used to concentrate sodium chloride of sea water. Electrodialysis can also be used to produce potable water from either sea water or brackish water. Ion exchange membranes are also used to manufacture caustic soda and chlorine by electrolysis of sodium chloride solution.

CONCLUSIONS

This chapter deals with industrial membrane separation processes. There are other separation processes in which membranes are used as membrane contactor, membrane extractor and membrane reactor. They are not included in this chapter because of their rather limited applications. Medical applications of membranes such as artificial organs, hemodialysis and drug release are also out of the scope of this chapter.

Membrane technologies emerged in the beginning of the 1960s. Having forty years of history, they have established themselves as important separation processes in various branches of industries. They are, however, still young and growing like the readers of this chapter. The coming millennium has to be awaited for these technologies to be in full bloom. I hope many of the readers will join this field which has fascinated me throughout my professional career.

REFERENCES

Binnig, R. C., Lee, R. J., Jennings, J. F. and Martin, E. C. (1961) *Ind. Eng. Chem.*, **53**, 45

Conserving the Global Environment (1991) Research Institute of Innovative Technology for the Earth, Kyoto, Japan

Henis, J. M. S. and Tripodi, M. K. (1980) Multicomponent Membranes for Gas Separations. U.S. Patent 4,230,463

Laverty, B. W. and O'Hair, J. G. (1986) Applications of membrane technology in the gas industry. In Proc. Fourth BOC Priestley Conf., pp. 291–310. London: Royal Society of Chemistry

Loeb, S. and Sourirajan, S. (1962) *Adv. Chem. Ser.*, **38**, 117

Matsuura, T. (1994) Synthetic Membranes and Membrane Separation Processes. Boca Raton, Florida: CRC Press

Membrane News (1999) *Membrane Quarterly*, **14**, 16

Mulder, M. (1991) Basic Principles of Membrane Technology, 1st edn. Netherlands: Dodrecht; Boston: Kluwer Academic

Schell, W. J. (1979) Gas Separation Membranes. U.S. Patent 4,134,742

Sourirajan, S. (1963) Nature, 199, 590

Tidball, R. A. and Tusel, G. F. (1986) In *Proc. First Int. Conf. on Pervaporation Processes in the Chemical Industry*, edited by R. Bakish. Englewood, NJ: Bakish Materials Corp

16. FOOD PRODUCTION AND PROCESSING

LARRY E. ERICKSON

Department of Chemical Engineering, Durland Hall, Kansas State University, Manhattan, KS 66506–5102

INTRODUCTION

Food production and processing is a major activity in the United States and elsewhere in the world. It is an essential activity which affects our health and welfare. Those who work as scientists and engineers in the food related industries include production oriented agricultural scientists (soil chemists, crop scientists, animal nutritionists, entomologists) and food processors (food scientists, chemical engineers, agricultural engineers, mechanical engineers). Chemical engineers are often found working for manufacturing companies that process agricultural raw materials into products such as ice cream, yogurt, instant coffee, and ketchup. They also work for other manufacturing companies which produce fertilizer and pesticides that are used in agricultural production or packaging materials for use in distributing food and agricultural products. Chemical engineers may also be found working for firms that design and build new manufacturing plants and for equipment manufacturing companies. Patent protection is often important in developing and marketing equipment and food products; chemical engineers with law degrees often work in patent and corporate law.

The products of agriculture may also be used for production of other non-food products such as ethanol. For example, starch has many non-food uses (Campbell *et al.*, 1996).

TYPES OF EMPLOYMENT

The opportunities in food production and processing range from highly technical to nontechnical, as shown in Figure 16.1. The highly technical areas include teaching, research, process development, and process engineering; a strong background in chemical engineering fundamentals is needed in these areas. Engineers who provide technical service need good engineering skills, but the work is not as technical as the highly technical areas listed above. Figure 16.1 only shows the extent to which technical skills are required by distributing the employment categories along a single line. Similar charts could be developed for other skills such as communication and supervisory skills.

Product quality is of great importance in the marketing of foods. Since processing can affect product quality, it is very important as well. While process development is shown separately from food product design and development, in some

Nontechnical | Highly Technical

College Teaching
Basic and Applied Research
Process Development
Process Design
Process Engineering
Technical Service
Technical Supervision
Quality Control
Food Product Design and Development
Equipment Design
Plant Supervision
Plant Engineering
Sales Engineer
Market Research
Market Development
Consulting Engineer
Process and Plant Economics
Food Preservation and Safety

Figure 16.1 Technical and nontechnical careers for chemical engineers in food production and processing.

cases they are very closely related because of the effect of processing on product quality. Examples include the need to retain volatile flavor compounds in products such as orange juice (King 1980) and the need for homogenization where fat globules are dispersed in food products such as ice cream. Since food preservation and safety are of great importance in the food industry, quality control receives considerable attention.

EDUCATIONAL NEEDS

Food chemistry and microbiology are important sciences for chemical engineers who work in the food industry. Since nutrition is receiving increasing attention, some background in nutritional chemistry is desirable as well. There are significant needs for both mechanical and chemical engineers in food production and processing. Food chemists and chemical engineers are needed at all degree levels. Many chemical engineers do not have any special food science education as they enter positions in the food industry; however, those who know they want to work in food processing are encouraged to build their background by taking food chemistry, food microbiology and other food related elective courses such as biochemical engineering and bioseparations which are offered in many chemical engineering departments. Interfacial phenomena affect the quality of many food products, especially those where emulsions are formed. Understanding the physical chemistry of food systems is often the key to developing a desirable product. Knowledge from courses in chemistry and biochemistry provides the background to understand and apply food chemistry principles to food systems.

FOOD PRESERVATION

The preservation of crops and foods has been a central concern for many centuries. Even before knowledge of microorganisms, there was an understanding of food spoilage. The principles of food preservation are based on knowledge from physical chemistry and microbiology as well as chemical engineering. It is well known that chemical reaction rates increase with temperature according to the well known Arrhenius activation energy model

$$k = A \exp(-E/RT) \tag{16.1}$$

where

- k represents the rate constant
- A is a frequency factor
- E is the activation energy
- R is the ideal gas constant
- T is absolute temperature

Since living microorganisms are supported through many chemical reactions, these rate concepts apply to microbial growth as well. Refrigeration and freezing reduce food temperature and provide methods to increase the shelf life of foods.

The second important principle in food preservation is to maintain the water activity of the food product below the critical value which might support microbial growth. Water activity is defined as the ratio of the fugacity at the conditions encountered to the fugacity at a chosen standard state (Erickson 1982). The standard state most widely used for water in foods is the fugacity of pure water (the solvent) at the temperature and pressure of the system. Since, at low pressures, the fugacity is essentially equal to the partial pressure of water, the standard state fugacity is the vapor pressure of pure water at the temperature of the system. When a food is in equilibrium with the gas phase,

$$a = p/P = ERH \tag{16.2}$$

where

a	is water activity
p	is the partial pressure of water in equilibrium with a food product
P	is the vapor pressure of pure water at the temperature of the food
ERH	is the equilibrium relative humidity at the temperature of the food

Sugar in a sugar bowl has a low water activity. It does not support microbial growth as long as it remains dry, while sugar in an aqueous solution is readily consumed by microorganisms. Products such as crackers and potato chips do not need to be refrigerated because the water activity is below the critical value which begins to support microbial growth. These products are packaged such that moisture does not diffuse into the product between production and when the product is opened for consumption.

Scott (1957) has determined the critical water activity required for microbial growth (Karel *et al.* 1975): most bacteria are unable to grow below $a = 0.91$ and molds do not grow below a water activity of 0.80. Moreover, minimum water activities required for production of toxins by microorganisms are often higher than those required for microbial growth.

Drying of foods to obtain products with water activities below the critical value is a common practice. Dried fruits, dry milk, instant tea, and dry breakfast cereals are examples of products that have had water removed prior to packaging. The water activity can also be lowered by adding humectants such as salt and/or sugar to the product (Erickson 1982). Preservatives which inhibit microbial growth such as benzoates and sorbates can also be added to foods.

The third method that is widely used to preserve food is to inactivate the microbial population. Heat is most commonly used for this purpose; however, irradiation is also being used. Sealing of foods in cans followed by thermal processing to inactivate all organisms has been a very successful food preservation technology. Thermal processing associated with canning affects the nutritional quality and sensory properties of many foods. Table 16.1 gives representative

Table 16.1 Representative activation energies for destruction of nutritional compounds and microorganisms. From Wang *et al.* (1979).

Substance	*Activation energy, kJ/mole*
Folic acid	70
Thiamine HCl	92
Bacillus stearothermophilus	283
Bacillus subtilis	318
Clostridium botulinum	343

Table 16.2 Examples of separation processes applied to food processing.

Separation Process	*Food processing example*
Adsorption	Sugar refining
Ion exchange	Purification of high-fructose corn syrup
Gas absorption	Adding carbon dioxide to beverages
Distillation	Concentration of ethanol in alcoholic beverages
Leaching	Separation of oil from soybeans
Extraction	Separation of active fatty acids in fish oil
Supercritical fluid	Decaffeination of coffee extraction
Precipitation	Soy protein purification
Filtration	Clarification of fruit juices
Ultrafiltration	Recovery of proteins from whey
Expression	Separation of apple juice from apples
Evaporation	Concentration of orange juice
Freeze drying	Freeze dried coffee
Spray drying	Powdered milk production

activation energies for destruction of nutritional compounds such as folic acid and thiamine HCl as well as common microorganisms in foods. Since the activation energies of the organisms are much larger, the concept of high temperature short time thermal processing has been developed and used in selected applications. This allows the microorganisms to be inactivated and minimizes the loss of nutritional compounds. For foods processed in cans, the canning time requirement is significant since the rate of heating and cooling is relatively slow. However, products such as milk have been processed as fluids using high temperature short time thermal processing (Garside and Furusaki 1994; Karel *et al.* 1975).

SEPARATION PROCESSES

Food processing often involves chemical separations to obtain desirable products. In his separation textbook, C.J. King (1980) describes the process for refining cane sugar in considerable detail. He identifies 11 different classes of separation processes associated with making raw sugar and refining cane sugar; these are settling, filtration, centrifugation, screening, expression, leaching, precipitation, evaporation, crystallization, adsorption, and drying. Table 16.2 gives examples of different separations applied to food processing. Some of these separations are commercialized because of the market for products with special attributes such as decaffeinated coffee, while others such as recovery of proteins from whey have environmental benefits as well. King (1980) identifies 54 separation processes in his book; many of these have applications in food production and food processing.

In the last 20 years, there has been a growth in the food industry in plants that process agricultural raw materials into a variety of products. These food refineries produce a variety of products such as corn starch, high-fructose corn syrup, and cooking oils. In other plants, cheeses, yogurt, frozen dinners, and extruded snack foods are prepared. As more ingredients are added to achieve nutritional goals and improve sensory qualities, the food processing plant more closely resembles a chemical plant. Separations are often essential in preparing some of the raw materials that are used in these manufacturing operations (Rizvi 1994).

The physical separation of fruit juices from the pulp by compressing the mixture is known as expression. This separation process is widely used in the food industry; however, theory regarding it is not ordinarily included in separation textbooks. This subject is considered in some detail in the Kirk-Othmer Encyclopedia of Chemical Technology in the section on Fruit Juices (Othmer and Kroschwitz 1994).

The growing importance of food additives and products with special functional properties is one of the driving forces for increased attention being given to the field of bioseparations (Belter *et al.* 1988; Asenjo 1990; Schwartzberg and Rao 1990). The health care industry is another driving force for advances in bioseparations. Some products are at the interface in that they have significant health benefits as well as nutritional value.

FERMENTED FOODS AND BIOCHEMICAL ENGINEERING

Fermentation has been used in the production of foods and beverages such as bread and wine for thousands of years. It is commonly used today in the production of beer, wine and alcoholic beverages, in bread making, in dairy products such as yogurt and cheeses, and in many other applications shown in Table 16.3. Food fermentations often help to preserve food products and to impart flavor to them. Most food fermentations are carried out under anaerobic conditions using mixed cultures (Erickson and Fung 1988). Brining and fermentation provided methods for preserving fruits and vegetables prior to the advent of canning and freezing. Some of these methods are still used today because they impart desired flavor and product quality to foods. In beverage alcohol fermentations, the ethanol in the product reduces the chance of microbial growth by spoilage organisms, while in the production of yogurt, the acidification of the product reduces the attractiveness of the mixture to other organisms.

Bacteria, yeasts, and molds are included in Table 16.3. Yeasts are widely used in bread making and in the production of alcoholic beverages. They may also be used in cheese production and sausage fermentations. Molds are used in making soy sauce, miso, tempe, and certain cheeses such as roquefort (*Penicillium roqueforti*) and camembert (*Penicillium camemberti*) (Bailey and Ollis 1986). Bacteria are used widely in lactic acid fermentations. In yogurt production, mixed cultures of lactic acid bacteria (*Lactobacillus bulgaricus* and *Streptococcus thermophilus*) are often employed with nearly equal numbers growing together.

Table 16.3 Examples of food fermentations and predominant microorganisms.

Raw Material	*Fermented product*	*Predominant microorganisms*
Cereals	Leavened breads	Yeasts, primarily *Saccharomyces cerevisiae*
	Crackers	Yeasts and lactic acid bacteria
	Beer	Yeasts, *Saccharomyces species*
Legumes, soybeans	Soy sauce	*Aspergillus* species
	Miso	*Aspergillus oryzae* and *Saccharomyces rouxii*
	Tempe	*Rhizopus oligosporus*
Vegetables	Pickles	*Lactobacillus* species
Fruits, grapes etc.	Wines	*Saccharomyces* species
	Vinegar	*Acetobacter aceti*
Milk	Yogurt	Lactic acid bacteria
	Cheeses	Lactic acid bacteria
Meats	Sausages	*Lactobacillus* species, *Pediococcus* species
Fish	Fish sauces	Lactic acid bacteria

Sources: Erickson and Fung (1988) and Steinkraus (1983).

Sauerkraut fermentations are carried out by several bacterial cultures; in the early stages of this cabbage fermentation, *Enterobacter cloacae* and *Erwinia herbicola* may be found together with *Leuconostoc mesenteroides* which produces lactic acid. About the second day of the fermentation the population of *L. mesenteroides* increases to about one billion/ml, and the acidity inhibits the enteric organisms. *Lactobacillus plantarum* succeeds *Leuconostoc mesenteroides* and produces more lactic acid. It also consumes the bitter mannitol produced by *Leuconostoc mesenteroides*. Small numbers of *Pediococcus cerevisiae* and *Lactobacillus brevis* may often be found during the middle stage of the fermentation (Erickson and Fung 1988).

Since yeast contain approximately 50% protein, they have been produced for use in foods and feeds (Blanch and Clark 1996). Yeast are consumed regularly in diets which include leavened bread that is made with yeast. The dry yeast commonly used in baking are produced by fermentation and dried in such a way that the yeast remain viable.

THERMAL PROCESSING

Thermal processing is widespread in agricultural production (grain drying) and in food processing (canning, evaporation, freezing, freeze drying, spray drying, extrusion cooking) and in food preparation (thawing, baking, roasting, boiling, broiling, frying) (Heldman 1975; Charm 1978; Watson and Harper 1988). Table 16.4 gives examples of thermal processes. Thermal processing often inactivates microorganisms which may be present in foods, and it usually improves taste, aroma, and texture. Mass transfer often occurs simultaneously with heat transfer in baking, roasting, and extrusion cooking. Water and juices in foods provide flavor and moistness which affect food quality and sensory characteristics. Thus,

Table 16.4 Thermal processes in food production and processing.

Process	*Example*
Canning	Fruits, soups
Freezing	Turkeys and fish
Thawing	Food preparation from frozen foods
Pasteurization	Milk and cream
Steam infusion heating	Soymilk production
Flash cooling	High temperature short time sterilization of milk
Microwave heating	Heating of prepared foods
Roasting	Coffee beans
Baking	Breads and cakes
Extrusion cooking	Snack foods containing starch

Please note that heat transfer is involved in many of the separation processes listed in Table 16.2.

thermal processing must be carried out such that microbial safety and sensory quality goals are both achieved. Heat transfer in solids in cans is by conduction; the time required increases with can size. Similarly, in roasting a turkey in a conventional oven, the cooking time is predicted based on heat conduction. Microwave cooking has provided a faster method to transfer heat to solids; however, no corresponding advances in heat removal have been reported for rapid freezing. Convection heat transfer is important in many thermal processing operations such as warming soup on top of a kitchen stove. Radiation heating is important in broiling.

In Table 16.4, canning often involves heat transfer by both conduction and convection. Conduction is the predominant mechanism of heat transfer in many freezing and thawing operations. Convection heat transfer provides rapid heat transfer in pasteurization, steam infusion heating and flash cooling. In soy milk production under steady state conditions, steam can be added continuously to a ground soybean slurry to achieve very rapid heating of the mixture (Tuitemwong *et al.* 1993). Slurries of liquids and solids can be heated under pressure to sterilization temperatures and maintained for the desired time by having an extended insulated pipe prior to a valve where pressure is reduced and the mixture is flash cooled as water evaporates and removes energy from the mixture. In this process it is important for the size of the solids in the slurry to be small because heat transfer within the solids is by conduction. The sterilization may not be complete if microbes are protected from the heat by being within the slurry solids.

In microwave heating, there is heat transfer directly to the food by converting electrical energy to heat. Electromagnetic radiation is generated and it is absorbed by the food product. Dipolar molecules contained in foods attempt to align themselves in this high frequency field, and their molecular motion results in the generation of heat (Karel *et al.* 1975). Because the dielectric properties of the food vary with water and fat content, heating is often not uniform in microwave ovens.

Roasting, baking and extrusion cooking are examples where mass transfer and heat transfer are both important. Conduction and convection are usually considered in modeling heat transfer in these processes.

BIOCHEMICAL REACTIONS IN FOOD PROCESSING

Many chemical transformations occur during food processing. Extrusion cooking of cereal products to produce snack foods involves disruption and melting of starch granules and reduction of molecular weight due to some breaking of the biopolymeric chain. Denaturation of proteins as well as some loss of the B vitamins thiamin and riboflavin and amino acids because of browning reactions has been reported (Karmas and Harris 1988).

Baking of cereal products denatures protein, which enhances protein digestibility. Nonenzymatic browning reactions involving reducing sugars and amino acids provide products which give odor and flavor to baked products. The yeast fermentation of dough has many biochemical reactions associated with it, including starch hydrolysis, glucose metabolism, and ethanol production.

Most prepared foods are transformed to some extent during the preparation process. Protein denaturation, browning reactions and nutritional losses of vitamins are commonly found.

Many compounds which are produced to be added to foods have been synthesized using chemical or biochemical processes. Immobilized enzymes are used to produce high-fructose corn syrup. Many vitamins and some other nutritional additives are produced by fermentation or by direct synthesis.

FOOD EMULSIONS

Many manufactured food products exist as emulsions, which are colloidal dispersions in a continuous phase. Some salad dressings are liquid droplets of oil dispersed in an aqueous phase, while butter and margarine are water-in-oil emulsions. Solids may be present in the dispersed phase and/or the continuous phase in many food dispersions. In ice-cream, crystalline material is present in the aqueous continuous phase while air and fat globules are present as dispersed phases. In both ice-cream and whipped cream, air bubbles are stabilized by smaller fat globules and milk protein. The smaller fat globules are found both at the air bubble surface and in the continuous phase that fills the void space between bubbles (Dickinson and Stainsby 1988).

Flavor ingredients only need to be soluble in either the lipid phase or the aqueous phase when an emulsion is marketed. Emulsifiers reduce the interfacial free energy and stabilize the dispersion by preventing coalescence. In order to stabilize an oil-in-water emulsion, one should use a stabilizer that has good solubility in the continuous or aqueous phase.

Some emulsions such as cake batters and ice cream mixes require stability of the emulsion for only a short period of time until they are stabilized by baking or freezing, while products such as butter and mayonnaise need to remain stable until they are consumed. Proteins often exist at the interface in food emulsions and act as emulsifiers and stabilizers. Droplet size ranges from 100 nm to 0.01 mm in food dispersions.

The high pressure valve homogenizer is a widely used method for making fine food emulsions. Droplets are disrupted by intense fluid flow forces as they pass through the homogenizer. Increased turbulence on the low pressure side of the valve favors the formation of finer emulsions. Adsorption of proteins and other surfactants at the newly created interface stabilizes the emulsion. Casein, the major protein of milk, is widely used in food emulsions as an emulsifier and stabilizer because of its backbone flexibility and hydrophobicity.

FLUID FLOW AND FOODS

One of the challenges of food process engineering is to create products with appropriate viscosity and fluid properties. Ketchup needs to flow from the bottle when it is first taken from the refrigerator, but it should also have similar flow properties after it has warmed to room temperature. Since viscosity is a strong function of temperature for many fluids, developing dressings, creams, and sauces with desirable flow properties is a challenge. In canning of foods, canning time is less for fluid foods compared to those heated by conduction through food solids. Thus, it is desirable to have fluid movement due to natural convection during canning. Because many foods are mixtures and present as dispersions and emulsions, rheological properties are often complex and non-Newtonian (Rha 1975).

CONCLUSIONS

A significant fraction of chemical engineers have careers related to food production and/or food processing. Chemical engineering educational programs provide the needed background for process engineering careers in the food industry. Many exciting applications of chemical engineering may be found in food processing and activities which support food production.

REFERENCES

Asenjo, J. A., ed. (1990) Separation Processes in Biotechnology. New York: Marcel Dekker

Bailey, J. E. and Ollis, D. F. (1986) Biochemical Engineering Fundamentals, 2nd ed. New York: McGraw-Hill

Belter, P. A., Cusler, E. L. and Hu, W. S. (1988) Bioseparations. New York: Wiley
Blanch, H. W. and Clark, D. S. (1996) Biochemical Engineering. New York: Marcel Dekker
Campbell, G. M., Webb, C. and McKee, S. L., eds. (1997) Cereals: Novel Uses and Processes. New York: Plenum
Charm, S. E. (1978) Fundamentals of Food Engineering, 3rd ed. Westport, Connecticut: AVI Publishing
Dickinson, E. and Stainsby, G., eds. (1988) Advances in Food Emulsions and Foams. New York: Elsevier
Erickson, L. E. (1982) Recent Developments in Intermediate Moisture Foods. *Journal of Food Protection*, **45**, 484–491
Erickson, L. E. and Fung, D. Y. C., eds. (1988) Handbook on Anaerobic Fermentations. New York: Marcel Dekker
Garside, J. and Furusaki, S., eds. (1994) The Expanding World of Chemical Engineering. Amsterdam, Netherlands: Gordon and Breach
Heldman, D. R. (1975) Food Process Engineering. Westport, Connecticut: AVI Publishing
Karel, M., Fennema, O. R. and Lund, D. B. (1975) Physical Principles of Food Preservation. New York: Marcel Dekker
Karmas, E. and Harris, R. S., eds. (1988) Nutritional Evaluation of Food Processing. New York: Van Nostrand Reinhold
King, C. J. (1980) Separation Processes, 2nd ed. New York: McGraw-Hill
Othmer, D. F. and Kroschwitz, J. I., eds. (1994) Kirk-Othmer Encyclopedia of Chemical Technology, 3rd ed. New York: Wiley
Rha, C. K., ed. (1975) Theory, Determination and Control of Physical Properties of Food Materials. Boston, Mass.: Reidel Publishing
Rizvi, S. S. H., ed. (1994) Supercritical Fluid Processing of Food and Biomaterials. New York: Chapman and Hall
Schwartzberg, H. G. and Rao, M. A., eds. (1990) Biotechnology and Food Process Engineering. New York: Marcel Dekker
Scott, W. J. (1957) Water Relations of Food Spoilage Microorganisms. *Advances in Food Research*, **7**, 83–127
Steinkraus, K. H., ed. (1983) Handbook of Indigenous Fermented Foods. New York: Marcel Dekker
Tuitemwong, P., Erickson, L. E., Fung, D. Y. C. and Tuitemwong, K. (1993) Effect of Processing Temperatures on Microbiological and Chemical Quality of Soy Milk Produced by Rapid Hydration Hydrothermal Cooking. *Journal of Food Processing and Preservation*, **17**, 153–175
Wang, D. I. C., Cooney, C. L., Demain, A. L., Dunnill, P., Humphrey, A. E. and Lilly, M. D. (1979) Fermentation and Enzyme Technology. New York: Wiley
Watson, E. L. and Harper, J. C. (1988) Elements of Food Engineering, 2nd ed. New York: Van Nostrand Reinhold

17. HOW CHEMICAL ENGINEERING HELPS TO SAVE LIVES – BIOMEDICAL ENGINEERING –

KIYOTAKA SAKAI

Chemical Engineering Department, Waseda University, Tokyo 169-8555, Japan

ABSTRACT

The body may be thought of as a small but finely-tuned chemical plant, incorporating a number of systems in which enzymatic processes and unit operations are taking place. Since artificial organs are intended to replicate these chemical processes, it stands to reason that knowledge of chemical engineering is essential in their design and optimization so that they will function with the maximum efficiency. For those not involved in specialities relating to medical engineering, it may come as a surprise to learn that the disciplines of physical chemistry and chemical engineering are actively applied in medical treatment. Here artificial kidney is described as a typical example of artificial organs including hemodialysis, dialysis membrane, dialyzer and computer-controlled dialysis, and further the other blood purification technique using adsorbents.

Keywords: chemical engineering, chemical plant, artificial organ, artificial kidney, dialysis, dialyzer, dialysis membrane, adsorbent

DREAMS FOR THE FUTURE OF CHEMICAL ENGINEERING

The countless contributions that chemical engineering has made to the development of the chemical industry are universally recognized. In spite of this we need, as never before, a transformation that will fire the imagination not only of chemical engineers, but of those outside the field as well. New ideas are germinating — some that will come to immediate fruition, others that will require deeper investigation. But just thinking about them can be a mind-expanding exercise.

Looking back on industrial advances we see that they have stood on two basic pillars — replacement of the natural by the artificial and of the manual by the automated. Consider diamonds or leather for example; their attractiveness is undeniable, but they are expensive and of limited availability. Creation of an artificial substitute is thus a praiseworthy endeavour. The long-standing impetus to create artificial replacements will doubtless remain with us indefinitely.

In one of his essays Dr. Torahiko Terada, a physicist active in the early years of this century, wrote "It seems that man's works are inevitably flawed, but whatever nature makes is of exquisite craftsmanship." Most artificial things that we have today represent attempts to replicate nature; artificial leather, artificial intelligence, synthetic textiles, synthetic rubber and synthetic diamonds. There are even artificial birds (aircraft), artificial humans (robots) and artificial life, organs, satellites and animals; the list is endless.

In an aircraft we see an artificial creation that outperforms its model, yet the imitation never quite measures up to the original. It is the original that sets the goal and our lives can be transformed if a comparable item is available cheaply and in abundance.

It is for this reason that the fullest knowledge of the natural model must be obtained. There are still countless things in nature of which we are ignorant or which as yet we are unable to replicate. Still further development of artificial objects would seem to be assured.

THE BODY: A FINELY-TUNED CHEMICAL PLANT

There are already many artificial organs in clinical use. Some have comparatively simple functions such as a bone, joint, oesophagus or the veins and arteries. Others are of greater complexity — coronary valves, pancreas, lungs or kidneys. It may come as a surprise to the reader to know that scientists and engineers play important roles in the development of such devices. Chemical engineering in particular is crucial in the design of those organs that are intended primarily to effect a transfer of mass, such as artificial lungs, kidneys or livers. The fact that artificial organs are not simply the concern of medical schools surely deserves wider recognition.

The body may be thought of as a small but finely-tuned chemical plant, incorporating a number of systems in which enzymatic processes and unit operations are taking place (see Figure 17.1). The heart pumping the blood through the body, the lungs absorbing and desorbing gas via a biological membrane, the stomach, other digestive organs and the liver — these are all bioreactors engaged in the decomposition or synthesis of various substances while the kidneys act as a filtration unit that extracts wastes and excretes them, together with water, in the form of urine. Since artificial organs are intended to replicate these chemical processes, it stands to reason that knowledge of chemical engineering is essential in their design and optimization so that they will function with the maximum efficiency.

THE CYBERNETIC ORGANISM

Extensive research and development is being carried out on artificial kidneys and other artificial organs and clinical applications are commonplace. It has been predicted that if these trends continue, practical replacements for almost all the organs of the body will be available in the near future. Thus, if any part of the

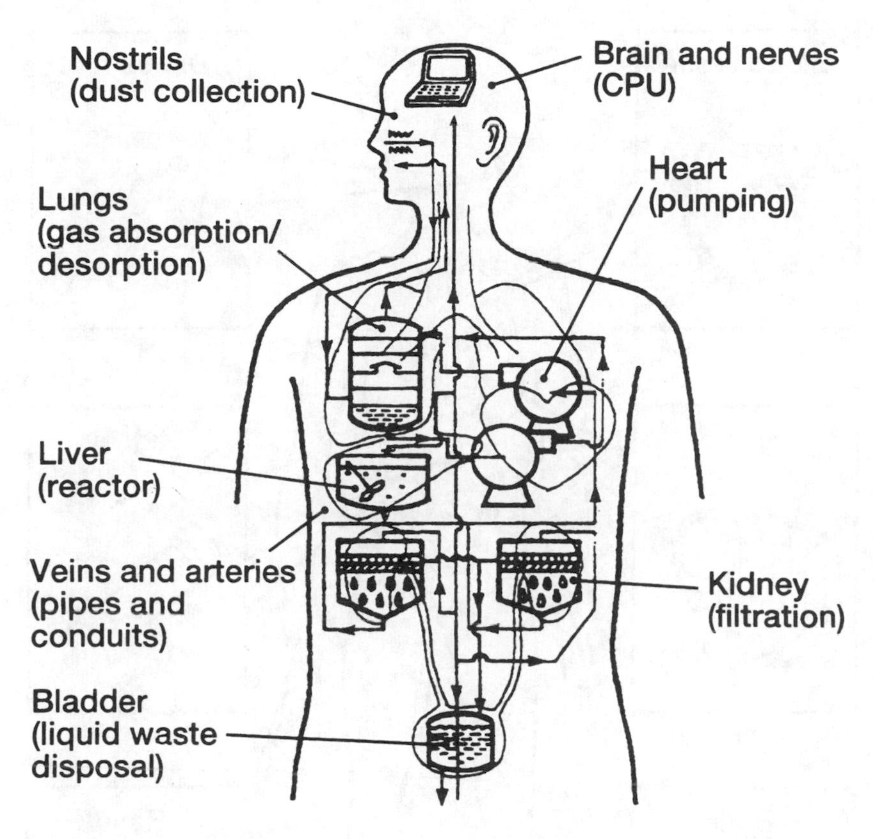

Figure 17.1 The body as a chemical plant — most of the better known organs are analogous to common chemical plant systems.

body fails, we will be able to use an artifitical replacement, just as we put a spare part into our car.

Some, however, have taken a critical position on the issue of whether the advent of such a cybernetic organism, or "cyborg" (see Figure 17.2), bodes well or ill for the future of humanity. Be this as it may, the fact remains that artificial organs offer an effective means of life support and chemical engineering will play a major role in their development.

THE ARTIFICIAL KIDNEY: APPLYING THE PRINCIPLE OF DIALYSIS

The most commonly used artificial organ is the artificial kidney, a machine that performs a treatment known as hemodialysis. This process cleanses the blood by dialysis and filtration, simple physicochemical processes. Dialysis can be defined

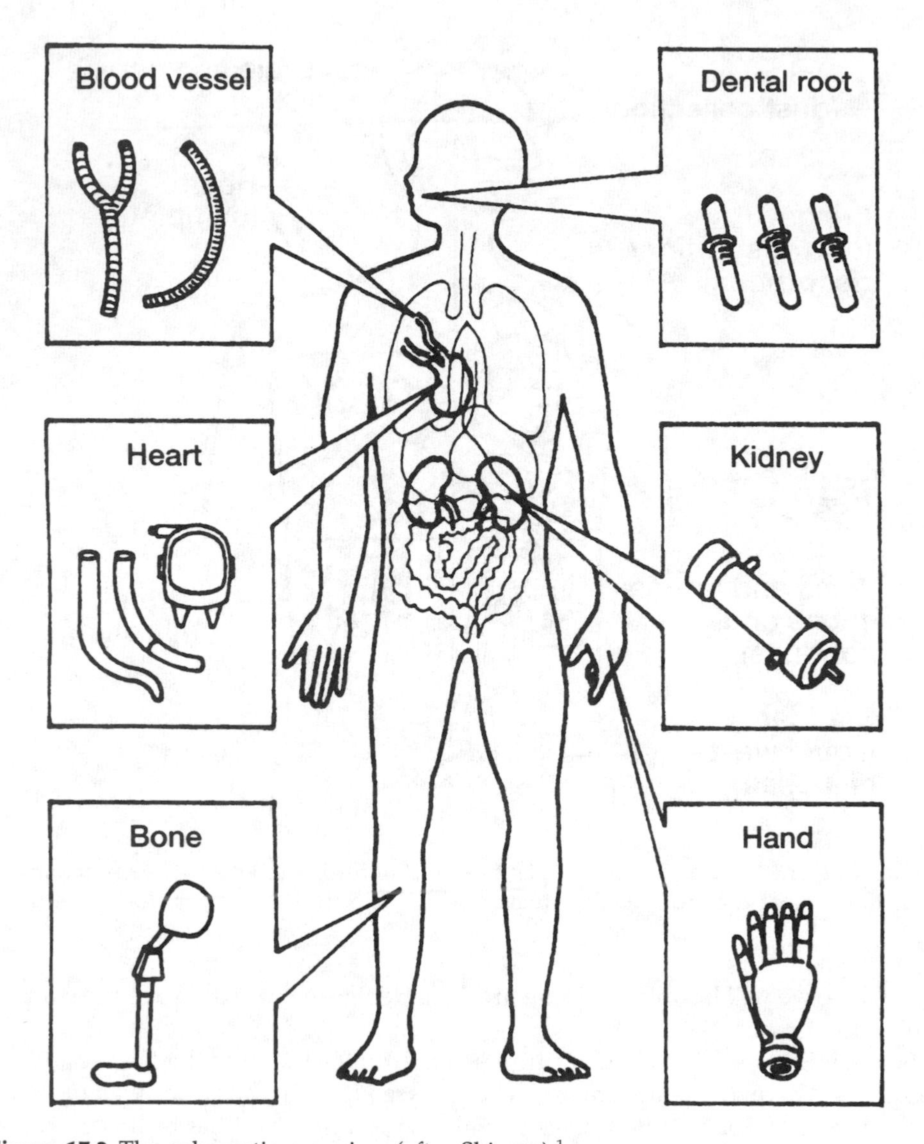

Figure 17.2 The cybernetic organism (after Shirane).[1]

as a transfer of mass between two solutions that have different solute concentrations. The solutions are kept apart by a semipermeable dialysis membrane and a difference in the rate of transfer of individual species in the two solutions results in the required separation (Figure 17.3). Filtration refers to the use of a pressure differential to force a solute through a semipermeable membrane, thereby separating the dissolved materials or the solute from the solvent.

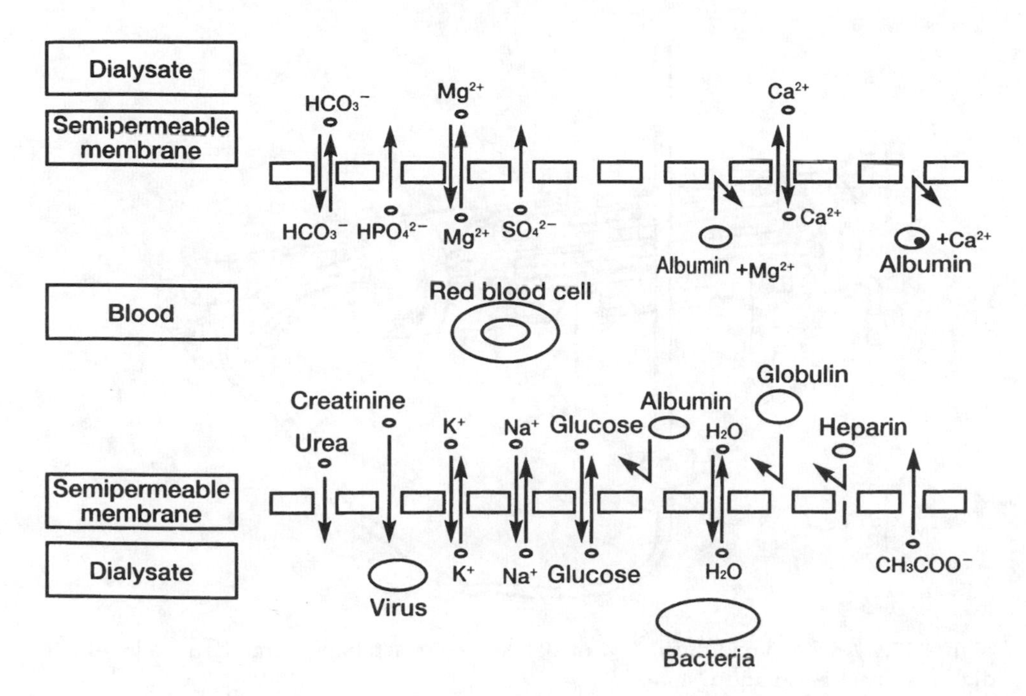

Figure 17.3 Principle of hemodialysis (from Ota).[3]

Patients suffering from renal failure are unable adequately to excrete metabolites and excess fluids and they exhibit abnormal concentrations of electrolytes and pH levels. This condition can be corrected by a dialyzer.

More than 50% of the dialysis membranes used in clinical dialyzers are composed of cellulose. Protein metabolites, electrolytes and other substances are driven by the difference in concentration through the molecular-level micropores in the membrane, the micropores being 4–7 nm diameter. However, only those solutes that have low molecular weights can pass through the micropores; albumin and other serum proteins, and of course viruses or blood cells, cannot. The dialyzers currently used in clinical situations perform filtration, making possible the removal of excess water.

THE CONTRIBUTIONS OF ABEL AND KOLFF

Figure 17.4 shows the first artificial kidney, which was composed of celloidin tubes. Proposed by John J. Abel of the United States in 1914, it was used successfully in the removal (or to use Abel's term, vividiffusion) of a drug (sodium salicylate) administered to anaesthetized dogs. In 1943, Willem Johan Kolff of the Netherlands used an enormous rotating-drum dialyzer (Figure 17.5) that was

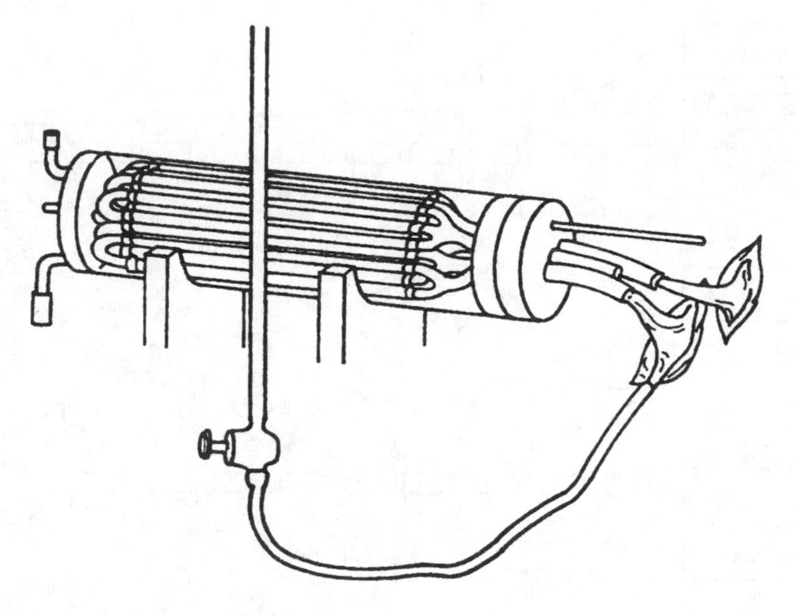

Figure 17.4 Abel's "vividiffusion" artificial kidney, resembling a modern hollow-fiber dialyzer (from Sakai and Oshima).[3]

wrapped with cellophane tubing taken from sausage skins and rotated in a physiological saline solution. He was able to remove uremic toxins (protein metabolites) from patients with acute renal failure, and in 1945 he succeeded in prolonging a patient's life in 1945. After moving to the United States in 1950, Dr. Kolff continued his work on artificial kidneys and artificial hearts. Despite the

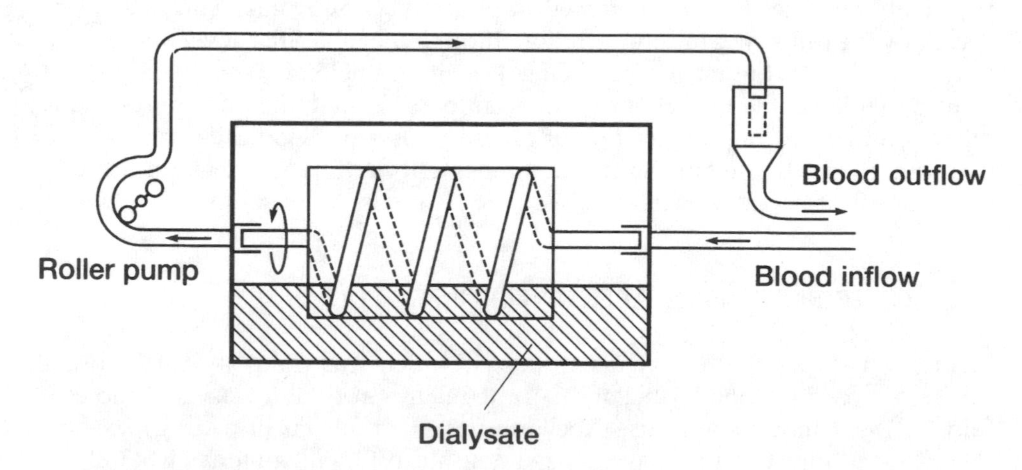

Figure 17.5 The rotating-drum kidney of Kolff.[4]

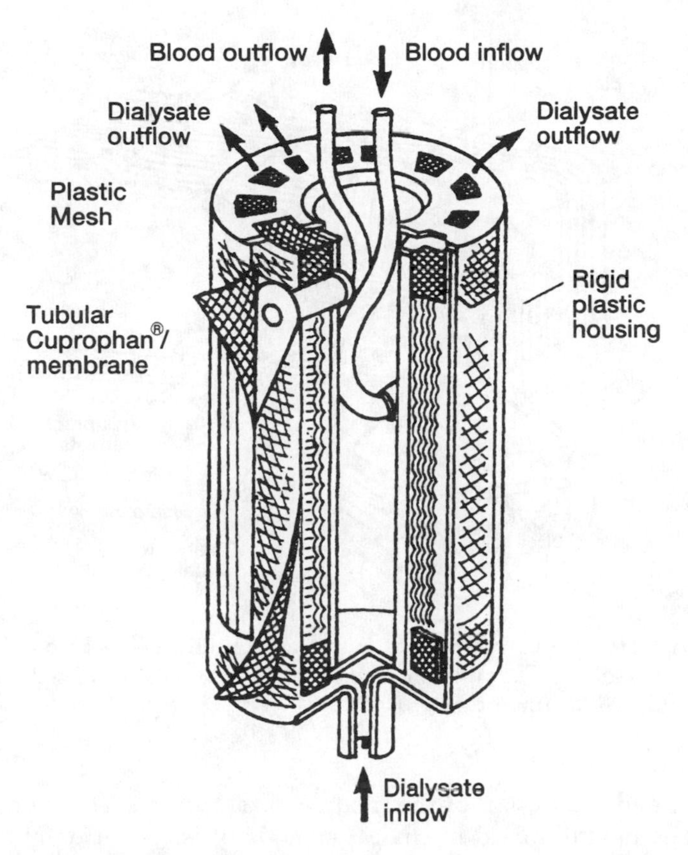

Figure 17.6 A coil-type dialyzer. The uremic blood flows inside tubular Cuprophan® membranes which are wrapped onto the drum with plastic mesh while dialysate flows between the membranes and the mesh.[3]

large size and limited performance of the equipment, Kolff's design marked the first step toward the advanced, disposable artificial kidney developed in 1956 by Travenol (now Baxter), which was called the twin coil-type dialyzer.

DIALYZERS

Although all dialyzers make use of a common principle, they can take a number of forms: coil-type (Figure 17.6), plate-type (Figure 17.7) or hollow-fiber devices (Figure 17.8). The standard Kiil dialyzer was developed in 1960 by Kiil of Norway. An improved and more compact disposable plate-type dialyzer was marketed by Sweden's Gambro starting in 1969. Since all use membranes in the separation process, they are all integrated into modules. Assembling these modules is a

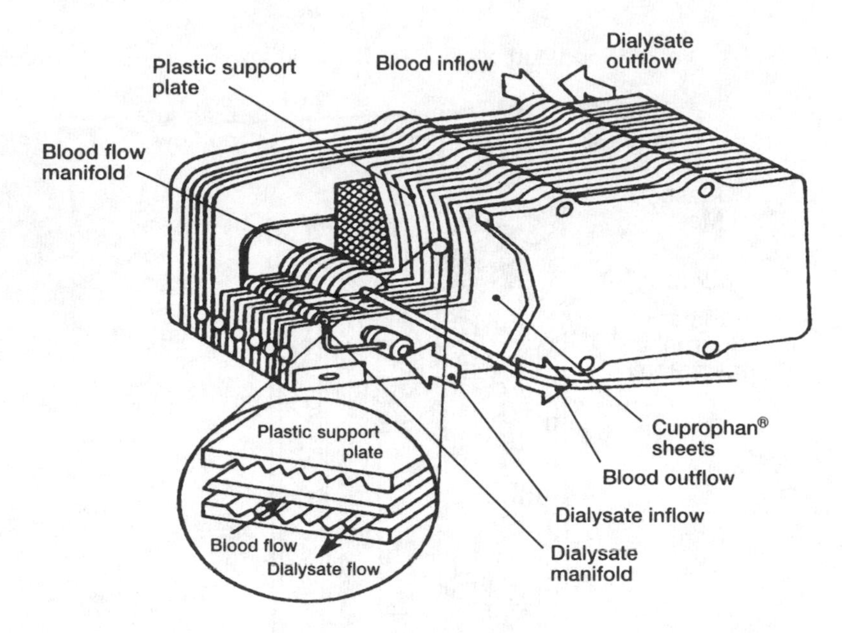

Figure 17.7 A plate-type dialyzer (from Sakai and Oshima). The blood flows from a manifold between sheets of Cuprophan® and the dialysate flows between the Cuprophan® membranes and plastic support plates.[3]

delicate task and the design of the modules has a crucial effect on separation efficiency. The modules used in clinical hemodialysis are virtually the same as those used in industrial-scale processes. The only real difference is in the size, which arises out of the different processing capacity requirements.

Some 98% of all clinical dialyzers in use today are of the hollow-fiber design proposed by Stewart in 1964, which was very similar to the Abel's "vividiffusion" artificial kidney of 1914. Cordis Dow marketed a hollow-fiber dialyzer using membranes of regenerated cellulose in 1971. Bundles of some 10,000 ultrafine capillary fibers with an inner diameter of only 200 μm are packed into a cylindrical plastic jacket. The blood flows inside the capillaries while the dialysate circulates outside. In more than 50% of these hollow-fiber dialyzers the fibers are made of cellulose; synthetic-fiber polymers are used in the rest. Because the fibers are so thin, an extremely large surface area can be contained in a compact space, resulting in high efficiency. Recently, fibers with walls only 6.5 μm thick (when dry) and 20–50 mm thick (when wet) have made their appearance.

Hollow-fiber dialyzers are somewhat more expensive than other types and they suffer from channelling of the blood and dialysate flows. These disadvantages are compensated for, however, by several factors; blood loss is minimal even if a fiber should rupture, the volume of residual blood is low, a large membrane area is contained in a compact, lightweight design, and the priming volume is low and does not change when pressurized.

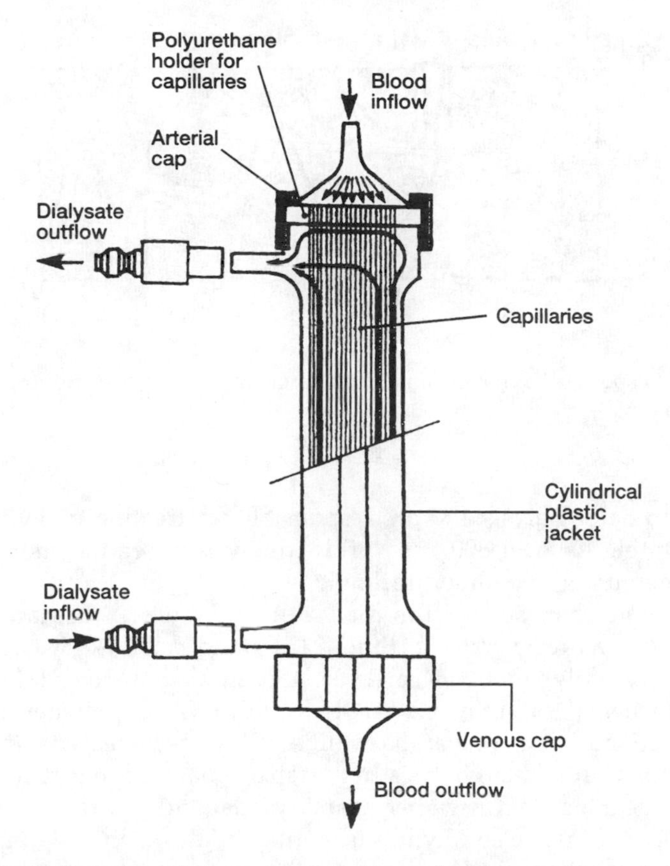

Figure 17.8 A hollow-fiber dialyzer (from Sakai and Oshima). Blood flows inside the capillary fibers and dialysate flows outside; the plastic jacket holds about 10,000 fibers with a surface area of approximately 10,000 cm^2.[3]

Much progress has been achieved through the efforts of industry in developing improved hemodialysis membranes. But despite constant improvements there are still many problems inherent in the design of the modules to hold the membranes. The attention of chemical engineers has been focused on creating ever-larger industrial plants but now their expertise is needed to design small-scale clinical equipment. In dialyzers there is a particular need to reduce size without compromising function. Experience shows that inadequate module design vitiates the advances made in membrane performance.

DIALYSIS MEMBRANES

Some 90% of all semipermeable membranes are used in therapeutic applications. When we realize that most of them — some 130 million square meters per year

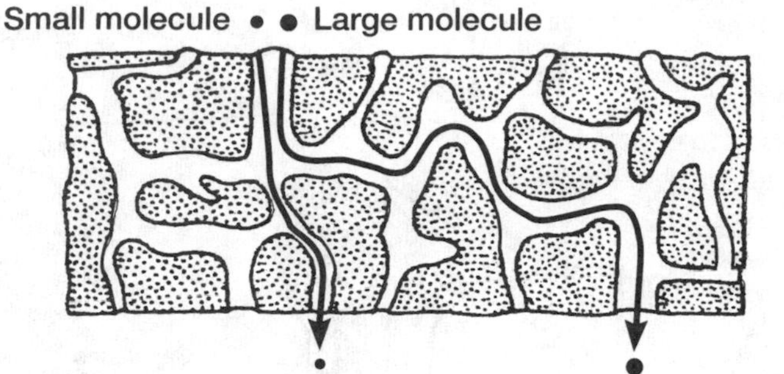

Figure 17.9 Transport behavior of small and large molecules in semipermeable membranes (not to scale).

— are used in hemodialysis and are responsible for treating patient with renal failure accounting for over 800,000 worldwide, we can see the enormous social benefits of membrane separation technology.

Although more than 50% of the dialysis membranes in use are still made of cellulose such as regenerated cellulose (RC) and cellulose acetate (CA), the use of synthetic polymer fibers such as poly(methyl methacylate) (PMMA), poly(acrylonitrile) (PAN), ethylene vinyl alcohol (EVA) copolymer, polysulfone (PS) and polyethersulfone/polyarylate (PEPA) alloy has increased in recent years. Membranes for use in hemodialysis must satisfy many stringent requirements. They must have a high diffusive permeability, adequate hydraulic permeability, balance between diffusive and hydraulic permeabilities, minimal degradation of the membrane over time, good biocompatibility, ample mechanical strength when wet, ease of sterilization, absence of eluted substances and low cost.

Let us assume that the pores of regenerated cellulose membranes are about 4.6 nm in diameter and open at right angles to the membrane wall which, when wet, is 20 μm thick. It may help us to realize the enormous relative distances that the solute must traverse if we liken this to a train passing through a tunnel which, if it were 8 m in diameter, would have to be approximately 35 km long. Even this assumes that the pores simply run in straight lines perpendicular to the membrane walls; in fact they form a maze requiring travel of 1.6 times that distance or more as shown in Figure 17.9.

The biological membrane that forms the capillary wall of the glomerulus in a living kidney is characterized by being highly permeable to water and electrolytes but impermeable to albumin and other serum proteins. Although the pores of these biological membranes are of adequate size to pass the albumin molecules, loss of albumin into the urine is prevented in the normal kidney by means of charge barrier. When the glomerulus is damaged the permeability of the capillary wall increases; serum proteins (mainly albumin) then appear in the urine.

The biological membrane with excellent selectivity and biocompatibility thus provides an excellent model and we must admit that its artificial counterpart lags far behind. New membranes that emulate these desirable properties will represent a major advance in membrane separation technology.

Recent achievements in the field have included thin membranes, charged membranes, membranes with huge pores, and biocompatible membranes. Thinner membranes are indispensable for the improvement of solute removal and they will also make it possible to reduce the size of dialyzers. Charged membranes allow control of serum electrolyte permeability, thereby giving membranes greater selectivity. Large-pore membranes promote the efficient removal of middle molecules (with molecular weights of 500–5,000) and β_2-microglobulin (molecular weight 11,800) which has been identified as a substance related to amyloidosis.

To improve biocompatibility, attention has been directed at the tendency for regenerated cellulose membranes to cause complement activation. The complement system mediates neutralization of the antigens to prevent infection and inflammation and in addition cell lysis, bacteria opsonization and histamine release from elements in the blood. The complement components are then activated via the classical pathway and the alternative pathway. This has been reduced by surface modifications involving the replacement of a very small number of hydroxyl groups with tertiary amino groups and coating with vitamin E or polyethylene glycol.

These are just some of the more significant improvements in dialysis membranes and advances continue. Less has been done on basic research into a relationship between permeability and structure of the membranes. This again is a field in which chemical engineers can shine.

COMPUTER-CONTROLLED DIALYSIS

In terms of technological sophistication, hemodialysis has attained a high level. It is still administered in a standardized way, however, which may not be optimally suited to individual patient needs. Thus patients may experience headache, nausea or decreased blood pressure and much staff time is spent in treating these symptoms. This has suggested the concept of a feedback control system in which sensors monitor the condition of the patient from minute to minute during the dialysis process, allowing treatment to be modified by computer in response to physiological changes. Thus, computerization promises a more positive and thorough treatment.

Figure 17.10 shows a specific example. Measurements are made of the concentrations of urea, creatinine and middle molecules in the dialysate and the corresponding blood concentrations are estimated. This enables the dialysis time to be adjusted and the dialysis conditions to be changed automatically during the treatment session. Water removal can also be controlled.

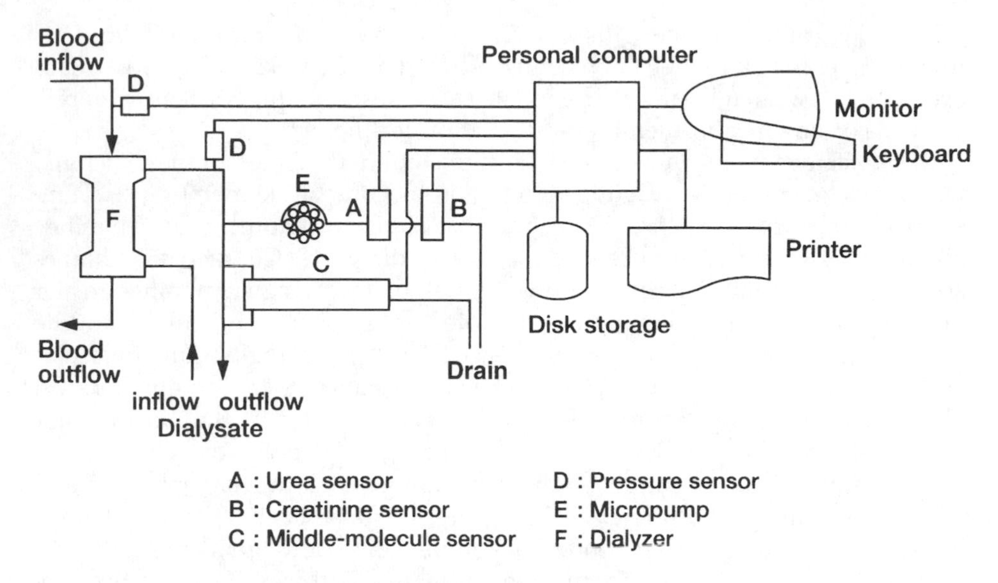

Figure 17.10 A computer controlled dialysis system.

Such a computerized dialysis system would have many advantages of reducing the workload of clinical staff, thus leaving more time for personal treatment and care. It would also make it possible to determine blood concentrations without taking a blood sample. It could provide a continuing picture of each hemodialysis session as well as a complete and permanent record — a major advantage of long-term patient care.

Despite the widespread use of computers in medical science, clinical applications have so far been limited to artificial pancreas. It may not be long before dialysis treatment can be administered in the home. When that day comes, computerized dialysis systems will play a major role. In developing high-quality systems that are accurate, safe and economical, chemical engineering will again make an important contribution.

MEDICAL APPLICATIONS FOR ADSORBENT AGENTS

Silica gel, used for dehumidification, and activated charcoal, used for deodorizing, are among the most commonly known adsorbent agents. Molecular sieves, too, are widely used in chemical plants. All of these have their individual properties and are also adsorbent agents that are used therapeutically.

One classic example is the system shown in Figure 17.11. This was intended to remove protein metabolites from the blood of patients with renal failure by direct hemoperfusion through an activated charcoal column. Despite its advan-

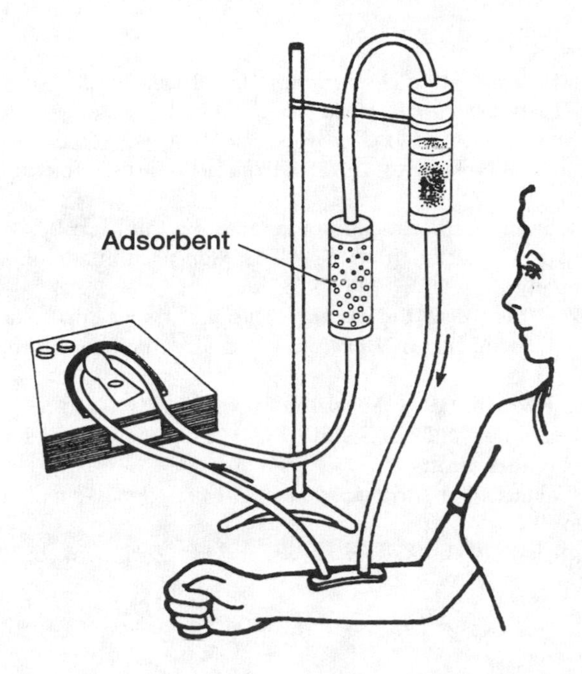

Figure 17.11 Device for direct hemoperfusion (from Sakai and Oshima), consisting of a cartridge containing adsorbent and a roller pump; easy to operate and convenient to carry.[3]

tages of compact design, the system was inferior in removing significant amounts of urea, adjusting electrolyte levels and removing excess water. Thus it was not a practical replacement for hemodialysis. Current clinical uses of this system, however, include treatment of drug addiction, hepatic coma and excess bilirubin in the blood.

Much attention has recently been focused on affinity chromatography, an adsorption technique for specific separation. The antigen-containing blood of a patient is introduced into a packed column containing the antibody which specifically removes the antigen. There are many who suffer from immune deficiencies and some cases are intractable. It is also more and more common to see autoimmune diseases in which a part of the body is the antigen and participates in an *in vivo* antigen-antibody reaction. Although not all of these antigens and antibodies have been identified, there is no doubt that treatment will involve specific separation techniques using such adsorbents.

For those not involved in these specialities, it may come as a surprise, or even a shock, to learn that the disciplines of physical chemistry and chemical engineering are actively applied in medical treatment. I experience great satisfaction at the life-saving role of this discipline. It is my fondest hope that there will be some among my readers who are led to greater interest in this less commonly known and idealistic aspect of chemical engineering.

REFERENCES

1. R. Shirane (1984) *"Fascinating Dictionary of High Technology."* Tokyo
2. K. Ota (1973) *"Facts about Artificial Kidneys."*, Tokyo: Nankodo
3. K. Sakai and M. Oshima (1979) *Gendai Kagaku*, **42**(4), 42
4. S. Koshikawa and S. Nakagawa (1972) *"Artificial Kidneys"*, Tokyo: Chugaigakusha

Other interesting references to artificial organs may be found in:

B. M. Brenner and F. C. Rector, Jr. (1981) *"The Kidney"*, Philadelphia, London: W. B. Saunders Company

D. C. Cooney (1976) *"Biomedical Engineering Principles"*, New York: Marcel Dekker

J. T. Daugirdas and T. S. Ing (1988) *"Handbook of Dialysis"*, Benston, Toronto: Little, Brown and Company

M. Mineshima and K. Sakai (1986) *Kagaku Kogaku*, **50**, 682

A. R. Nissenson, R. N. Fine and D. E. Gentile (1984) *"Clinical Dialysis"*, Norwalk, Connecticut: Appleton-Century-Crofts

Y. Sakurai and K. Sakai (1984) *"Recent Artificial Organ Techniques and Their Future Development"*. Tokyo: IPC

F. Yoshida and K. Sakai (1993) *"Kagaku Kogaku to Jinko Zoki"*, Tokyo: Kyoritsu Shuppan

18. DEVELOPMENT OF A MAN-MACHINE INTERFACE FOR TASK ASSIGNMENT AND TELEOPERATION OF AN INTELLIGENT MULTIROBOT SYSTEM IN A BIOPLANT

TSUYOSHI SUZUKI[1], YOSHITO SAKINO[2], TAKESHI SEKINE[3], TERUO FUJII[4], HAYATO KAETSU[5], HAJIME ASAMA and ISAO ENDOS

[1]*Advanced Technology Center, The Institue of Physical and Chemical Research (RIKEN), Saitama, 351-0198, JAPAN*

[2]*Osaka Prefectural Police, Osaka, 540-8540, JAPAN*

[3]*Ajinomoto Co., Inc., Kanagawa, 210-8681, JAPAN*

[4]*Institute of Industrial Science, Univ. of Tokyo, Tokyo, 106-8558, JAPAN*

[5]*Biochemical Systems Lab., The Institute of Physical and Chemical Research (RIKEN), Saitama, 351-0198, JAPAN*

INTRODUCTION

Recently, expert systems for diagnosing bioreaction processes have been studied and developed enormously in the world. But there are very few systems which consider maintenance of a bioplant. The maintenance of the plant means inspection of malfunctioning elements like sensors and/or valves and agitators, and repairing them as well.

In our previous papers the authors (von Numers *et al.* 1994, Endo *et al.* 1996, Suzuki *et al.* 1997) have presented a novel man-machine interface for a bioprocess expert system by taking into account the mutual cooperation between the operator and the system. This interface aims to aid the operator's precise decision making and correct operation on the basis of ergonomics.

Adding to the functions of this expert system, the authors are considering the introduction of an adaptive scheduling ability for multiple intelligent robots to maintain a bioplant. The intelligent robot means that it can move autonomously, recognize its own position in the plant, acquire the environmental information and communicate with the plant operation systems as well as with other robots. Asama (1994) has named this robot system as Distributed Autonomous Robotic Systems (DARS) and developed them since 1987. The principal aims of DARS is to maintain an atomic or chemical plant, where human operators can not access freely. Although the environment of a bioplant is not as dangerous as an atomic plant, it should be kept clean to avoid contamination to microorganisms. Thus, operation access should be limited. We can not enter nor get out the working site freely. The operation of the bioplant also requires inconvenient shift work. In other words, there are many discrete works in the bioplant like inoculation, washing inside bioreactors, attachment and detachment of sensors and so on. Furthermore, among industrial countries like the U.S., the E.C. and Japan, labor costs are very high. Considering these conditions, the DARS is obviously powerful in the bioplant, particularly for maintenance.

Efficient utilization of DARS requires the optional assignment of tasks for routine and emergency responses to multiple robots which include various kinds of functions. Namely, an adaptive multirobot scheduling system for maintenance operation should be established. In the present paper, the authors present this scheduling system and evaluate the performance of the system by numerical experiments.

On the other hand, a human operator should somehow take over the operation of the robotic system according to the task requirements when the robot cannot execute it. We have already proposed a teleoperation system for mobile robots (Suzuki *et al.* 1996, Suzuki *et al.* 1998, Kawabata *et al.* 1997, Ishikawa *et al.* 1998). But, there are some problems based on time delay. Hirukawa *et al.* (1997) proposed WWR (World Wide Robotics). Sekimoto *et al.* (1997) reported a design of driving device for a mobile robot. They used the real feedback image with a touch panel. But, it would have the problem in case of long distance (Sekimoto *et al.*, 1997). Moreover, teleoperation systems that can be used by nonspecialists from any Internet site are discussed (Inagaki *et al.* 1995, Mitsuishi *et al.* 1997, Masson and Fournier 1997, Adams *et al.* 1995). But, it is difficult for these teleoperation systems to have communication infrastructure with stable data transfer speed.

Therefore, we discuss the problems of teleoperation systems of a mobile robot and utilization of the virtual world, in which a human-robot collaboration based on the teleoperation system is provided as comfortable human interfaces. Even if the communication link shows unstable performance, we could compensate the control or monitoring signals by virtually computed ones. Performance of the developed system is examined through the experiments with an actual mobile robot and network.

ADAPTIVE MULTIROBOT SCHEDULING SYSTEM

We call here an adaptive multirobot scheduling system which applies DARS as the local agent and maintains a bioplant as an Adaptive Multirobot Scheduling System (AMSS). The meaning of "adaptive" is that the system is able to assign the robots adaptively based on the conditions in the plant and the operating robots, as well as adapting to emergency conditions. In the following we explain the scheme and performance of the AMSS on the basis of several working hypotheses.

Working Hypotheses

Firstly, there are many autonomous robots distributed in the bioplant, which have different tools; camera, forklift and manipulators. The camera installed robot can

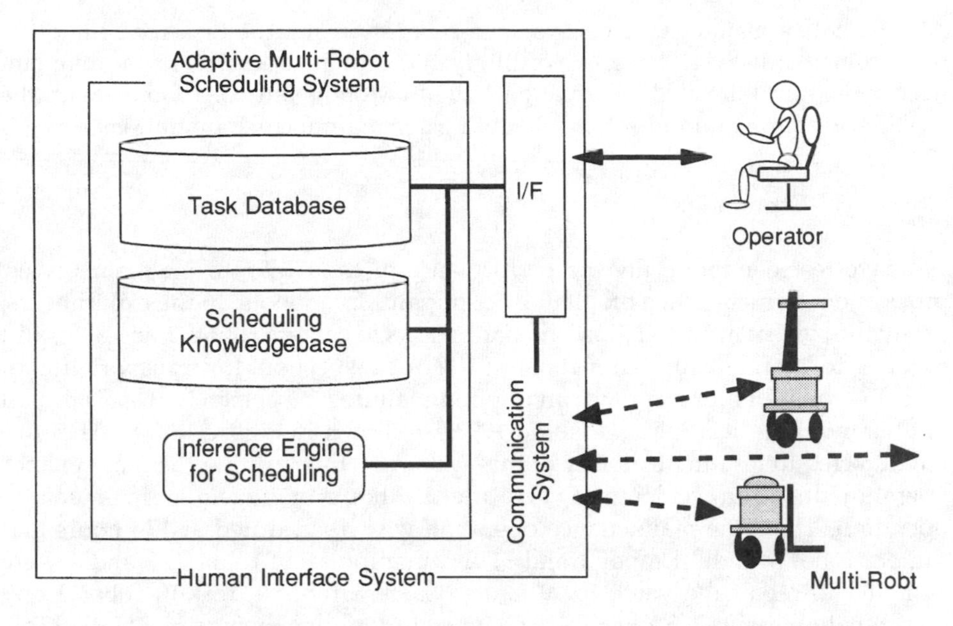

Figure 18.1 Adaptive multirobot scheduling system.

inspect parts of the plant in detail, the forklift robot can transport heavy materials here and there in the plant and the manipulation robot can adjust plant elements, like sensors and valves skillfully. Secondly, these robots which have the same tool are originally equal in their abilities, but they differentiate their skillfulness according to the experiences. Thirdly, there are several maintenance operations in the plant but every operation is executed as a sequence of unit operations. Fourthly, a unit operation is defined as a group of actions which is executed by a unit command.

Scheme of the AMSS

The AMSS is composed of a task database, a scheduling knowledge base and an inference engine for scheduling as shown in Figure 18.1. The task database controls task recipes and a unit operation database. It is written in a task recipe that describes the operated procedures and sequences for every maintenance operation. An operator can change the operation manual freely according to the urgency. In this way, the scheduling for maintenance operations become flexible and adaptive. Meanwhile, the unit operation database installs the data which define the specific robots used, the numbers of robots and the operation time. The scheduling knowledge base stores the knowledge about specifying robots and their limited abilities. On the basis of this knowledge base, an operator of the plant can decide a strategy for specifying robots easily. The inference engine in the

AMSS allots a maintenance task to each robot by using the task recipe and the scheduling knowledge base. Accordingly the human operator in the bioplant need not control detailed information issued to robots but collaborates with the AMSS for adaptive, multirobot scheduling for maintenance operations.

Numerical Experiments

We have tested numerically the performance of the AMSS to maintain valves, pipes and a fermenter in a bioplant by using multiple robots. In the experiments, firstly the system has allotted maintenance tasks to camera robots 1 and 3, forklift robots 1 to 4, and manipulator robots 1 to 6 for inspection, for transportation of machine tools and for examination of piping failures, respectively. This is drawn in a gaunt chart (Kuroda and Ishida 1993) as shown in Figure 18.2(a). Then we have started to test the flexibility of this system by interrupting the maintenance operation due to an accident in a valve after 127 hours of the regular maintenance operation. Then, the maintenance operation was rescheduled at 136 hours and camera robot 3, forklift robot 1 and 3, and manipulation robot 1, 3 and 4 were assigned to repair the valve for 9 hours. Camera robot 1, forklift robot 2 and manipulation robots 2, 5 and 6 were assigned to continue their original tasks. This shows one of the flexibilities of the AMSS. Then, trouble with the manipulation robot 1 (thick line in Figures 18.2(b) and 18.2(c)). The remaining robots continued their tasks, but 78 hours after the trouble, forklift robots 1 and 3 began to help manipulation robots 2, 3 and 4 for the relief of the manipulation robot 1 as shown in Figure 18.2(c) (at 225 hours). From these experimental results, we could confirm that the scheduling system has allotted maintenance tasks to each robot without overlap. Obviously sound robots could continue their original task without being bothered by the troubled robot. It was proved, too, that this scheduling system was adaptive even when maintenance procedures were interrupted obligedly by accidents or by other problems.

Strategy of the Schedule

If we describe the total time for any robot which has tool Dj to accomplish a maintenance unit operation Ui as $t(R_{Dj}, Ui)$ and the time for robot k which has the same tool to accomplish the same task as $t(R_{Djk}, Ui)$ then occupied ratio of the robot k, $Eo(R_{Djk})$ is expressed as follows;

$$Eo(R_{Djk}) = \frac{\sum_i t(R_{Djk}, U_i)}{\sum_i t(R_{Dj}, U_i)} \qquad (1)$$

Namely, $Eo(R_{Djk})$ means an occupation ratio of robot k which has tool Dj for the accomplishment of unit maintenance operation U_i.

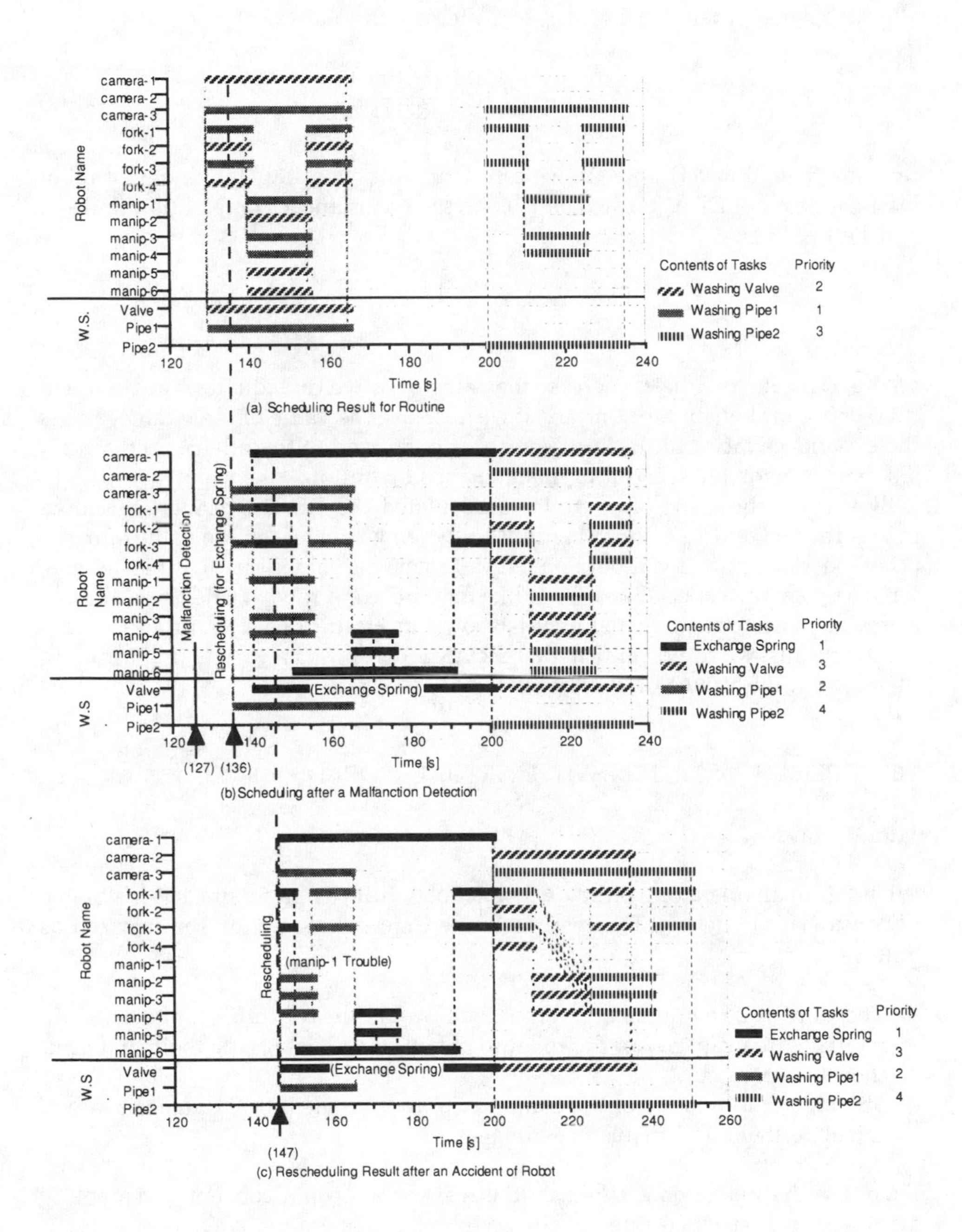

Figure 18.2 Results of scheduling.

In the next, we define an experience factor for robot k, Ee(R_{Djk}, Ui) as a ratio of time that it has spent for maintenance operation Ui to the time originally allocated for it. Then, Ee(R_{Djk}, U_i) is expressed as follows;

$$Ee(R_{Djk}, U_i) = \frac{t(R_{Djk}, U_i)}{t(R_{Dj}, U_i)} \tag{2}$$

By using equations (1) and (2), we can define the working state index of robot k which has tool Dj, as S.I. which is expressed as a linear combination Eo(R_{Djk}) and Ee(R_{Djk}, U_i);

$$S.I. = \alpha\,(1 - E_o(R_{Djk})) + (1 - \alpha)\,E_e(R_{Djk}, U_i) \quad (0 \le \alpha \le 1) \tag{3}$$

where, α means that if we increase the value of α, we can equalize the occupied ratio of each robot, on the contrary, if we decrease the value of α we can evaluate the experience factor of the specified robot compared to the others. So that, α is a strategy factor for scheduling maintenance operation.

By varying the value of α, we have calculated the S.I. The results are shown in Figures 18.3 and 18.4. From Figure 18.3, when we increased the value of α, the occupied ratio of every robot converges to certain value, say, 0.3. Meanwhile, when α is small, a specialized robot like manipulation robot 4 often was applied to maintenance operations Ui as it is shown in Figure 18.4. From these experimental results, we could confirm that the AMSS proposed a variety of scheduling plans when the values of α were varied.

TELEOPERATION SYSTEM WITH VIRTUAL WORLD

Time Delays

When a human operator operates a mobile robot from a remote site, it is inevitable to consider time delays. The reasons of the time delays can be summarized as follows.

- There are large amounts of information flow and heavy traffic in the network.
- Shortage of a capacity of the robot to obtain image data, save them and send them to operator.
- Shortage of a capacity of the computer to receive the image data from robot, calculate them and display the images.

Therefore the image data taken by the real mobile robot can not be smoothly displayed, e.g. at video rate, on the screen of the operation site.

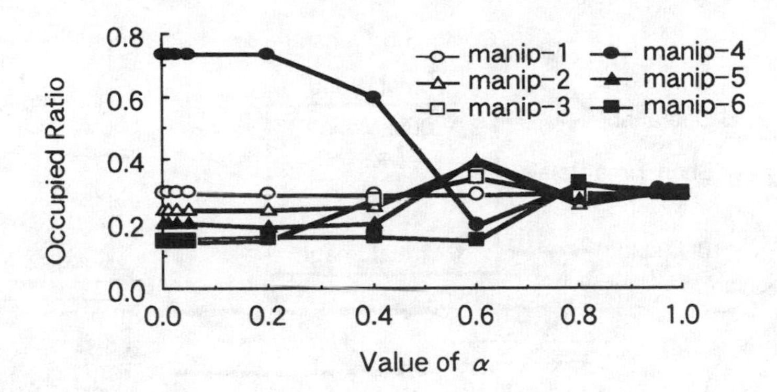

Figure 18.3 Relationship between occupied ratio and a.

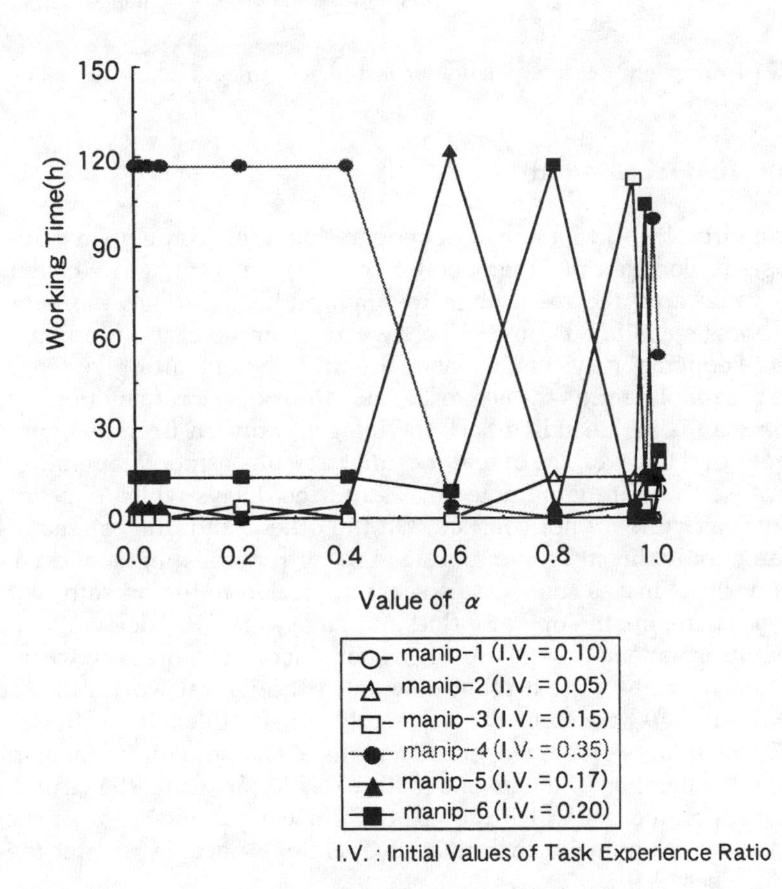

Figure 18.4 Working time of the robot "manip-4".

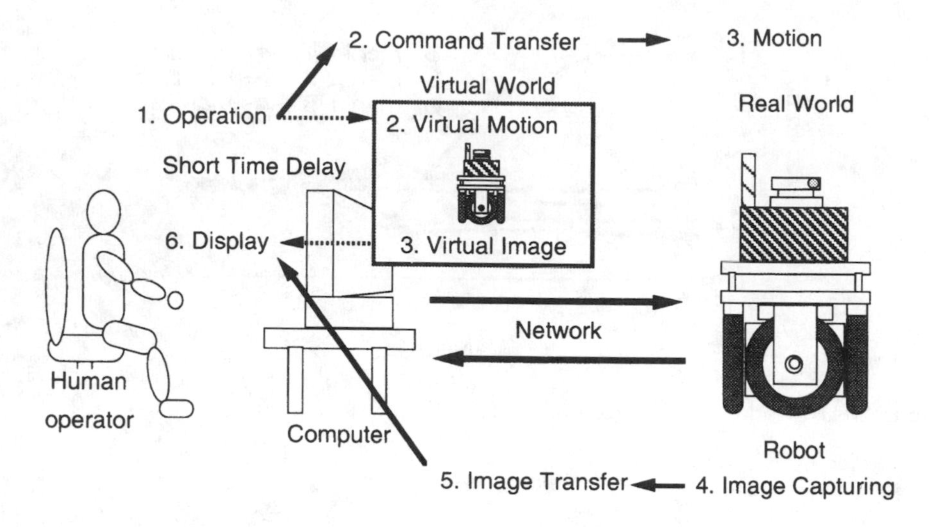

Figure 18.5 Example of the interpolation of insufficient images.

Advantage of Virtual World

By using a virtual world for a real robot operation, there come many advantages in the system. For example, the operator often uses real images which are sent from the camera on the robot or in the environment through the network to operate robots smoothly. Figure 18.5 shows an example of an information flow in such a teleoperation system. In the real world, the operator gives some commands to the robot through the network. Then, the robot starts a motion according to the commands with sending back the images captured from its own camera to the operator. However, the operator cannot execute a smooth operation in this cycle, because there are large time delays and the delays which depend on the state of the network are not constant. On the other hand, the commands from the operator could directly communicate to the virtual robot in the virtual world. Then, the virtual images could be also communicated to the operator with little time delay. Therefore, the operator does not have to feel the delay of communicating the information and operates the robot smoothly. Moreover, by using the image taken from the virtual robot moving in the virtual world, the intervals between the image data from the real world can be filled in by those virtual images. These images give the operator a realistic and smooth visual display for a comfortable operation of the mobile robots. Besides, by using the virtual world, the human operator can watch the object from various view points such as a bird's-eye view, etc. from which the human operator is not able to watch the image in the real cases.

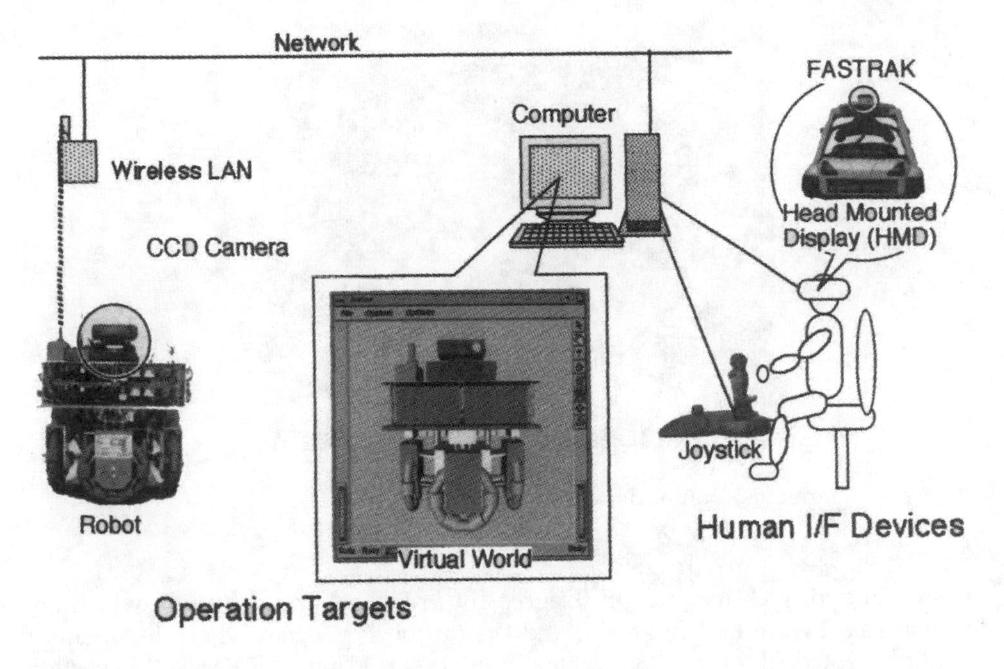

Figure 18.6 Proposed teleoperation system.

OVERALL STRUCTURE OF THE TELEOPERATION SYSTEM

We have developed a prototype teleoperation system for mobile robots utilizing the virtual world for better image display to human operators. Figure 18.6 shows the overall structure of the developed system. The system consists of human I/F (interface) devices, operation targets and a computer. This system uses TCP/IP (Transmission Control Protocol/Internet Protocol, generally used communication protocol in the Internet), and each part is connected to the network. In particular, between the robot and the computer are linked via the Internet through a wireless LAN (Local Area Network).

Human I/F Devices

Figure 18.7 shows the input and output devices. In order for the operator to operate the robots easily, joystick is used as an input device for the movement of the robot, where as pan and tilt motions of the camera are controlled by the output from the FASTRAK™ motion tracking system attached to a HMD (Head Mounted Display). The real image captured by the real camera and the virtual image captured by the virtual camera on the virtual robot are displayed on the

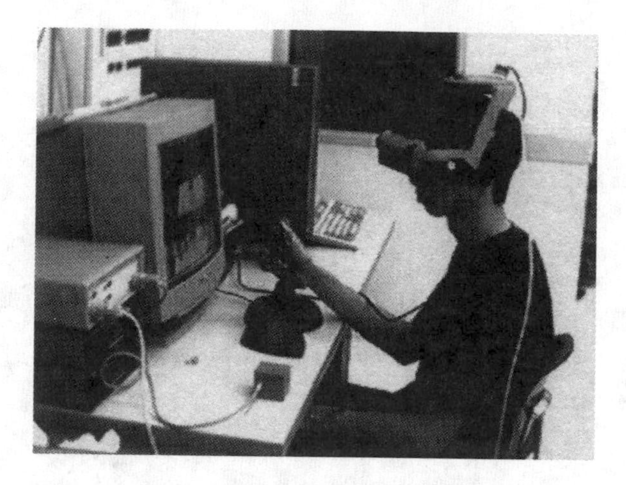

Figure 18.7 Input and output devices.

HMD. The joystick has 2 DOF (Degrees of Freedom) and a button, which are forward-backward and left-right, and the button is pushed when the operator wants to rotate the robot (Kawabata *et al.* 1997). When the FASTRAK motion tracking system detects a change of magnetic field, it converts the change into the numerical value of an angle. The motions of the camera on the real robot and the virtual robot are managed by the value of the angle.

Operation Targets (Robots)

In the Real World

The real world consists of a mobile robot, a CCD (Charge Coupled Devices) camera mounted on the robot and an environment where the mobile robot moves. Figure 18.8 shows the omnidirectional mobile robot. The robot has specially designed wheels and a drive mechanism. These mechanisms realize omnidirectional motion and decoupled control of 3 DOF movement in a horizontal plane (Asama *et al.* 1995). The robot moves bases on deadreckoning control method, and corrects its error in attitude by using own gyroscope data. The robot has its own IP address in the network. It is a kind of a physical agent on the Internet. The robot carries a CCD camera with pan (it can move from +40 degree to –40 degree), tilt (it can move from +20 degree to –20 degree) and zoom functions.

In the Virtual World

The virtual world consists of an environment model, a robot model and a process for virtual robot. The environment model is made by computer graphics based

Figure 18.8 Real omni-directional mobile robot.

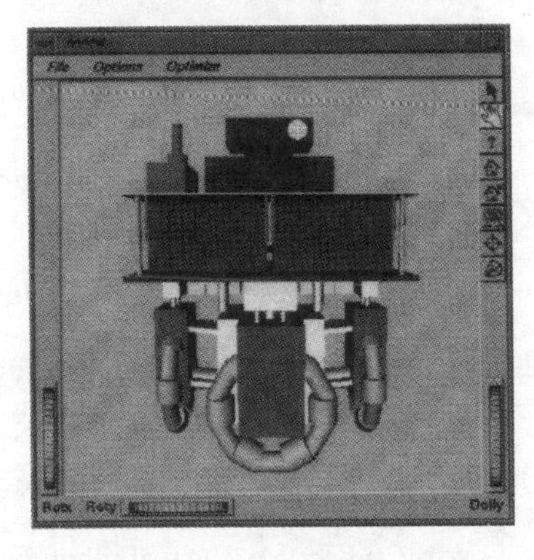

Figure 18.9 Virtual robot.

on the real environment where the real robot behaves. Similarly, the robot model is also made by computer graphics based on the real robot (Figure 18.9).

Figure 18.10 shows the structure of the overall processes for generation of the virtual world. The roles of the processes for virtual robot are data acquisition and data processing for operation of the robot in the virtual world. In a graphical workstation, the process which displays animation is the parent process, and processes which receive the angle of the camera and calculate the coordinates of the robot positions are child processes.

The data flow of the process for virtual robot is as follows:

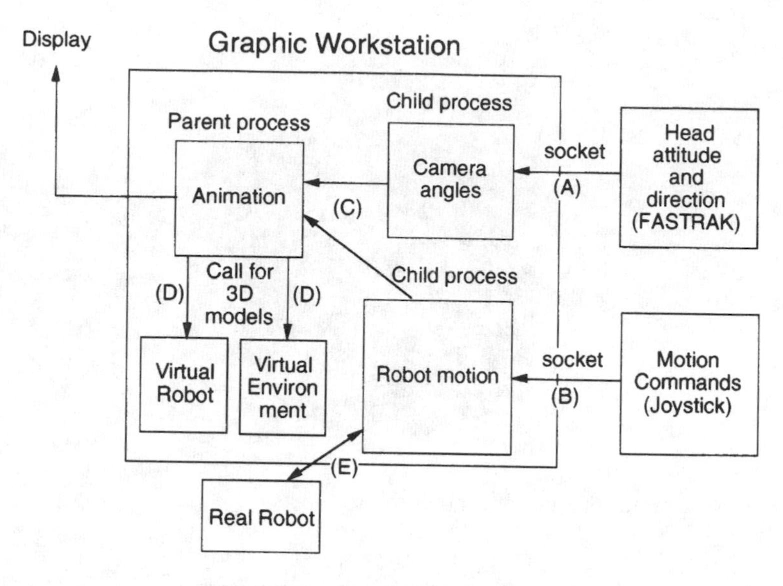

Figure 18.10 Process for virtual world.

(A) The camera control commands which are calculated on the basis of the head attitude and direction detected by FASTRAK are sent to graphical workstation through socket communications.

(B) Similarly, the motion commands which are calculated on the basis of the joystick are sent.

(C) Camera and position data are sent to the parent process for animation.

(D) The parent process calls for 3D models to construct animated image to animate.

(E) The position of the real robot is not always controlled precisely. So we have to correct the deviation of the position between the real robot and the virtual robot. For that purpose the position of the real robot detected by the onboard rotary encoder and the gyroscope is sent to the child process for robot motion. Figure 18.11 shows the image of the virtual world from the birds-eye view and the image from the virtual robot's view. The operator can operate the robot easily with watching these virtual images.

TELEOPERATION OF MOBILE ROBOT IN A VIRTUAL WORLD

We examined the performance of the developed teleoperation system to see whether or not similar motion of the robots can be displayed to the operator compensating the deviation of the position between the real robot and the virtual robot. Figure 18.12 shows a real working environment for experiment. The

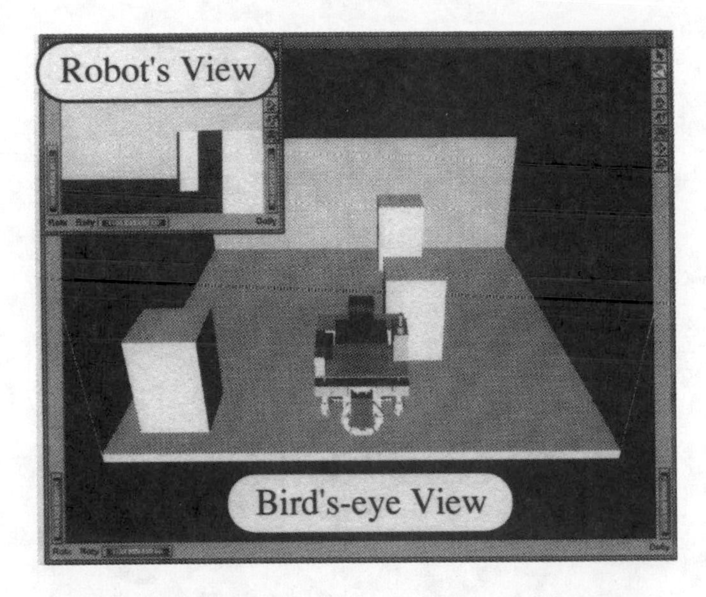

Figure 18.11 View of virtual world.

Figure 18.12 Real environment for experiment.

operated robot trajectory is shown in Figure 18.13. The task of the robot is to watch the objects A and B. The operator controlled the real robot and the virtual robot simultaneously using the joystick and the FASTRAK.

The examination has confirmed that the actual motion of the real robot and the virtual robot is almost linked. Figure 18.14 shows the position of the real and

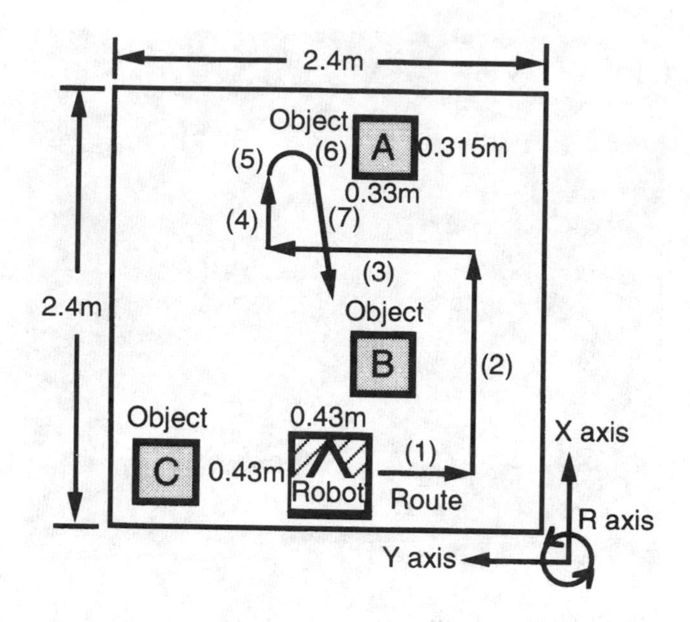

Figure 18.13 Environment and robot's route.

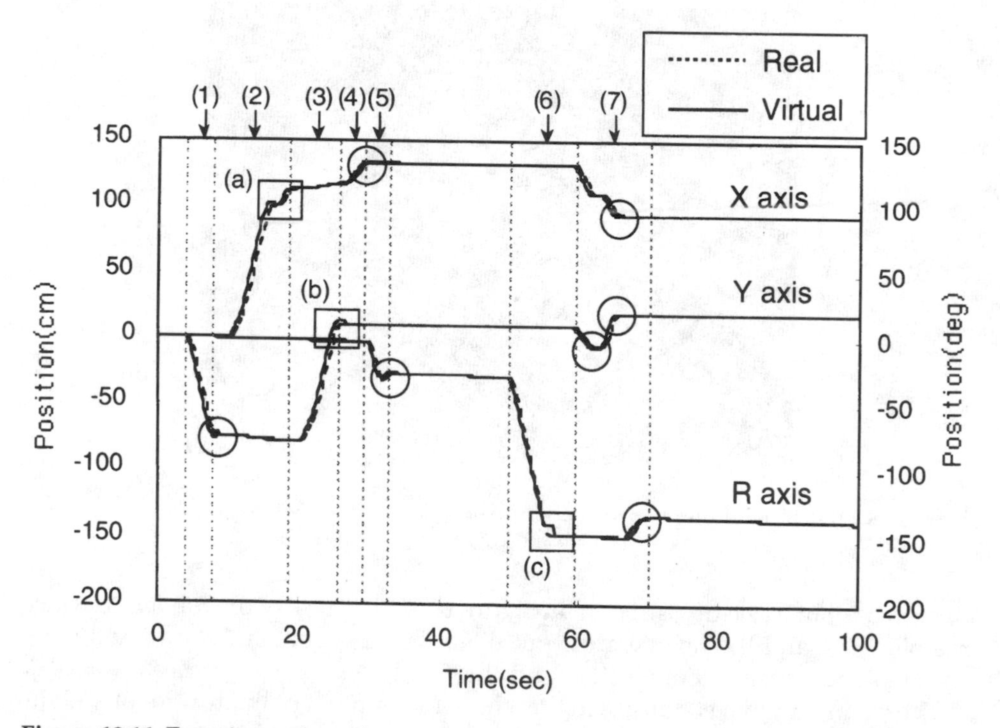

Figure 18.14 Experimental result.

virtual robots where it is toward the operator's order in each time. In Figure 18.14, X-axis, Y-axis, and R-axis indicate the X direction, Y direction, and rotation on the world coordinate, respectively. The number (1) to (7) in Figure 18.14 indicates a position in the route of the robot also in Figure 18.13. Between (5) and (6), the operator is watching the object, and also after (7). The initial positions and postures of the real robot and the virtual robot are denoted as (0.0, 0.0) [cm] and 0.0 [degree] in the world coordinate system, respectively. Due to time delay and/or motion error, a real robot does not position correctly while a virtual robot moves correctly according to the command by an operator. So that, an operator stops the command about one second, the virtual robot communicates with the real one and can adjust its position to that of the real one. When the operation is paused, the position of the virtual robot is corrected using the position and the posture data which is sent from the real robot. Figure 18.15 shows an enlargement of (a) in Figure 18.14. These figures show each deviation of the position between the real robot and the virtual robot, but the next moment the position of the virtual robot is corrected according to the position data from the real robot. We can confirm that compensation between the position of the real robot and the virtual robot has been also realized at each circle in Figure 18.14.

CONCLUSION

We have proposed in this paper a novel man-machine interface for maintenance of a bioplant, in particular, an adaptive multirobot scheduling system (AMSS).

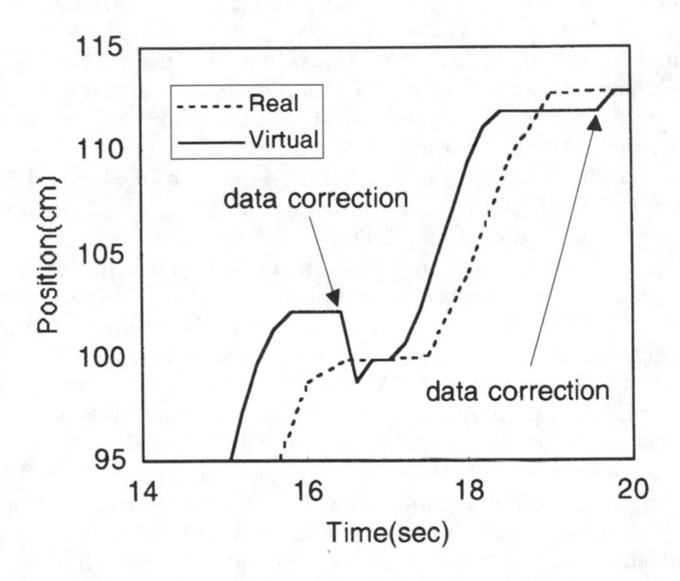

Figure 18.15 Enlargement of part (a) in Figure 18.14.

From numerical experiments we have confirmed that the AMSS could perform its original idea of adjusting the scheduling of maintenance operation even when an accident occurred in the plant or when a robot was disabled.

A framework of human intertace systems for teleoperation of a mobile robot was presented in this paper, too. The teleoperation system was constructed by using the virtual world as an operation interface. A human operator could simultaneously operate a robot in the real world as well as in the virtual world which could compensate the incomplete data transfer to the real robot. We could confirm the correction of the deviation of the position between the real robot and the virtual robot through the experiment. Although this experiment was executed in an experimental room, this system might be applicable in the real bioplant.

REFERENCES

Adams, J. A., Bajcsy, R., Kosecka, J., Kumar, V., Mandelbaum, R., Mintz, M., Paul, R., Wang, C., Yamamoto, Y. and Yun, X. (1995) Cooperative Material Handling by Human and Robotic Agents: Module Development and System Synthesis. *Proceedings of IEEE/RSJ International Conference on Intelligent Robots and Systems*, 200–205

Arai, Y., Fujii, T., Asama, H., Fujita, T., Kaetsu, H. and Endo, I. (1996) Self-Localization of Autonomous Mobile Robots using Intelligent Data Carriers. *Distributed Autonomous Robotic Systems 2*, pp. 401–410. Springer-Verlag

Asama, H. (1994) Trends of distributed autonomous robotic system. In: *Distributed Autonomous Robotic Systems*, edited by Asama, H., Fukuda, T., Arai, T. and Endo, I., pp. 3–8. Tokyo: Springer Verlag

Asama, H., Sato, M., Bogoni, L., Kaetsu, H., and Endo, I. (1995) Development of an Omni-Directional Mobile Robot with 3DOF Decoupling Drive Mechanism. *Proceedings of International Conference on Robotics and Automation, Vol. 2*, 1925–1930

Endo, I., Sakino, Y., Suzuki, T., Nakajima, M., Asama, H., Fujii, T., Sato, K. and Kaetsu, H. (1996) A novel man-machine interface for a bioprocess expert system constructed for cooperative decision making and operation. In: *The 5th World Congress of Chemical Engineering*, pp. 428–431. San Diego

Hirukawa, H., Matsui, T. and Hirai, S. (1997) A Prototype of Standard Teleoperation Systems on an Enhanced VRML. *Proceedings of IEEE/RSJ International Conference on Intelligent Robots and Systems*, 1801–1806

Inagaki, Y., Sugie, H., Aisu, H., Ono, S. and Unemi, T. (1995) Behavior-based Intention Inference for Intelligent Robots Cooperating with Human. *Proceedings of the International Joint Conference of the Fourth, IEEE International Conference on Fuzzy Systems and the Second International Fuzzy Engineering Symposium*, 1695–1700

Ishikawa, T., Kawabata, K., Ueda, Y., Asama, H. and Endo, I. (1998) Graphical User Interface for Collaborative System of Human and Mobile Robots with Sensors. *Distributed Autonomous Robotic Systems 3*, pp. 319–328. Springer.

Kawabata, K., Ishikawa, T., Fujii, T., Noguchi, T., Asama, H. and Endo, I. (1997) Teleoperation of Autonomous Mobile Robot under Limited Feedback Information. *Proceedings of the International Conference on Field and Service Robotics*, 158–164

Kuroda, C. and Ishida, M. (1993) A proposal for decentralized cooperative decision-making in chemical batch operation. *Engng Appl. Artif. Intell.*, **6**, 399–407

Masson, Y. and Fournier, R. (1997) EVEREST: A Virtual Reality Interface to Program a Teleoperated Mission. *Proceedings of IEEE/RSJ International Conference on Intelligent Robots and Systems*, 1813–1817

Mitsuishi, M., Chotoku, Y. and Nagao, T. (1997) Tele-guidance system using 3D-CG and real-time tracking vision. *Proceedings of 5th IEEE International Workshop on Robot and Human Communication*, 172–175

Sekimoto, T., Tsubouchi, T. and Yuta. S. (1997) A Simple Driving Device for a Vehicle-Implementation and Evaluation. *Proceedings of IEEE/RSJ International Conference on Intelligent Robots and Systems*, 147–154

Suzuki, T., Fujii, T., Asama, H., Yokota, K., Kaetsu, H., Mitomo, N. and Endo, I. (1996) Cooperation between a Human Operator and Multiple Robots for Maintenance Tasks at a Distance. *Distributed Autonomous Robotic Systems 2*, pp. 50–59. Springer

Suzuki, T., Sakino, Y., Nakajima, M., Asama, H., Fujii, T., Sato, K., Kaetsu, H. and Endo, I. (1997) A novel man-machine interface for a bioprocess expert system constructed for cooperative decision making and operation. *J. Biotechnol.*, **52**, 277–282

Suzuki, T., Fujii, T., Asama, H., Yokota, K., Kaetsu, H. and Endo, I. (1998) A multi-robot teleoperation system utilizing the Internet. *Advanced Robotics*, **11**(8), 781–797

von Numers, C., Nakajima, M., Siimes, T., Asama, H., Linko, P. and Endo, I. (1994) A knowledge based system using fuzzy inference for supervisory control of bioprocesses. *J. Biotechnol.*, **34**, 109–118

19. DIRECT USE OF SOLAR ENERGY: SOLAR CELL

MANABU IHARA[1] AND HIROSHI KOMIYAMA[2]

[1]*Institute for Chemical Reaction Science, Tohoku University, Sendai, Japan*
[2]*Department of Chemical System Engineering, University of Tokyo, Tokyo, Japan*

CHARACTERISTICS OF SUNLIGHT

The total energy of sunlight irradiated to the earth is equivalent to 10000 times the total energy consumed over the entire world. Sunlight is absorbed by the air, and irradiated to the surface of the earth. The strength of the incident sunlight on the surface of the earth varies by latitude, the time of day and season, and the weather. The amount of air that sunlight passes through is called the air mass (AM). The sunlight at AM-1 corresponds to the sunlight incident at a right angle to the zenith. The sunlight at AM-1.5 is nearly equivalent to the sunlight in Tokyo on a clear day at noon in winter. Figure 1 shows the solar spectrum at AM = 0, and 1.5, which indicates that sunlight is in the UV and the visible region, both of which have relatively high energy. At AM = 1.5 the energy density of sunlight is $1 kW/m^2$, which is low compared with other energy sources such as fossil fuels.

Despite the large total incident solar energy on the earth, solar energy has a relatively low energy density. This implies that a large area is required to capture

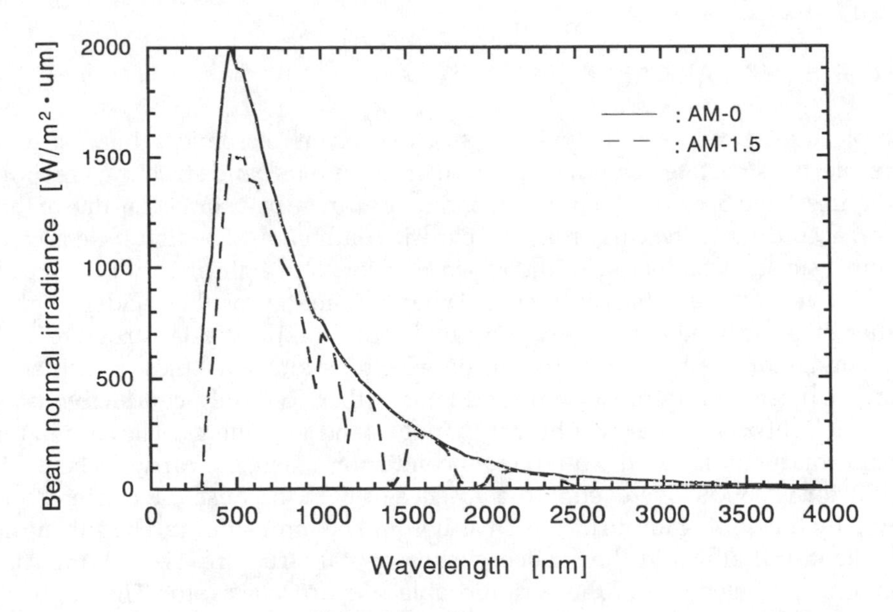

Figure 19.1 Solar spectrum at AM-0, 1.5.

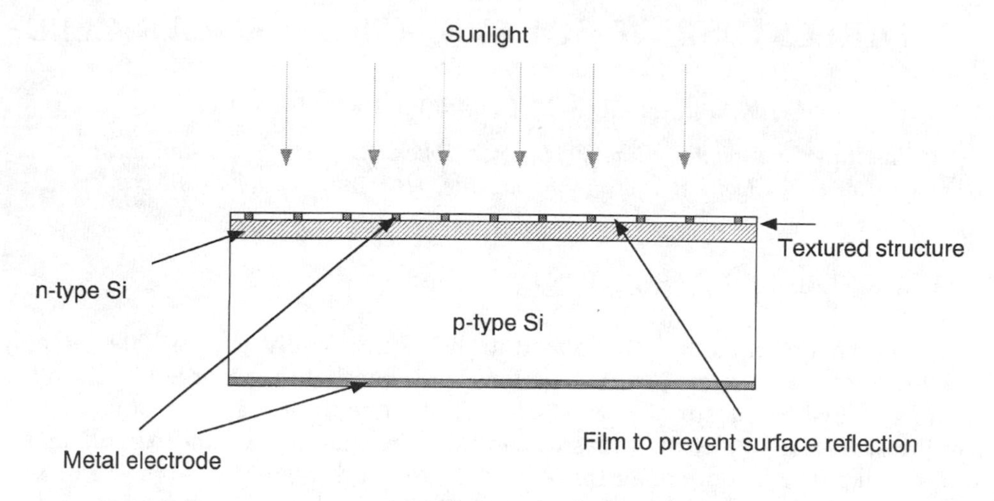

Figure 19.2 Structure of a Si p-n homojunction solar cell.

sufficient sunlight for practical applications. On the other hand, solar energy is indirectly responsible for most energy reserves on earth, such as fossil fuels, wind, and water power. Although these are indirect solar-energy sources, two promising direct solar-power technologies are solar-heat and solar-cell electric power generation. In this chapter, electric power generation by solar cells will be discussed as a technology for reducing global warming.

WHAT IS A SOLAR CELL?

A typical solar cell essentially consists of two layers of semiconductors. As an example, the structure of a p-n junction solar cell and its band structure are shown in Figures 2 and 3, respectively. Semiconductors can be categorized as one of three types, according to the carrier for the electric conduction. The first type is called an intrinsic semiconductor. In the absence of incident light it is not electrically conductive because it has no carrier. When the energy of the incident light is higher than the band-gap energy, the atomic bonds in the crystal structure of the semiconductor are broken, releasing the electrons and the holes, which act as charge carriers. An intrinsic semiconductor is therefore only conductive under incident light whose energy is higher than the band-gap energy. The second type of semiconductor is called a p-type semiconductor, where the carrier is hole. This conductor is conductive even in the absence of light, because the carrier is generated by the doping impurities such as boron. The impurities can be substituted for silicon (Si) atoms in the semiconductor crystal structure. For Si, impurities from group III elements of the periodic table take up an acceptor. The impurities accept the valence electrons. It means the generation of the holes. The hole is

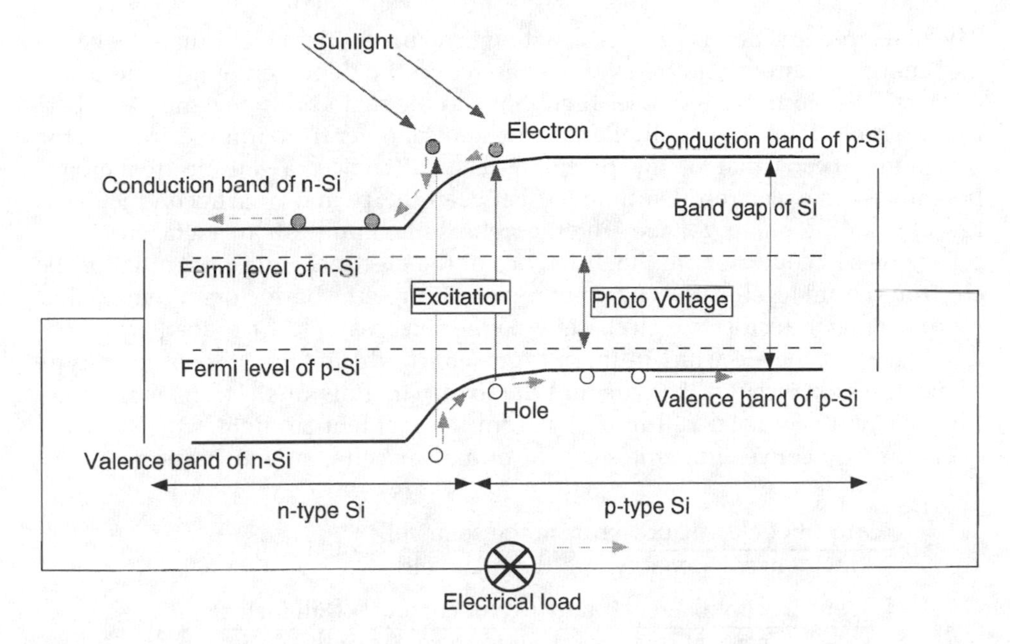

Figure 19.3 Principle of power generation in a Si p-n homojunction solar cell.

thermally generated at room temperature as a carrier. The third type of semiconductor is called an n-type semiconductor, where the carrier is an electron. Similar to a p-type semiconductor, an n-type semiconductor is conductive even in the absence of light, because the carrier is generated by impurities such as phosphorus. The impurity can be substituted for silicon (Si) atoms in the semiconductor crystal structure. For Si, impurities from group V elements of the periodic table are used to generate electrons.

Typical solar cells essentially consist of a combination of n- and p-type semiconductors, which is called as p-n junction, as shown in Figure 2. The p- and n-type semiconductors generate many holes and electrons at room temperature, respectively. Electrical contacts of these two different types of semiconductors transfer electrons from n- to p-type semiconductors along the gradient of the electrochemical potential of the electron (the Fermi level), and transfer holes from p- to n-type semiconductors along the gradient of the electrochemical potential of the hole. Through this electrical transport, the electrochemical potential of the electron (Fermi level) of the two semiconductors is equalized. The electrochemical potential of the hole of two semiconductors is also equalized through the transport. At the interface, the transfer causes a depletion of electrons and holes in the n and p regions respectively. This region is called the depletion layer. The n region in the depletion layer becomes positively charged and the p region in the depletion layer becomes negatively charged.

When a p-n junction is irradiated with light with energy higher than the band-gap energy, electrons and holes are generated in the depletion region and act as carriers. The electrons and holes then move to the n- and p-type regions, respectively. Through this transport, the electrochemical potential of the electrons in the n region exceeds that of the p region. The difference of the electrochemical potential of the electron (Fermi level) between the p- and n- semiconductors is revealed as the photo voltage. The theoretical maximum photo voltage of solar cells is nearly equivalent to the difference of the electrochemical potential of the electron (Fermi level) between two non-contacting semiconductors. When n- and p-type semiconductors are electrically connected to an electrical load by a conducting wire, the electrons in the n-type semiconductor transfer to the p-type semiconductor through the wire and the load. The flow of electrons is the electrical current of a solar cell resulting from the incident sunlight.

The energy conversion efficiency [η] of a solar cell can be expressed as

$$\begin{aligned} \eta &= \frac{[\text{Output of electric power from the solar cell}]}{[\text{Incident solar energy into the solar cell}]} \times 100\ [\%] \\ &= \frac{[\text{Open circuit voltage}] \times [\text{Short circuit current}] \times [\text{Fill factor}]}{[\text{Incident solar energy into the solar cell}]} \times 100\ [\%] \end{aligned} \quad (1)$$

The energy conversion efficiency of commercial Si solar cells ranges from 15 to 20%, and research solar cells have achieved a maximum of about 24.4% (Green *et al.*, 1998). The incident solar energy is lost by many factors, such as surface reflection, incomplete absorption of sunlight, and the recombination of electron and hole carriers. Also, with respect to the band-gap energy, there are two competing effects. If the difference between the energy of the incident solar energy and the band-gap energy is increased, the number of holes and electrons generated increases. However, as this difference increases, heat generation in the semiconductor increases, causing an overall loss of energy. There is, therefore, an optimum band gap. The maximum conversion efficiency of semiconductors with a band gap is shown in Figure 4, considering only absorption losses. For p-n homojunction semiconductor solar cells, the optimum band gap ranges from 1.0~1.5eV at AM = 1.5.

Solar cells are expected to contribute to a reduction in global warming. Global warming is caused partly by an increase of carbon dioxide (CO_2) emissions. Many types of solar cells have been developed, such as Si and compound semiconductors. The desirable conditions for increased commercial use of solar cells, such as residential power generation, are (a) availability of a large amount of required natural resources, (b) nonpoisonous materials, and (c) high cost-performance of the solar cells. (Lower conversion efficiency at lower cost need larger area of solar cell, which cause higher cost, to generate a unit of the electricity.) Considering these three desirable conditions, among the solar cells currently being studied, Si solar cells seem to be the most promising technology for commercial applica-

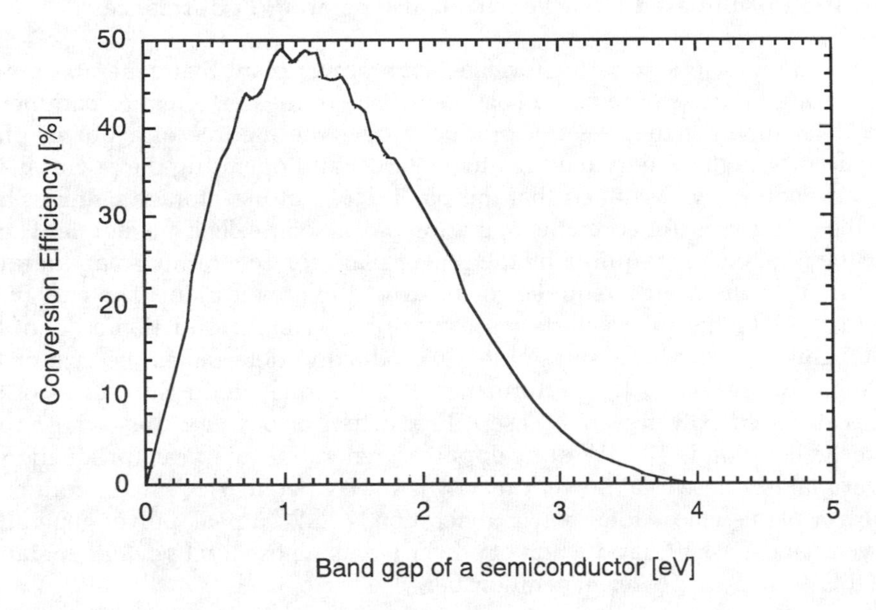

Figure 19.4 Calculated conversion efficiency of a p-n homojunction solar cell vs. band gap.

tions. In the next section "Evaluation of Si p-n junction solar cells", we estimate the economic and energy cost to make Si solar cells and discuss the potential reduction in global warming by using solar cells.

EVALUATION OF SI P-N JUNCTION SOLAR CELLS

The categorization of Si p-n junction solar cells is usually based on the crystallinity of Si as single-crystal, polycrystal, and amorphous Si. Although single-crystal Si solar cells have the highest conversion efficiency, their fabrication requires more energy than either polycrystal or amorphous Si. Furthermore, it is difficult to make the large surface area required for solar cells out of a single crystal. Making many small cells and connecting them is another method for making solar cells, but this also increases both the energy and economic costs because of the large number of chips that must be cut from ingots. Single-crystal solar cells are therefore not appropriate for commercial applications, although their high conversion efficiency makes them appropriate for space applications, where conversion efficiency is more important than cost. Therefore, in this section we evaluate energy and economic costs for making both polycrystal and amorphous Si solar cells. The evaluation method was first reported in "Evaluation of the technology of the solar power generation II", Komiyama *et al.* (1995).

Evaluation of Si-based Solar Power Plant Operating Performance

We evaluated the operating performance of a power plant that generates power only through the use of Si-based solar cells. One of the performance parameters is to determine the total energy produced over the lifetime of a power plant compared with the energy required to construct and operating the power plant. For our analysis we assumed that the plant does not use storage batteries and that the energy consumed by the operation of the solar cells is negligible. Therefore, the total energy required by the power plant for generating solar energy is equivalent to the energy required to construct the power plant. The energy required to make the solar cell decreases with increasing production scale of the solar plant per year because the rate of production determines the size of the facilities. We evaluated the performance at three total production scales of the solar cell based power plants: Case I, Production of power plants whose total power generation is 10 MW at peak power generation per year (production of power plants whose total power capacity is 10MW per year); Case II, Production of power plants whose total power generation is 1GW at peak power generation per year; and Case III, Production of power plants whose total power generation is 100 GW at peak power generation per year.

Performance Parameters for Solar Power Plants

Although we classify solar power plants by their peak power generation, the actual power generation capacity is less than this peak value and can be expressed as

$$WH = Wc \cdot Ty \cdot Ed \cdot Ei \cdot Eb \tag{2}$$

where WH is the total amount of generated electricity per year, Wc is the total power capacity of the solar cell based power plants produced per year (i.e., for Case I, 10 MW), Ty is the total hours of sunlight per year, Ed is the efficiency of the DC controller, Ei is the conversion efficiency of the inverter, and Eb is efficiency of the battery. Because an analysis is for a power plant without batteries, we assume that Eb = 1. The values for the other parameters are shown in Table 1. For Case III, the generated electrical power is equivalent to about 10% of Japan's annual electrical energy consumption of about 8×10^{11} KWh/yr. We assumed continual improvements in the required technologies and the efficiencies of each process (scale-up factor) with increasing scale of production. Specifically, we assumed year-on-year increases in single-cell energy conversion efficiency, increases in the area of single cells, reduction in the cover-glass thickness (this reduces energy loss), and reduction in the weight of the aluminum frame of the solar cell module. For calculating the energy payback time (EPT), we also assumed year-on-year improvements in the power generation efficiency (Ef in the equation (6)) of fossil-fuel based power plants. The calculation method of EPT will be explained afterward.

Table 19.1 Suppositions to evaluate.

	Polycrystal-Si			Amorphous-Si		
Case	*I*	*II*	*III*	*I*	*II*	*III*
Total power capacity of solar based power plants produced per year: Wc[MW]	1.0E+01	1.0E+03	1.0E+05	1.0E+01	1.0E+03	1.0E+05
Parameters related to the calculation of the amount of the power generation						
Energy density of incident sunlight: Cs [kW/m2]	1	1	1	1	1	1
Time of irradiation: Ty [h/yr]	1200	1200	1200	1200	1200	1200
Efficiency of control: Ed [–]	0.8	0.8	0.8	0.8	0.8	0.8
Inverter Efficiency: Ed [–]	0.9	0.9	0.9	0.9	0.9	0.9
Total amount of the power generation: WH [kWh/yr]	8.64E+06	8.64E+08	8.64E+10	8.64E+06	8.64E+08	8.64E+10
Parameters related to production of the cell						
Conversion efficiency of a single cell: Ec [–]	0.15	0.17	0.2	0.08	0.13	0.16
Transparence of the cover glass: Eg [–]	0.96	0.96	0.96	1	1	1
Size of a single cell: [m × m]	0.1 × 0.1	0.15 × 0.15	0.2 × 0.2	0.3 × 0.4	0.4 × 1.2	0.4 × 1.2
Area of a single cell: [m2]	0.01	0.0225	0.04	0.12	0.48	0.48
Total surface area of the cells: SYc [m2/yr]	6.94E+04	6.13E+06	5.21E+08	1.25E+05	7.69E+06	6.25E+08
Total number of the single cells: NYc [/yr]	6.94E+06	2.72E+08	1.30E+10	1.04E+06	1.60E+07	1.30E+09
Parameters related to production of the module						
Packing factor: Ep [–]	0.82	0.86	0.88	0.89	0.92	0.92
Conversion efficiency of a module: Em [–]	0.118	0.14	0.169	0.0712	0.12	0.147
Total surface area of the modules: SYm [m2/yr]	8.47E+04	7.12E+06	5.92E+08	1.40E+05	8.36E+06	6.79E+08

The *Wc* can be expressed as

$$Wc = Cs \cdot Ec \cdot Eg \cdot Sc \tag{3}$$

where Cs is the energy flux of the incident sunlight, Ec is conversion efficiency of the solar cell, Eg is the transparency of the cover glass, and Sc is the total surface area of the single solar cells. We assumed that Cs = 1kW/m^2, which is the energy flux of sunlight at AM-1.5.

The total surface area of solar-cell modules, Sm, is larger than Sc because a module consists of some single solar cells, and the placing the single cells need the area between the cells. The ratio of the Sc to Sm is defined as the packaging factor, Ep. Therefore, we can express the definition of the energy conversion efficiency of the solar-cell modules, *Em*, on the *Sm* as

$$Wc = Cs \cdot Em \cdot Sm \tag{4}$$

The parameters related to the assembly of the module are shown in Table 1.

The energy consumption to construct the solar power plant was divided into the energy required to produce raw materials, such as glass and aluminum; to construct solar-cell fabrication equipment, such as chemical vapor deposition (CVD) reactors; and to operate this fabrication equipment.

The energy consumed in the construction of a solar power plant is typically compared with its annual power generation and is expressed as the energy payback time, or EPT:

$$\text{EPT [year]} = \frac{\text{[Sum of the consumed electric power to construct the power plant]}}{\text{[Amount of the power generation per year]}} \tag{5}$$

If ETP is shorter than the life of the solar plant, the system is useful for reducing global warming. The consumed energy is composed of two types: electrical and chemical (fossil fuel). The fossil-fuel energy can be converted to electrical energy by a factor of 1.164 ($1[\text{cal}] = 4.186[\text{J}] = 4.186 \times 1/3600[\text{J} \times \text{h/s}] = 1.164\text{E-3}[\text{Wh}]$). Therefore, expressing the fossil-fuel chemical energy as an equivalent amount of electrical energy yields the total amount of electrical energy required to construct the power plant as

$$\begin{aligned}&\text{[Equivalent electric power to construct the power plant](KWh)}\\&\quad = \text{[electric energy](KWh)} + \text{fossil energy(Mcal)} \times 1.164 \times \text{Ef}\end{aligned} \tag{6}$$

The power-generation efficiency of fossil-fuel based power plants, Ef, for Case I, Case II, and Case III was assumed to be 0.35, 0.37 and 0.40, respectively.

The economic cost to construct the solar plant was estimated as the sum of the material costs to construct the factory that produce the solar-cell modules, the construction cost of the factory, the cost of the raw materials to make the solar-cell modules, the cost of the inverter and other control equipment, the operating cost of the utilities in the factory, and the labor cost.

Fabrication of Polycrystal Si Solar Cells

Figure 5 schematically shows the fabrication of polycrystal Si solar cells. The Si used for LSI (large-scale integration) (SEG-Si) must be refined with a gasi-

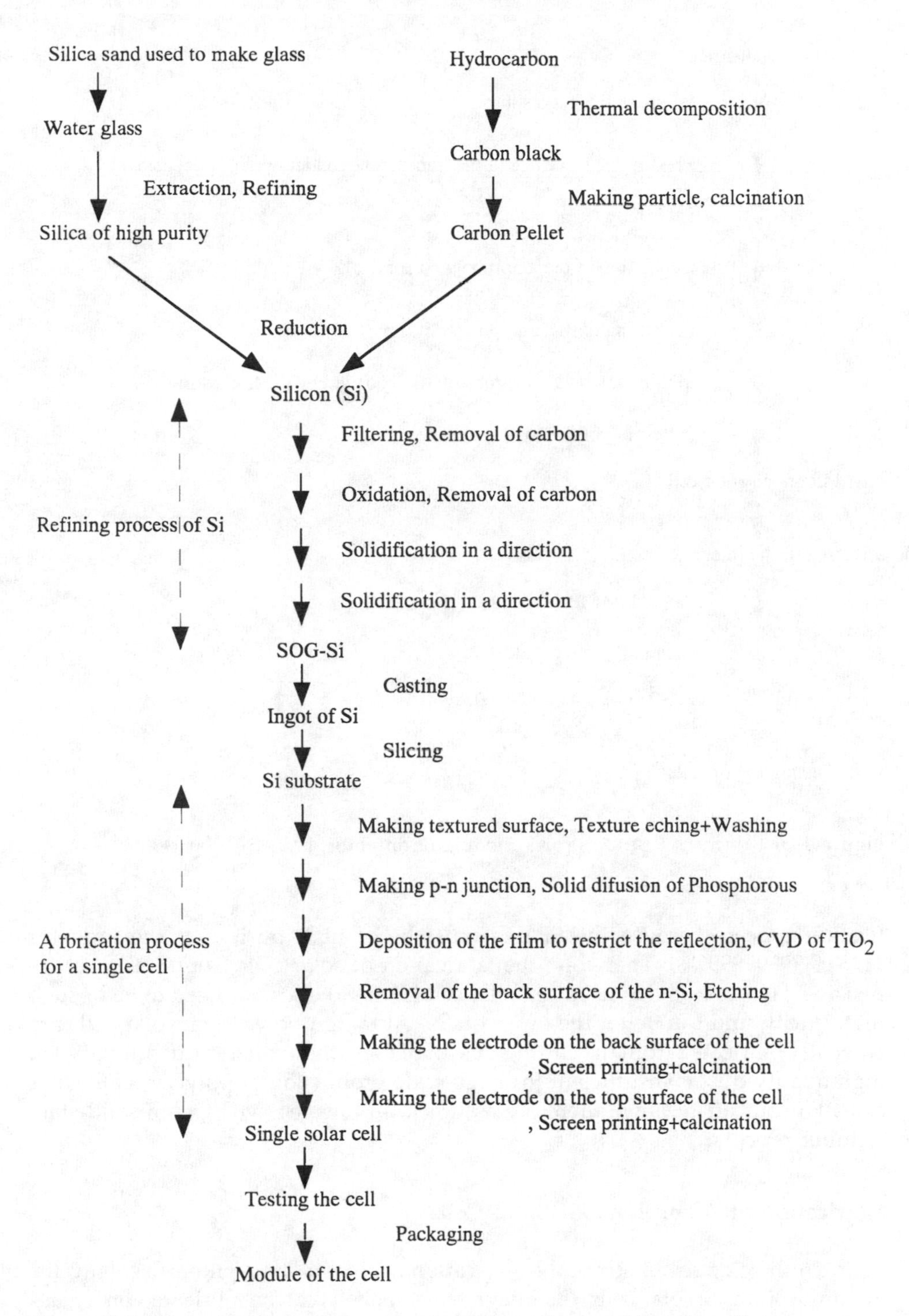

Figure 19.5 Fabrication process for a module of a polycrystal Si solar cell.

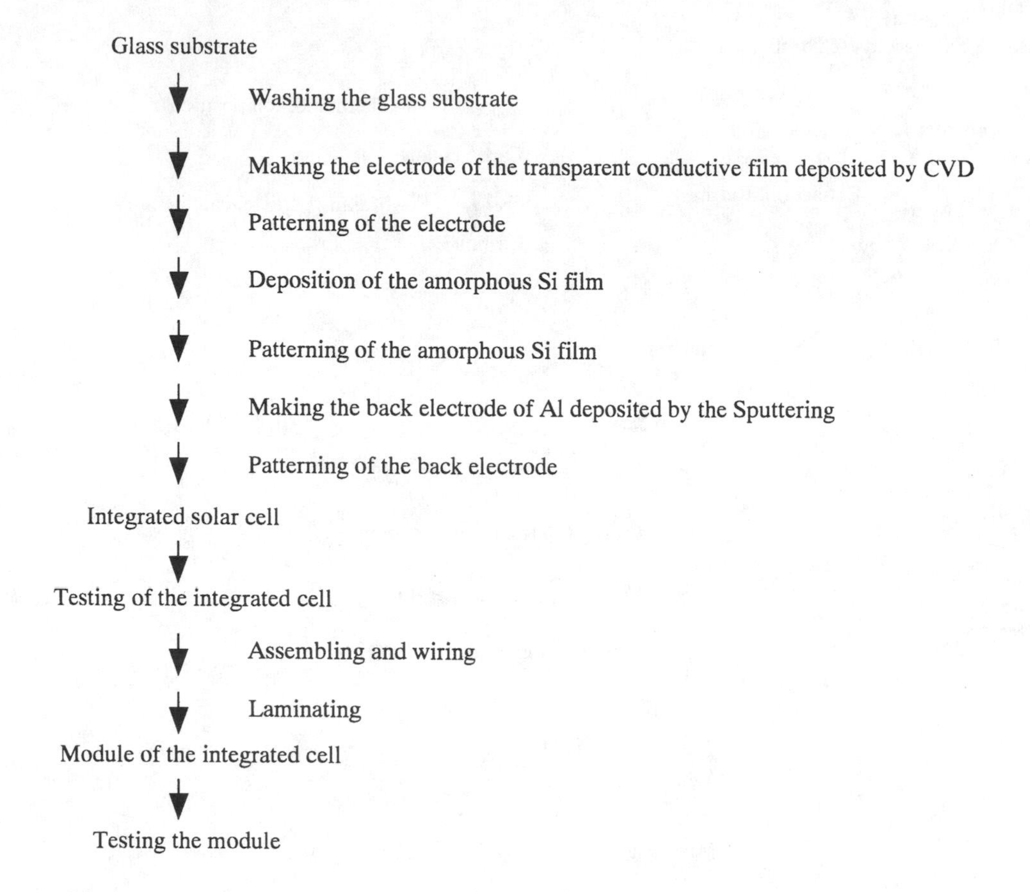

Figure 19.6 Fabrication process for a module of an amorphous Si solar cell.

fication process because the SEG-Si must be of high purity, on the order of 11N(99.999999999%). Therefore, the fabrication of solar cells from SEG-Si is expensive. However, the Si used for solar cell (SOG-Si) does not need to be of such high purity, and can be on the order of 7N. Although polycrystal solar cells are currently fabricated from the cutting dust of SEG-Si, the available quantity of such high-quality dust is insufficient for large-scale production of polycrystal Si solar cells. For our analysis we therefore assumed an inexpensive SOG-Si solid-state refining process.

Fabrication of Amorphous Si Solar Cells

Figure 6 shows a schematic of the fabrication process of amorphous Si solar cells. For amorphous Si solar cells, the integrated solar cells can be fabricated on a glass

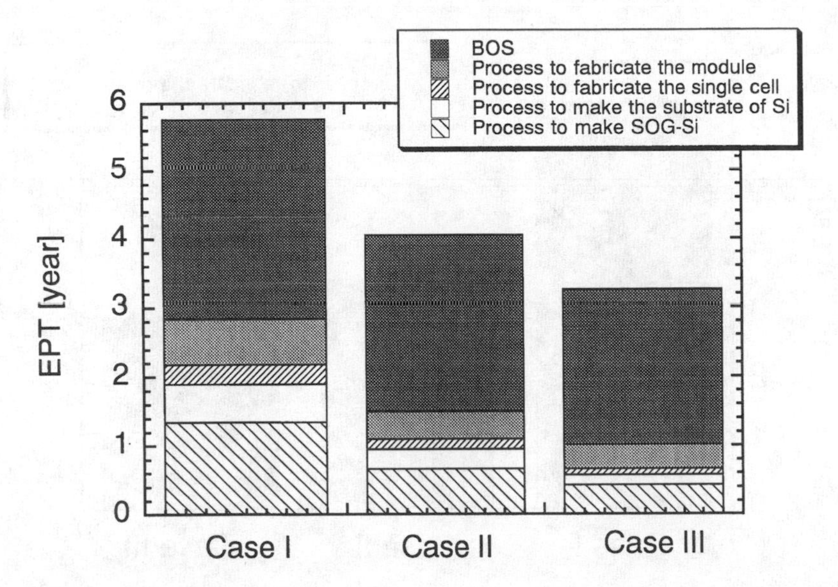

Figure 19.7 Energy Payback Time (EPT) of polycrystal Si solar cell based power plant in three total production scales of plants.
Case I: Power plants (whose total power generation is 10MW at peak power generation)/year.
Case II: Power plants (whose total power generation is 1GW at peak power generation)/year.
Case III: Power plants (whose total power generation is 100GW at peak power generation)/year.

substrate by using laser patterning without the assembling process because the Si was deposited as the film by the CVD (Chemical vapor deposition). The integrated cell has the structure that the single cells are connected in series on a glass substrate. Therefore, the integrated cell can generate the voltage higher than the single cell voltage. The characteristics of the fabrication process of amorphous Si solar cells are (a) small number of the fabrication processes, (b) low-energy consumption because of low-temperature processing, (c) small amount of raw materials required because of thin-film solar cells and (d) the process is suitable process for mass production because the larger area of a integrated cell can be fabricated by CVD and sputtering, and the processes easily can be made automatic.

Energy Payback Time (EPT)

The EPT of polycrystal Si solar-cell based power plants is shown in Figure 7. The BOS (balance of system) are all facilities to control the power generation such as

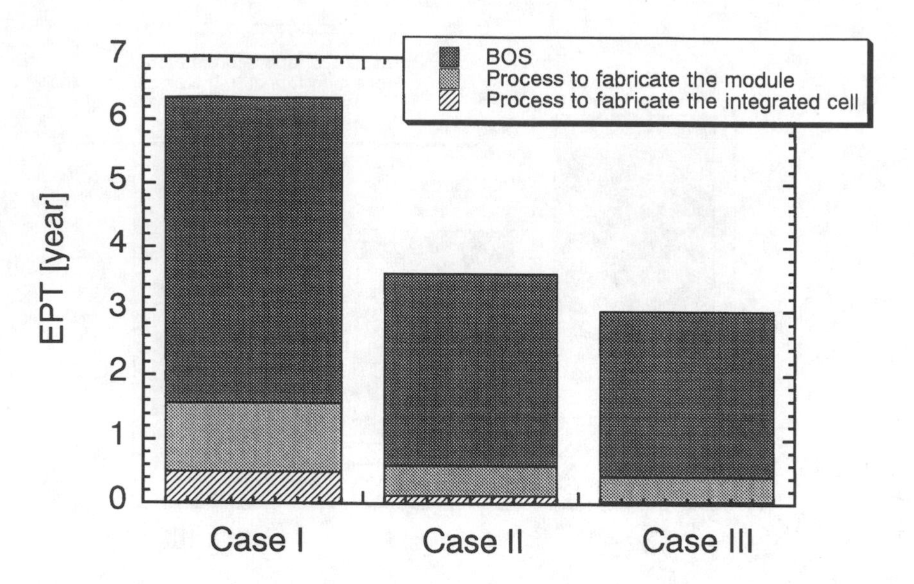

Figure 19.8 Energy Payback Time (EPT) of amorphous Si solar cell based power plant in three total production scales of plants.
Case I: Power plants (whose total power generation is 10MW at peak power generation)/ year.
Case II: Power plants (whose total power generation is 1GW at peak power generation)/ year.
Case III: Power plants (whose total power generation is 100GW at peak power generation)/ year.

the inverter and the batteries. The EPTs for Cases 1, 2, and 3 were calculated to be 5.8, 4.1 and 3.3 years, respectively. The EPT decreased with increasing production size, thus reflecting the economy of scale possible with large power plants. This indicates that solar-power based power plants are effective for reducing global warming because the expected EPT is much shorter than the expected 15~20-yr life of the power plant. For all cases, the ratio of the EPT to the BOS to the total EPT is large, and the ratio increases with increasing scale of the power plant.

The EPT of amorphous Si solar-cell based power plants are shown in Figure 8. The EPTs for Cases 1, 2, and 3 were calculated to be 6.4, 3.6 and 3.0 years respectively. The fabrication energy cost per unit area of amorphous Si solar cells was smaller than for polycrystal cells. However, the EPT of the amorphous Si solar cell was not significantly different than for the polycrystal cells because of the lower conversion efficiency of the amorphous Si solar cells. Similar to the polycrystal cells, the EPT for amorphous Si solar cells was much shorter than the expected life of the solar power plants. For amorphous solar cells, the calculated ratio of the EPT for the BOS to the total EPT is larger than that for the polycrystal Si solar cells.

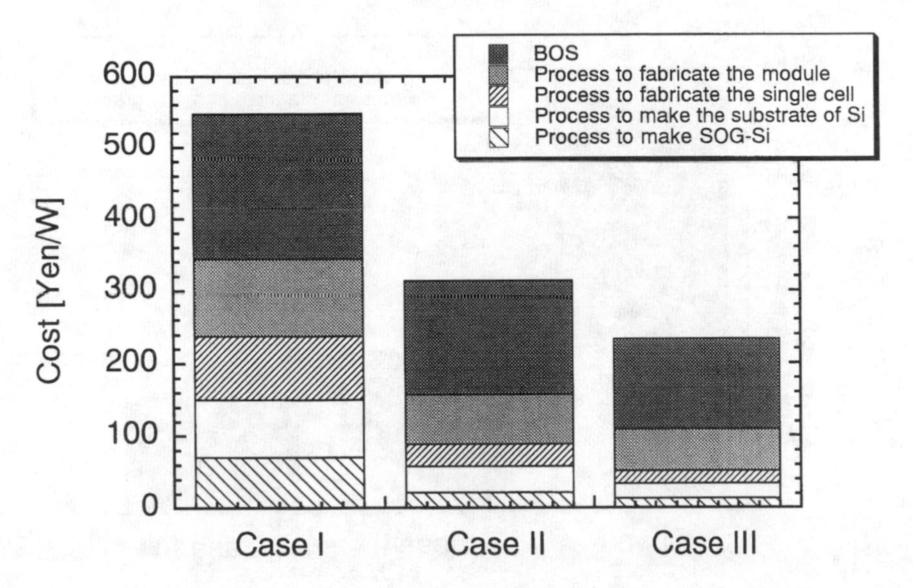

Figure 19.9 Fabrication cost of polycrystal Si solar cell based power plant in three total production scales of plants.
Case I: Power plants (whose total power generation is 10MW at peak power generation)/year.
Case II: Power plants (whose total power generation is 1GW at peak power generation)/year.
Case III: Power plants (whose total power generation is 100GW at peak power generation)/year.

Cost to Construct a Solar Power Plant

The cost to construct either a polycrystal or an amorphous Si solar-cell based power plant is shown in Figure 9 and Figure 10, respectively. The estimated costs of the polycrystal Si solar cell for Cases 1, 2, and 3 are 547, 314, and 233 Yen/W, respectively, and for amorphous cells are 790, 288, and 210 Yen/W, respectively. The larger decrease in the cost for an amorphous Si solar-cell based power plant with increasing scale of production is due to the assumption that the conversion efficiency improves with the power plant scale. Because of their lower conversion efficiency, amorphous Si solar-cell based plants require more surface area than do polycrystal-based Si power plants. However, in our estimation, the cost of the land was not considered. The ratio of the BOS cost to the total cost was high for both types of solar cells and for all cases considered.

The power generation costs were estimated by using the assumptions ; (a) life of the solar cell is 20 yrs, (b) interest on money is 6.0%/yr, (c) the other economical assumptions. The power generation capacity of a plant was needed to calculate the power generation cost. The capacity of a plant supposed to 1MW, 10MW and 100MW. Table 2 show the estimated power generation costs. The lowest costs for

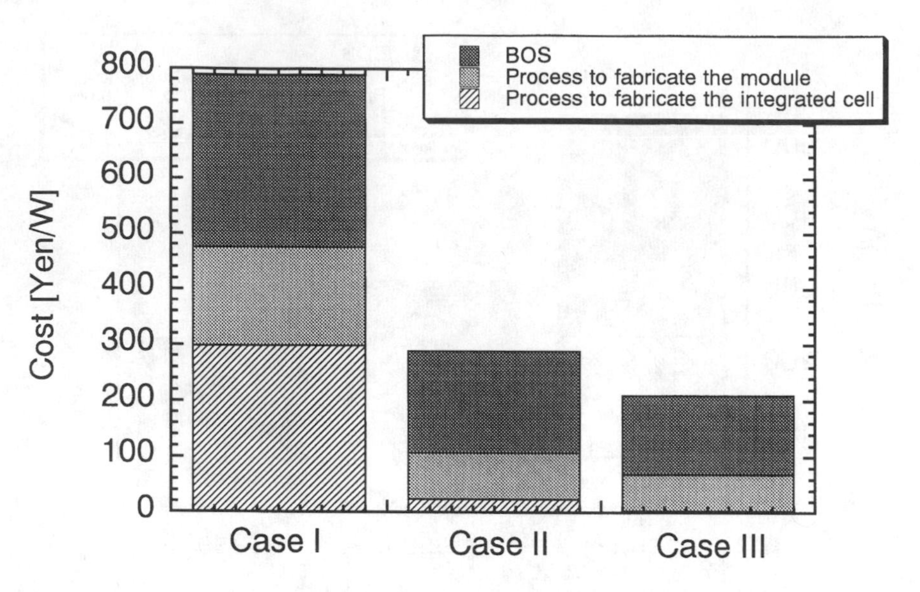

Figure 19.10 Fabrication cost of amorphous Si solar cell based power plant in three total production scales of plants.
Case I: Power plants (whose total power generation is 10MW at peak power generation)/year.
Case II: Power plants (whose total power generation is 1GW at peak power generation)/year.
Case III: Power plants (whose total power generation is 100GW at peak power generation)/year.

Table 19.2 Power generation cost.

Power capacity of a plant [MW]		*1*			*10*			*100*	
Case	*I*	*II*	*III*	*I*	*II*	*III*	*I*	*II*	*III*
Total power capacity of solar based power plants produced per year = Wc [MW]	1.0E+01	1.0E+03	1.0E+05	1.0E+01	1.0E+03	1.0E+05	1.0E+01	1.0E+03	1.0E+05
Polycrystal Si solar cell Power generation cost [Yen/kWh]	61.75	38.41	30.53	55.42	32.52	24.79	54.91	32	24.27
Amorphous Si solar cell Power generation cost [Yen/kWh]	55.33	32.9	27.64	49.12	27.12	21.96	48.6	26.6	21.44

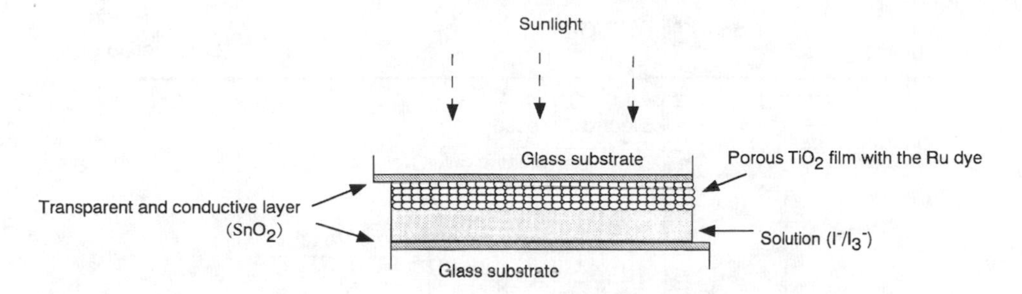

Figure 19.11 The structure of the dye sensitized solar cell.

the polycrystal-based 100 MW power plants for Cases 1, 2, and 3 were 55, 32 and 24 Yen/KWh, respectively. This cost is higher than that for thermal power generation, which is about 8~10 Yen/KWh.

Summary of the Evaluation

The EPTs for both polycrystal and amorphous Si solar cells were much shorter than the expected life of the solar-cell based power plant. This means that solar power generation is useful for reducing global warming. Although the costs of solar power plants is high, it is expected to decrease with increasing scale. To continue to decrease power-generation costs, the solar-to-electrical energy conversion efficiency must be increased, solar-cell fabrication costs must be decreased, and the cost of inverters and control equipments must also be decreased.

The development of new fabrication process is desired.

DYE-SENSITIZED SOLAR CELL: NEW TYPE OF SOLAR CELL

At the current technological level, the Si solar cells appear to be the most promising technology for developing solar power plants. In this section a new type of solar cell will be introduced. It is a developing solar cell and is not commercialized yet. For a nanocrystalline dye-sensitized solar cell using a cis-$(NCS)_2$ bis(2,2′-bipyridyl-4,4′-dicarboxylate) ruthenium (II) dye, the solar-to-electric energy conversion efficiency was reported about 10% (O'Regan *et al.*, 1991), which is quite high for solar cells fabricated using the inexpensive wet method. The cost-performance of efficiency is expected to be high. Figure 11 shows the structure of such cells. The principle of generating electricity with dye-sensitized solar cells is shown in Figure 12. Sunlight is absorbed by the dye and generates electrons and holes, and the separated electrons are then injected into a TiO_2 nanocrystalline film while the holes are transported to the I^-/I_3^- solution. In this cell, the dye absorbed the sunlight instead of the TiO_2 film because the band gap of TiO_2 of

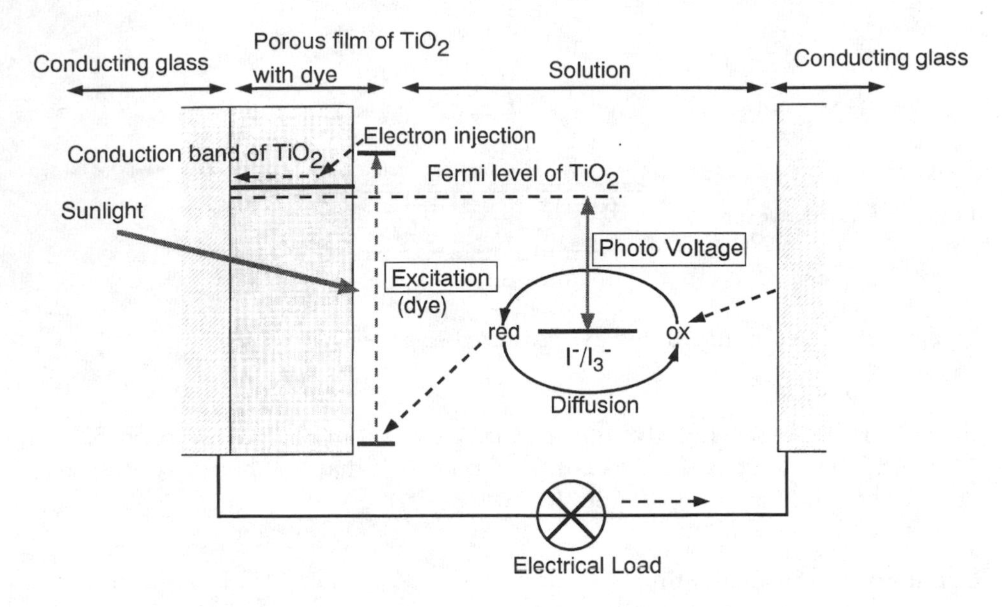

Figure 19.12 Principle of the power generation by the dye sensitized solar cell.

3.0 eV is much higher than the appropriate band gap as the solar cell (shown in Fig. 4). The requirements of the dye sensitized solar cell are improvement in the solar-to-electrical conversion efficiency, minimal leakage of the I^-/I_3^- solution, and minimal degradation of the solar cell.

REFERENCES

M. A. Green (1998), the internet homepage of the Centre for Photovoltaic Engineering UNSW, http://www.pv.unsw.edu.au

B. O'Regan and M. Grätzel (1991) a low cost, high-efficiency solar cell based on dye-sensitized colloidal TiO_2 films. *Nature*, **353**, 737–740

20. RESOURCES AND RECYCLING OF MATERIALS

TOSHINORI KOJIMA* and KYOICHI SAITO**

**Department of Industrial Chemistry, Seikei University,*
3-1 Kichijoji-kitamachi 3-chome, Musashino-shi, Tokyo 180-8633 Japan
***Department of Materials Technology, Chiba University, 33 Yayoicho 1-chome,*
Inage-ku, Chiba-shi, Chiba 263-8522 Japan

RESOURCES

The word "resource" is used in many ways: metal resources, wood resources, water resources, inorganic resources, energy resources, good resources, human resources and so on. In the present chapter, we would like to discuss the issue of resources from the viewpoint of chemical engineers.

Various Resources and Their Reserves

Among the various resources, the amounts of uranium, coal and petroleum are surely finite. After using them, we cannot recover and re-use them because they are changed into other materials. They are classified as "non-renewable" resources.

On the other hand, metals are manufactured from metal ores using various metallurgical processes by which the metals are concentrated. The metallic products are used in various manufactured articles, sold as products and are dispersed to various places. The metals, however, do not change into other materials and they are thus classified as "diffusive " resources. They are not going to disappear, the resources will not be exhausted, but their grade or quality will gradually decrease. Wastes that include these materials can also be considered as resources since they can be "recycled".

Resources such as metal ores, fossil fuels and uranium are mostly buried in the ground. When we want to use them, we usually have to mine them and transport them to consumers using energy, manpower, production facilities and so on. The mined materials are sometimes pre-treated *in-situ* to enhance their value. When the price at which the materials are sold to consumers is higher than the cost of production, the resources are deemed to be usable. Although "diffusive" resources are defined as those that are not exhaustible, the amount of resources that can be economically obtainable is limited.

Improvement in Recovery Processes

Chemical engineers can contribute solutions to the resource problem. When new applications such as electronic and superconducting materials are found, and when new and low cost processes of mining, separation and purification are developed, some resources that were previously uneconomic to mine become economic.

The recovery of resources from various wastes undergoes the same economic change. Thus, as chemical engineers, we can help convert a pile of rubbish into a mountain of treasure. The application of chemical engineering to these processes will be illustrated in the following sections by describing some typical processes.

LOW GRADE ORES

Metal extraction has traditionally employed pyrometallurgical techniques in which the ores are usually smelted and purified. In such a process all materials in the ore, including impurities, are heated to a high temperature and so have to be supplied with large quantities of heat. The cost of smelting low grade ores by pyrometallurgy is therefore very high. In order to recover metals from low-grade ores, completely different processes from pyrometallurgy have been developed and applied.

Hydrometallurgy

Hydrometallurgical processes can substitute for pyrometallurgical processes: indeed some hydrometallurgical processes have already been commercialized. At present, their application is limited to metals such as uranium, rare metals, nickel, copper and cobalt but it is possible that most pyrometallurgical processes will be replaced by hydrometallurgical techniques in the 21st Century.

The overall hydrometallurgical process includes a number of operations — leaching of the target valuable metals in the ore into acidic, alkaline or other media as ionic or complex species, mutual separations of metal ions, their concentration and reduction into metals from their ions. Hydrometallurgical processes save energy because they are operated at or near ambient temperatures, whereas pyrometallurgical processes are energy intensive since they operate at high temperature. In particular, hydrometallurgical processes have advantages when the resources originally include large amounts of water, the grade of ore is extremely low, or valuable materials need to be recovered from waste water. Various types of extractants and ion exchanger resins having a high selectivity have been developed. The role of the chemical engineer is to design the optimum process using these reagents.

Application of Microorganisms in Hydrometallurgy

The large-scale operation of the hydrometallurgical process in the copper industry had started in Spain even before the eighteenth century. The leached copper ion was usually replaced by metallic iron and metallic copper recovered as precipitated copper. A few decades ago, new commercial chelating extractants were developed and the solvent extraction process is now the most common of the hydrometallurgical routes. The concentrated copper ion is reduced electrolytically to metallic copper.

Sometime after 1950, a microorganism called *Thiobacillus ferrooxidans* that consumes ferrous ion and sulfur was identified in mines and was found to be playing an important role in the process of leaching. Two mechanisms have been suggested to elucidate the observation that the leaching rate of copper is accelerated by the microorganism. The first is an indirect mechanism by which ferric ion, which accelerates the leaching rate of copper from the ore, is produced by the microorganism. Experimentally it has been found that the existence of the microorganism increases the production rate of ferric ion by around a million fold. The second mechanism is a direct one where the microorganism directly oxidizes the copper ore into cupric ion using oxygen from the air. Both mechanisms may work simultaneously although their relative contributions have not been determined.

Identification of the optimum conditions to activate the microorganism is one task of the chemical engineer; the most advantageous values of pH and temperature must be determined and devices with high contact efficiency between ore and leach liquor must be developed.

The most attractive feature in employing the microbiological mining process is that metals can be recovered from mines that were once thought exhausted, from extremely low grade ores, and from waste ores. In the dump leaching process, soils or low grade ores with a low content of valuable metals are accumulated in a huge dump, sulfuric aced is sprayed onto the top and the liquor containing metal ions is collected from the bottom of the dump. What is more, the amount of waste is reduced by applying these techniques.

Liquid Membranes

Use of liquid membranes is typical of new techniques developed by chemical engineers. The liquid membrane process combines simultaneous extraction on one side of the membrane and stripping on the other side. Extractants in the membrane act as mobile carriers for the valuable ions (see Figure 20.1). Dramatic reductions in the amount of extractant are expected, but the mechanism of permeation is complicated when compared with conventional extraction processes. To develop new processes using liquid membranes, chemical engineers should elucidate the mechanism, predict the performance in terms of the permeation rate and selectivity and present the best process system for the target.

RESOURCES IN SOIL

It is usually thought that Japan is lacking in natural resources, but this is not totally true although most fossil fuels and metal ores are imported. Soft water flows in the Japanese rivers, the sea surrounds Japan and huge amounts of

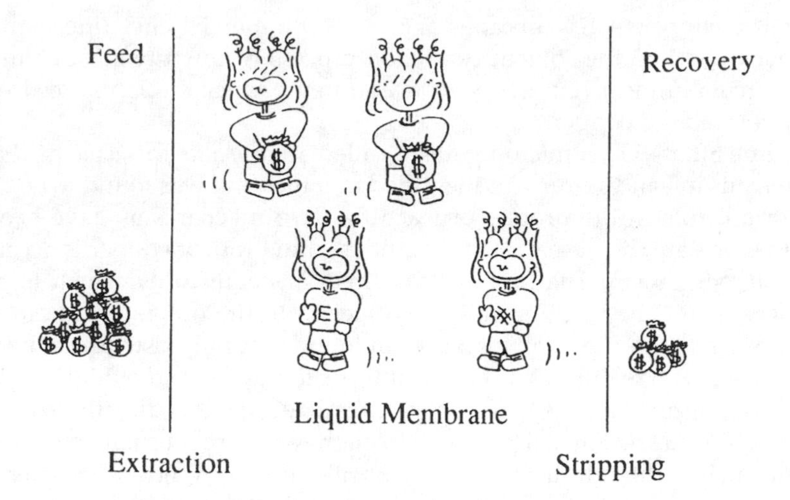

Figure 20.1 A liquid membrane in action.

limestone deposits exist in the mountains. And soil is possibly a bonanza of inorganic resources.

Acid Clay

The Japanese cement industry has grown very large because of these rich limestone resources. But there is another resource abundant in Japan, "acid clay", and a unique chemical company just two hours away from Tokyo by the Shinkansen bullet train.

The company produces various inorganic chemicals from acid clay. From the viewpoint of chemical composition, acid clay is a kind of soil but it consists of very fine particles. By treating with sulfuric acid, part of the aluminium and other elements, except for silica, are leached out as mother liquor. The remaining solid, called "activated bleaching earth", can be used as a refining agent for fats and oils, while activated silica is obtained by further acid treatment of the activated bleaching earth. The remaining component of the mother liquor, aluminium sulfate, is recovered and used directly as a flocculant for wastewater treatment. Activated alumina is also produced from the mother liquor. Finally a synthetic zeolite that is used as a detergent builder and is an alternative to sodium phosphate, whose use is restricted from the environmental viewpoint, is made from silica and alumina. Figure 20.2 shows the whole flowsheet by which these various inorganic materials are produced from acid clay. Most of the processes have been developed in Japan to make efficient use of these abundant resources. This is an example of the development of new processes that give additional value to existing resources.

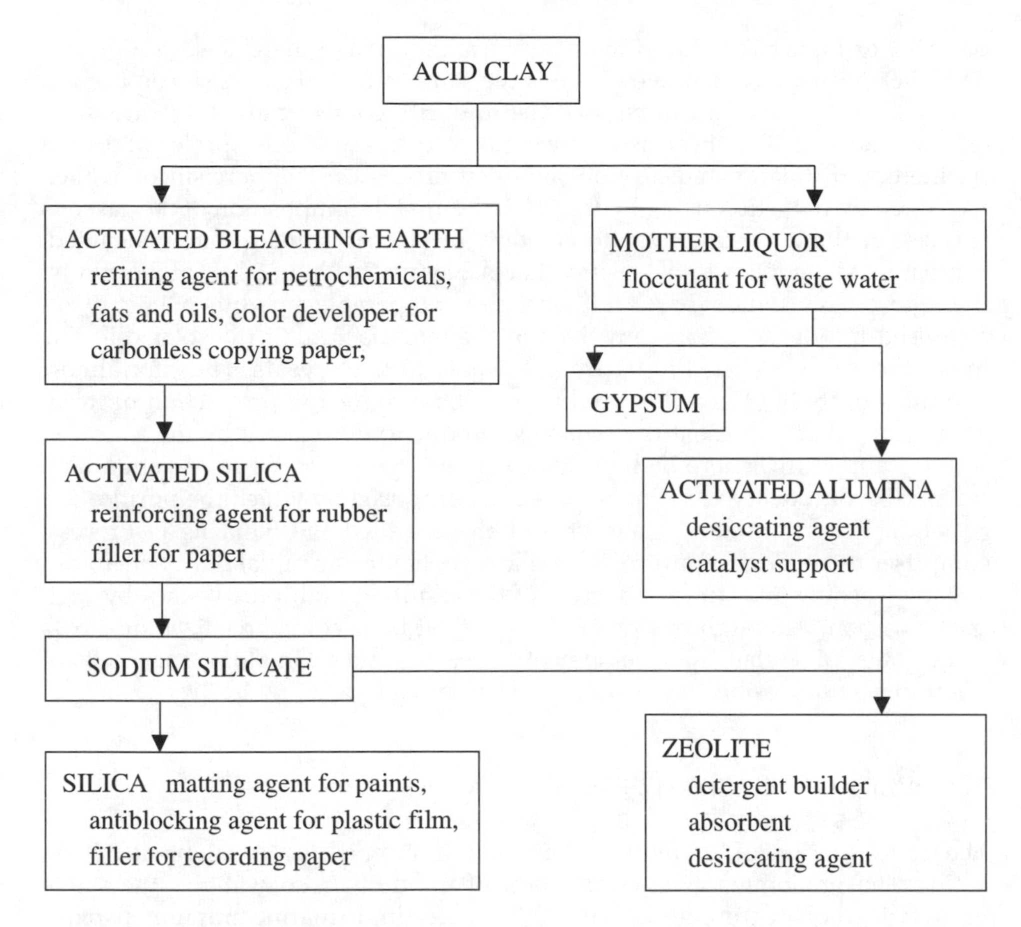

Figure 20.2 Product routes from acid clay.

Solar Cells from the Soil

We are now living with too large an energy consumption and we mostly depend on fossil fuels for energy resources. Fossil fuel consumption inevitably causes an increase in the atmospheric concentration of carbon dioxide, the main greenhouse-effect gas (see next chapter). Engineers, including chemical engineers, are now making every effort to develop new soft energies.

It has been said that our reserves of natural gas will be consumed in thirty years, oil in forty years and coal in two hundred years. On the other hand, solar energy will last almost forever. The best way to utilize solar energy as electrical power is by employing a solar cell. The efficiency of such cells is now more than 10% and will reach 20% in the near future. Silicon is now normally the principal material of the cell. To produce high purity silicon, silica sand having a purity of 99.5% or higher is reduced in an arc furnace to make metallic silicon of 98% or higher purity. This metallic silicon is the same chemical as the silicon used in solar cells, but before it can be used the impurity content must be reduced. To do this, the metallic silicon is chlorinated to form a gaseous species and then purified by distillation, adsorption and other processes. The pure silicon is then recovered by reduction using hydrogen or thermal decomposition of the gaseous species. In these processes, large amounts of electrical energy are consumed, comparable in amount to the energy that is eventually obtained from the sun by the solar cell for some years. Additional and comparable amounts of energy are also used in other processes involved in the manufacture of the solar cell. The life of the cell is expected to be around twenty to thirty years. Thus, for almost one third of its life the cell is producing electricity for the production of itself. This means that at present the economic production of energy by the solar cell is uncertain. Furthermore high purity silica sand resources are not so abundant.

Some revolutionary techniques have been proposed by which the purification is conducted in the solid state. One of these, called the Siemens D Process, comprises removal of impurities from silica ores having an initial purity of around 95% by leaching and direct reduction of the resulting high purity silica by high purity carbon. Although this process has not yet been commercialized, it shows great potential so that by development in science and technology, solar cells of the future will possibly be manufactured from soil.

RECOVERY OF URANIUM FROM SEAWATER

The concentration of uranium in seawater is remarkably constant at 3.3 mg U/m^3. The predominant dissolved form of uranium in seawater is the stable uranyl tricarbonate complex $UO_2(CO_3)_3^{4-}$. Interestingly marine uranium displays no detectable deviation from the normal terrestrial U-235/U-238 isotope ratio. The total uranium content of 4.5×10^9 tons, dissolved in the world's oceans, is almost 1000-fold larger than terrestrial resources of reasonable concentrations. Thus, the ocean is a virtually limitless reservoir of dissolved uranium in a well-defined chemical environment. Atomic power plants continuously require uranium resources; therefore, the 4.5 billion tons of uranium in seawater will be essential for atomic power utilization.

Trial to Recover Uranium

A recovery program was begun in England in the early 1960's. Extensive research on the recovery of uranium in seawater has been conducted to replace uranium locally deposited as terrestrial ore with uranium uniformly dissolved in seawater. Many methods of recovery have been suggested: coprecipitation, adsorption, ion floatation, and solvent extraction. Adsorption using solid adsorbents is promising with regard to economic and environmental impacts.

The molar concentration of uranium, 1.4×10^{-5} mol/m^3, is about 1 part in 4×10^6 of that of magnesium, which is a representative bivalent cation in seawater. Extensive efforts have been exerted to develop an adsorbent capable of separating uranium from the other elements. Hydrous titanium oxide was identified as the most promising inorganic adsorbent. Organic adsorbents were classified into three types: a chelate-forming resin, a macrocyclic compound bound to a resin, and a cellulose resin immobilized by polyphenol compounds. At present, a resin containing an amidoxime group ($-C(=NOH)NH_2$), which can be prepared by reaction of cyano groups ($-CN$) with hydroxylamine (NH_2OH), is promising in view of adsorption rate, capacity, durability, and production cost.

Recovery Process

The recovery process of uranium from seawater consists of three stages: (1) adsorption from seawater using an amidoxime resin, (2) purification of the eluate with another chelating resin, and (3) further concentration of uranium using an anion-exchange resin. Since an adsorbent is required to contact a tremendous volume of seawater in the first step, various effective contacting systems have been suggested and evaluated.

The adsorption system in the recovery process of uranium from seawater can be classified with respect to three factors: (1) the shape of the adsorbent, i.e., spherical or fibrous; (2) the mode of the adsorption bed, i.e., fixed or fluidized; and (3) the method of moving the seawater, i.e., by pumping or ocean current. The system consists of a combination of the three factors.

Preparation of Amidoxime Fibers

A recovery system for uranium in seawater using amidoxime (AO) fibers is promising in that the ocean current forces seawater to easily move through an adsorption bed charged with AO fibers (Figure 20.3). A research group at the Shikoku National Industrial Research Institute has developed bundled AO fibers based on commercially available polyacrylonitrile fibers; the fibers exhibited a uranium adsorption rate of 2 mg U per g of fiber for 60 days of operation in a flow-through mode. A research group at the Japan Atomic Energy Research Institute has proposed a method of preparing AO fibers based on polyethylene and polypropylene fibers by radiation-induced graft polymerization of acrylonitrile ($CH_2=CHCN$, AN) onto the fiber, followed by amidoximation.

Hydrophilization of AO adsorbents is effective in improving the uranium adsorption rate; the diffusion of uranyl tricarbonate ion, which is the predominant species of dissolved uranium in seawater, governs the overall adsorption rate of uranium. Cografting of methacrylic acid ($CH_2=CCH_3COOH$, MAA) with AN onto polyethylene (PE) fibers was suggested as a method of enhancing the ad-

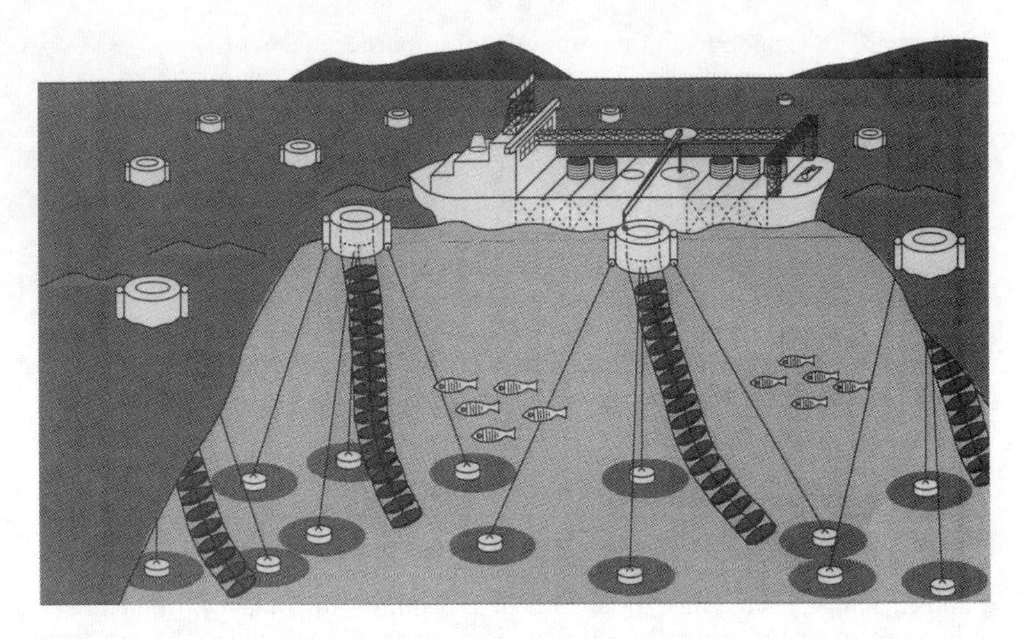

Figure 20.3 Schematic illustration of a recovery system of uranium from seawater.

sorption of uranium on AO adsorbents. The preparation of hydrophilic AO fibers based on PE fibers by radiation-induced cograft polymerization and subsequent chemical modifications is illustrated in Figure 20.4. PE fibers of about 30 μm diameter were used as the trunk polymer for grafting. First, a combination of MAA with AN was cografted onto the PE fiber. Second, the AN/MAA-cografted fibers were immersed in a 3 (w/v)% solution of hydroxylamine hydrochloride to convert the produced cyano groups to AO groups. Finally, the AO/MAA fiber was immersed in KOH solution.

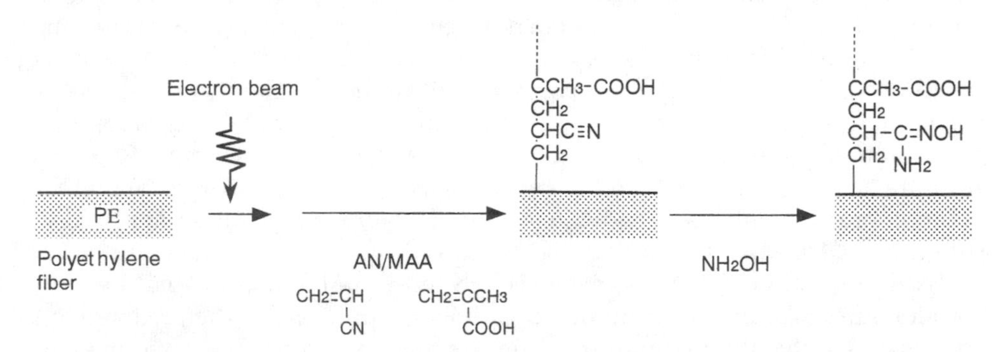

Figure 20.4 Preparation of hydrophilic amidoxime adsorbents by cografting acrylonitrile and methacrylic acid and subsequent amidoximation.

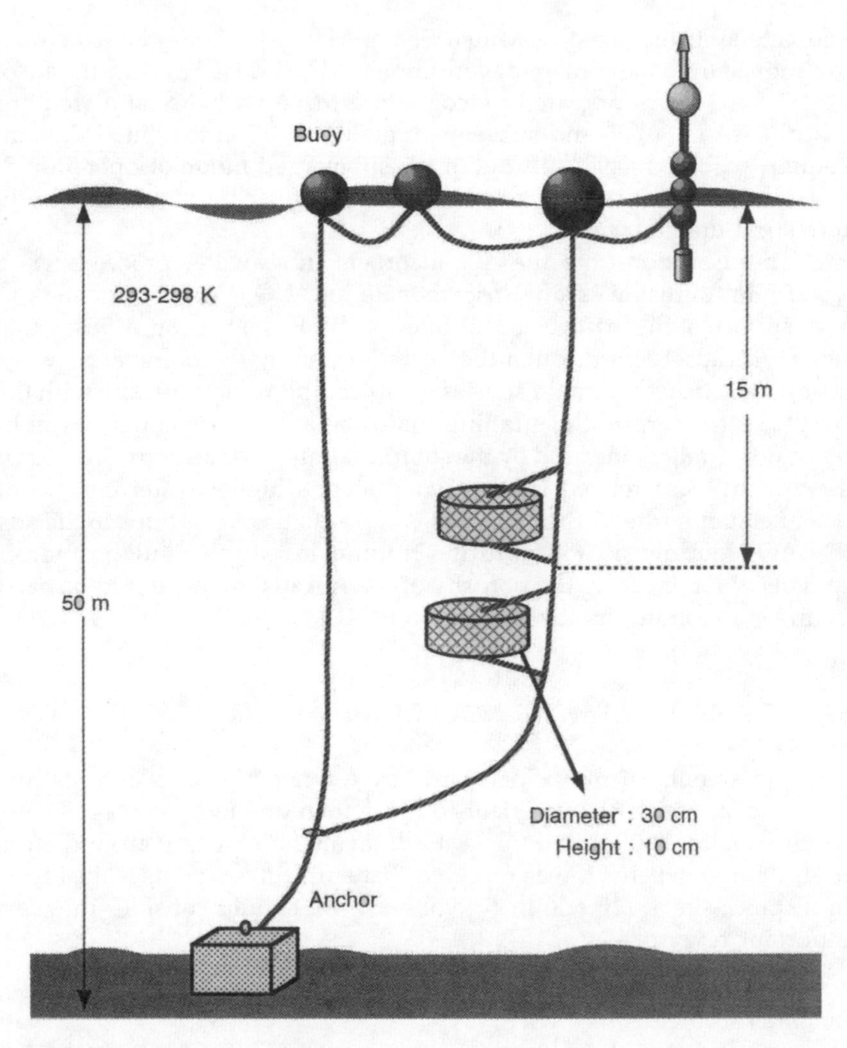

Figure 20.5 Experimental apparatus for uranium recovery from seawater in the submerged mode of operation at the ocean site.

Evaluation and Demonstration

The uranium adsorption rate of the hydrophilic AO fibers was evaluated in a submerged mode of operation in the ocean far from the coast of Japan; about 0.5 g of the AO/MAA fiber was charged in a container made of plastic mesh, which was attached to the outside of a frame (adsorption unit) made of stainless steel 30 cm in diameter and 10 cm in height, as illustrated in Figure 20.5. The adsorption units were submerged for 20 days 15 m below the surface of the sea located about six kilometers off the coast of Mutsu Sekine-Hama in Aomori Prefecture. The fibers removed from the container were immersed in 1M HCl to elute the adsorbed metals. The AO fibers prepared by cografting MAA with AN at a weight ratio of AN to MAA of 60/40 and subsequent amidoximation exhibited the amount of uranium adsorbed, 0.90 g U/kg, in the submerged mode of operation for 20 days of contact. This uranium content of the AO fiber is comparable to that of the terrestrial uranium ore.

Uranium adsorption onto the AO adsorbent in seawater proceeds via three steps. (1) Film diffusion: uranyl tricarbonate ion, $UO_2(CO_3)_3^{4-}$, transfers to the external surface of the adsorbent. (2) Interior diffusion: the ion diffuses into the interior of the adsorbent through the pores formed by the polymer network. (3) Intrinsic adsorption: the uranyl species forms complexes specifically with the AO group. Overall adsorption of uranium onto the adsorbents continues until the concentration gradient induced by the complexation becomes zero. The cografting of MAA with AN is related to the second step; a higher water content of the adsorbent enhances the diffusion of $UO_2(CO_3)_3^{4-}$ to the AO groups in the adsorbent. The AO group density for capturing uranium in seawater and the hydrophilic group density for ensuring the porosity of the AO adsorbent should be balanced to maximize the uranium adsorption rate.

WASTE AS AN IMPORTANT RESOURCE

In 1996, the amount of municipal waste from our daily life was to 50 million ton/y in Japan, which is equivalent to more than one kg/day/cap. The waste from industry was 400 million ton/y. Each Japanese is casting around ten kg of wastes off. But to where? It was reported that harmful materials leaked from the waste disposal site at Hinode in Tokyo. Now, we should consider the waste as an important resource.

Clean Japan Center

The Clean Japan Center is an organization which promotes recycling of waste materials for further use. It has, for example, proposed a plant to produce 48 tons of solid fuel from every 80 tones of waste. After crushing the waste, combustible materials are selectively collected and valuable fuels produced by drying and granulation. These processes such as drying and crushing are almost the same unit operations as are employed in many parts of the chemical industry.

Recycling of Mercury from Wastes

In 1984 production of dry batteries in Japan was approximately three billion pieces and around 60,000 tons per year of potentially valuable materials were used in their manufacture. The Clean Japan Center constructed a demonstration plant for recycling mercury-containing wastes in the northern island of Hokkaido in July 1985 and experimental operation continued for two years until September 1987. The demand for mercury is around 300 ton/year and most of this has to be imported to Japan. The plant capacity is around 6,000 ton of batteries per year, only one tenth of the Japanese consumption. Nevertheless this plant is unique and all the waste batteries collected by municipalities, electrical shops and so on are collected there. The plant consists of three major processes — pretreatment, roasting, separation. Waste batteries are classified into five shapes according to the Japanese Industrial Standard. An electromagnetic separator together with vibrational and load-cell separators are first employed to identify and separate batteries according to their size and type. The iron jackets of the batteries are detached for use as iron scrap and then mechanically crushed. The pre-treated batteries are fed to a furnace where they are roasted at between 600–800°C. The mercury vapor in the flue gas is introduced to a condenser after electrostatic precipitation which first separates any dust. In the condenser, droplets of liquid metallic mercury are recovered and refined to purity of 99.99% or above, while the uncondensed gas from the condenser is cleaned by chemical cleansing, wet electrostatic precipitation and gaseous mercury adsorption.

A mercury-stabilizing agent is added to the wastewater from the various process steps before it is filtered and the dried solid residue returned to the roasting furnace. Overall recovery of mercury is 98%. The calcined solids from the furnace are crushed and magnetic and non-magnetic substances separated from each other, the former being used as iron scrap and the latter as zinc residue.

In most rubbish treatment processes, mercury in the batteries is left in the soil or is combusted, vaporized and partly dispersed into the air. The Clean Japan Center process should be installed in many places from the viewpoint not only of resource recovery but also based on environmental considerations. At the present time, however, the income obtained by selling the products of the process and paid by municipalities for waste disposal does not cover the operating cost. Of course we chemical engineers should make great efforts to improve this process, but more cooperation from consumers and back-up from the government are also needed. Such social and political pressures are vital to make a success of recycling processes.

Coal Ash and Boiler Slag

A large amount of coal ash is produced in power plants, most of which is either thrown away or used for land reclamation. Although the concentration of either harmful or valuable elements in the ash is not so high, their total amount is large and increasing. At present the ash is partly utilized as cement, building materials, soil conditioner or fertilizers and so on, but the amount so used is not large. Development of a process to utilize the ash and recover valuable elements in the ash is required. Analysis of the behavior of any harmful elements present in the ash and of finding ways to prevent them reaching the environment are also important tasks for chemical engineers. Fairly large amounts of vanadium are contained in boiler slag. They originate from the fuel oil, having been concentrated in the heavy fraction through the distillation process. After combustion of the heavy oil, they are concentrated in the slag. Vanadium is one of the rare metals and investigations to recover it from boiler slag are underway. Hydrometallurgical processes are better than the conventional pyrometallurgical processes when they are applied to the recovery of elements from boiler slag and coal ash. Precious or harmful elements are leached out and the residues can be used as light building materials. The development and application of new separation technologies, such as liquid membrane techniques and development of new metal absorbing materials is also a promising line of research for concentrating the elements from the leach solution.

21. GLOBAL ENVIRONMENT AND CLIMATE CHANGE

TOSHINORI KOJIMA

Department of Industrial Chemistry, Seikei University,
3-1 Kichijoji-kitamachi 3-chome, Musashino-shi, Tokyo 180-8633 Japan

Environmental problems have so far been topics of great concern throughout the world, including chemical engineers. The present chapter first reviews the history of environmental problems. The global warming/climate change mainly caused by CO_2 emission due to fossil fuel burning is then focussed on. For more details, please refer to my previous publications (Kojima 1998, Kojima 1999).

HISTORY FROM POLLUTION TO GLOBAL ENVIRONMENT

An awareness of pollution problems significantly developed during the period of post-war growth, and nowhere was this more true than in Japan. Now, global climate change is a topic of great concern. Let us first review the history of environmental problems from the age of pollution to recent global environment problems.

The Age of Pollution

The most infamous incidents in Japan were the appearance of Minamata disease (due to mercury poisoning) in 1953, *itai itai* disease (a painful bone disease caused by cadmium poisoning) in 1955, Yokkaichi asthma (in Mie prefecture, due mainly to the emission of sulfur oxides) in 1959 and the second outbreak of Minamata disease that occurred in Niigata prefecture in 1964. The major sources of pollution were traced to chemical plants and petrochemical complexes. In order to combat pollution, various regulations were enacted and a range of procedures were implemented to ensure adherence to the regulations. Naturally, these measures included regulations concerning the recovery and removal of pollutants. In some cases, which occurred in the soda industry, the process itself was changed (from the mercury method to the diaphragm method, and then to the ion exchange method). Yes, in the course of combating pollution, chemical engineers have played a significant role.

Japan After the Age of Pollution

Even though the uproar which accompanied the early major pollution incidents has died away, serious problems still exist with matters such as photochemical smog and red tide. Fresh problems have also arisen in various parts of Japan, such as the contamination of the water table due to the use of agricultural chemicals at golf courses, and the recent problem in which harmful materials leaked from the waste disposal site at Hinode in Tokyo. However, it is clear that many of these cases do not stem from pollution directly originating from the manufacturing processes themselves. Instead, the main source of pollution has either been ordinary people or sites closely associated with everyday life.

One example of pollution that is not restricted to manufacturing plants is the problem concerning nitrogen oxides, which are a source of air pollution; these pollutants are also emitted from sites at which electricity is generated for household consumption etc. (for example, thermal power plants). In Japan, however, there has been a great reduction in the quantity of emitted nitrogen oxides and in this respect, the standards in the country are now at the highest level in the world. At the present time, it is estimated that most of the nitrogen oxides is emitted from transportation that is closely linked with everyday life, such as diesel trucks and buses.

This is not pollution whereby a certain small number of companies cause harm to many unspecified people; human beings destroy the human environment, which then produces a threat to people's daily lives. It is not only the sufferers who remain unspecified, but the same applies to the emitters of pollution. In 1972 the United Nations Conference on the Human Environment established the United Nations Environment Program; with this, attention turned away from pollution itself to the environment in general, thus perhaps marking the start of a new era. However, in essence, the specific program discussed at the conference still addressed regional (as opposed to global) problems.

The Limits of Growth

Another landmark event also took place in 1972: the publication of "The Limits of Growth" by the Club of Rome (Meadows *et al.*, 1972). This book was the one that most accurately portrayed the trend away from a regional perspective to a global viewpoint. The book pointed out that the earth's resources (and energy resources in particular) were not infinite, and painted the scenario of human destruction in convincing fashion. The wasting of resources is a double-edged sword for the human race. Even if the environment can recover, the devouring of resources would still leave human beings on the road to ruin.

The scenario seemed to have become reality when the fourth Middle East war resulted in the oil crisis in 1973, and when that was then followed by the second oil crisis in 1979. However, although countries such as Japan, which were overdependent on oil referred to the situation as an energy crisis, from a global perspective, it was simply an oil crisis. Certainly, the situation was not caused by a shortage of resources.

Resources and Energy

Among the various resources, most of them, such as metals will not disappear but will be simply dispersed (Kojima and Saito, 2000). However, after dispersion, recovery becomes difficult, and this task will necessitate the expenditure of considerable amounts of manpower or energy. Fortunately, this is where technological advances are expected to provide a contribution, although at the present time, the actual amount of energy required for separation is many times the theoretical amount.

The only situation that is different is the one concerning energy resources, such as uranium, coal and petroleum. Strictly speaking, energy itself is being conserved. It is merely being transformed and dispersed; entropy, however, is increasing.

However, given that there is a finite limit, how extensive are the world's ultimate reserves? The world's energy reserves are listed in Table 1. The ultimate reserves include those that have already been used, but it is possible that the estimates may rise substantially since they contain errors, and because new resources may be discovered which are completely different in origin from those already known.

How long will fossil fuel resources last? Also, will the scale of consumption continue to increase? Even though various factors are involved, for simplicity, let us consider the values produced by dividing the estimated ultimate reserves by the amount of production. According to Table 21.1, the total reserves of oil and natural gas should both last for 100 years. Of the fossil fuel consumption, oil accounts for 43% and natural gas for 23%, with the other one-third accounted for by coal. It was thought that if people exhausted the supplies of oil and natural gas, it would be acceptable to replace them by using coal. In the end this would mean that total fossil fuel supplies would last for 700 years; the figure would be over 800 years if oil sand and oil shale were also included. As a result, it was believed that even if energy resources are considered to be finite, the situation would be satisfactory if new forms of renewable energy could be developed within a few hundred years.

Global Environment

There have been many significant global environmental incidents and conferences since the publication of "The Limits of Growth" (Table 21.2). Various effects of the global environmental problem have become apparent in desertification, marine pollution, nuclear waste contamination, acid rain, destruction of the ozone layer, and the transfer of harmful substances over international borders. These effects have stemmed from the activities of human beings; development and over-consumption mainly by advanced nations with population problems in developing

Table 21.1 Comparison of various fossil fuels; CO_2 emission characteristics and amount of reserves (Kojima, 1998).

Characteristics and reserves	*Carbon*	*Coal*	*Oil*	*Natural gas*	*Hydrogen*	*Total (oil equivalent)*	*(Uranium) (LWR basis)*
* Higher heating value (kcal/kg)	7,800	7,000	10,000	13,000	34,000	–	
H/C ratio (by number of atoms)	–	0.9	1.8	3.9	–	–	
H/C ratio (by weight)	–	0.08	0.15	0.33	–	–	
Amount of CO_2 emitted (gC/kcal)	0.13	0.11	0.078	0.058	0	–	
Estimated ultimate reserves (E) (10^{12} t)	–	9.9	0.27	0.15	–	7.4	(15 Mt)
Proven recoverable reserves (R) (10^{12} t)	–	0.73	0.12	0.08	–	0.74	(2 Mt)
Annual production (P) (Gt)	–	3.5	3.0	1.4	–	T=7.3	(50 kt)
Remaining years of supply (R/P, y)	–	200	40	60	–	100	(40 y)
Remaining years of supply (E/P, y)	–	2,800	90	100	–	1,000	(300 y)
Remaining years of supply (E/T, y)	–	950	37	27	–	–	(24 y)

T = Total annual fossil fuel production, oil equivalent
Revised and enlarged from The Carbon Dioxide Problem, Table 3.3, p. 77, T. Kojima, Copyright © 1998 OPA (Overseas Publishers Association) N.V., with permission from Gordon and Breach Publishers.

countries who have also exhibited the desire to follow the same "road to prosperity" as the advanced nations.

The first category of the global environmental problem is that harmful or waste products are left untreated, dispersed or transported elsewhere. The effects are seen in worldwide contamination, various forms of marine pollution, nuclear waste contamination, and the transport of harmful materials across international borders. The next case is an environmental problem which used to be a localized one but which then developed into a global problem; examples include acid rain, and some forms of marine pollution. Unlike the issues discussed so far, the question of the destruction of the ozone layer is characterized by the fact that scientists can only clarify the causal relationship. Freons are chemical substances which are harmless to human beings and which used to be regarded as extremely useful substances.

Resolution of these problems would be possible if existing technology could be applied (or further developed), or if the production and use of certain materials were ceased and other materials used as a substitute. Chemical engineers are

Table 21.2 Significant events affecting global environmental issues (Kojima, 1998).

March 1972	"The Limits to Growth" published by the Club of Rome
June 1972	United Nations Conference on the Human Environment, Stockholm
December 1972	Establishment of United Nations Environment Programme (UNEP)
August 1977	United Nations Conference on Desertification, Nairobi (Kenya)
May 1984	World Commission on Environment and Development (WCED)
March 1985	Vienna Convention for the Protection of the Ozone Layer (by United Nations Environment Programme, UNEP)
June 1985	Tropical Forestry Action Plan (Food and Agriculture Organization of the United Nations, FAO)
July 1985	Helsinki Protocol: Protocol on the reduction of sulfur emissions or their transboundary fluxes by at least 30 percent (Helsinki, Finland)
April 1986	Nuclear accident at Chernobyl, Ukraine, Soviet Union
February 1987	Final meeting of World Commission on Environment and Development (WCED)
April 1987	Publication of "Our Common Future" (WCED)
September 1987	Montreal Protocol on Substances that Deplete the Ozone Layer ("Montreal Protocol")
June 1988	Toronto Conference on "The Changing Atmosphere" produced a 'call for action' on energy and environmental questions
October 1988	Sofia Protocol: Protocol concerning the control of emissions of nitrogen oxides or their transboundary fluxes (Sofia, Bulgaria)
November 1988	Establishment of the Intergovernmental Panel on Climate Change, IPCC in Geneva, Switzerland (by UNEP and WMO, World Meteorological Organization)
March 1989	Basle Convention on the Control of Transboundary Movements of Hazardous Wastes and Their Disposal, Basle, Switzerland
November 1989	Noordwijk Declaration (Netherlands). Ministerial-level participants agreed a declaration on "atmospheric pollution and climactic change".
Jan.–Feb. 1991	Gulf war (crude oil spills, torching of oil fields)
June 1992	United Nations Conference on Environment and Development, Rio de Janeiro, Brazil. Commonly referred to as the 'Earth Summit'. (Rio Declaration on Environment and Development, Agenda 21; included agreements on forest preservation, biodiversity, climate change and global warming).
March 1994	Framework Convention on Climate Change came into force (following ratification of agreement at UNCED, Rio, 1992)
December 1997	The third session of the Conference of the Parties to the Framework Convention on Climate Change (Kyoto Protocol)

again expected to play a significant role in developing and improving the technology for treating them.

In contrast, the destruction of tropical forests and desertification are in many cases rooted in the overpopulation problem and North-South dialog, which in turn produces the problem of inadequate food production. It is also a problem that people only evaluate forests in terms of their traditional economic value rather than in terms of their true non-economic value (i.e., the value of their environmental role in flood prevention and in support of ecosystems, and their role in the carbon dioxide problem). Even so, the causal relationship is clear, and if human beings seriously appreciate that value, it is unlikely that anyone would oppose countermeasures to combat the threat.

Global warming results from the actions of so-called "greenhouse gases". These gases are typical of a global environmental problem in that their sources are widely dispersed, virtually all of them are harmless to humans, and their causal relationship with global warming remained obscure for a long time. Furthermore, in the case of carbon dioxide, in addition to the large quantities involved and the difficulty in specifying the sources of emission was the contribution of a range of other factors such as the North-South problem, and the problems of overpopulation and food production.

GLOBAL WARMING AND CLIMATE CHANGE

The most essential point is that energy sources are intrinsic to growth and the majority of the energy used today is not sustainable. However, it was believed that even if energy resources are considered to be finite, the situation would be satisfactory if new forms of renewable energy could be developed within a few hundred years. However, the emergence of the global warming problem means that the above conclusion was probably over-optimistic. The times, and indeed the earth's environment, are demanding a halt to fossil fuel use.

Intergovernmental Panel on Climate Change

One notable gathering was the inaugural meeting of the Intergovernmental Panel on Climate Change (IPCC), which took place in 1988. It seems to have been after this meeting that the recognition of the earth's environment being finite and the problem was expected to develop not in a few centuries but rather during our own children's lifetime began to spread.

The readers may think that global warming itself is not such a bad thing. Yes, the higher temperature is not bad, or rather, it may make our life more comfortable. However, the problem is the rate of temperature rise. If the rate is slow, human and other ecosystem can adapt themselves to the new environment.

First, let us consider the question of a rise in sea levels. It has been postulated that the first effect that would be seen if the sea temperature rises would be an increase in volume. However, it was predicted that the rise in sea level during these several decades would be not several meters but only about one-tenth of that. Notwithstanding this, the effect has been clearly visible to the naked eye.

The effects of global warming are also clearly evident in climatic changes, due to their aggravation of various natural phenomena. A rise in atmospheric temperature increases water vaporization, resulting in more frequent and more ferocious typhoons. An increase in precipitation might, at first, appear to be beneficial from the point of view of food production, but if there should be an increase in torrential downpours, this would create other worries such as land being washed away. The reverse scenario would be that a rise in atmospheric temperature would also increase the evaporation, which is a cause of land becoming parched. It is actually unclear as to which areas would experience greater precipitation and which would suffer desertification. However, it would be a tragedy if an area that had just been converted into agricultural land was later turned back into a desert.

Rapid global warming will also damage the ecological system. It may also result in an increase in respiratory diseases and infections. Furthermore, it is possible that the large amount of methane trapped in the ice fields of Siberia would be released, which results in a further increase in temperature.

Mechanism of Global Warming

The balance of the input/output of energy determines the temperature of the earth. Short-wavelength light is radiated from the sun and it can readily pass through the earth's atmosphere. However, after this radiation is absorbed by the earth, it is radiated back into space in the form of long-wavelength (infrared) radiation. The greenhouse gases, such as water vapor, CO_2, CH_4, N_2O, and chlorofluorocarbons (CFCs, floen) in the air absorb the infrared rays and partly re-radiate them back to the earth surface. Thus, the greenhouse gases act as a resistance to heat transfer from the earth's surface into space. This mechanism keeps the earth surface temperature at 15°C while the earth temperature looks like –18°C from space. However, we should remember that the earth's temperature is primarily controlled by the sun activity, and secondarily by the existence of materials which prohibit the introduction of sunlight (e.g., volcanic activity) and lastly by the existence of greenhouse gases.

The calculations would not be very difficult if the actions of only the gases conventionally regarded as greenhouse gases were taken into account. However, a rise in atmospheric temperature would lead to other complicating factors such as an increase in water vapor, which also produces clouds as a barrier to solar radiation. It has been calculated by IPCC that an increase in temperature at the end of 21st century would range from 1.0°C to 3.5°C.

The greenhouse effect per unit concentration of carbon dioxide is small compared with that of other trace gases, which can be deduced from the fact that the present concentration of carbon dioxide in the atmosphere is already more than 300 ppm, and since considerable absorption already occurs at its characteristic wavelength. In addition, the length of time for which each gas remains in the atmosphere varies. Consequently, the degree of the observed effect varies with the period selected for evaluation. The situation is often referred to as the calculation of carbon dioxide concentrations, but even this is not simple to determine.

Kyoto Protocol

In December, 1997, COP3 (the third session of the Conference of the Parties to the Framework Convention on Climate Change) was held in Kyoto. An outline of the agreement is as follows:

a) By the period of 2008 to 2012 compared with 1990 levels (compared with 1995 levels for alternative gases to freon; HFC, PFC, SF6).
b) Reduction in greenhouse gas emissions at least 5% totally for developed countries; e.g., –8% for EU, –7% for USA, –6% for Canada and Japan, +8% for Australia.
c) In terms of CO_2-equivalent emissions; x21 for methane, x310 for N_2O, x1300 for HFC, x6,500 for PFC and x23,900 for SF6 according to Global Warming Potentials by IPCC.
d) Inclusion of the effects of land-use changes; CO_2 absorption by forests established after 1990, though its details will be discussed in the near future.
e) Introduction of trading of emission reductions and joint implementation between advanced countries, and clean development mechanism between advanced and developing countries.

Nevertheless, the realization of these figures is thought to be difficult, if any innovative technologies are not introduced. For example, Japan emitted around ten percent more of these gases in terms of CO_2-equivalent in 1996 compared with the 1990 level. In the next section, various ways to mitigate the climate change, especially the CO_2 problem are classified and critically evaluated in terms of their possibility, scale, merits and demerits.

CO_2 MITIGATION TECHNOLOGIES AND THEIR EVALUATION

To evaluate the measures, the most important point is the term during which the problem is to be mitigated. How long should we consider? If we considered just "the present", the evaluation would be too simple; we need not reduce the CO_2 emission. We should recognize that the fossil fuel which is the main cause of the CO_2 problem will last around several hundreds years. However, several hundred years are still thought to be too long for us. Here, we will consider what we should do now to realize the ideal direction of technology or society of several decades later, taking into account the fact that fossil fuels will be exhausted within several hundred years.

We will here consider the global, not the local problem. The south-north problem is possibly one of the most important and essential aspects of the CO_2 problem. Some other environmental effects may also occur when some of these measures are undertaken. These factors should also be included in the present evaluation.

Classification and Evaluation Maps of Technologies

The measures for global warming are first classified into four categories: cooling the earth, symptomatic measures against possible effects of global warming, reduction in non-CO_2 gas concentrations and that of CO_2. As for the CO_2 problem, they are conventionally classified as follows.

a) Primary energy: changes from fossil fuel resources to renewable or nuclear energies and from higher carbon content fossil fuels to lower ones.
b) Secondary energy system: energy conservation, material and energy recycle, change in life style, and improvement in efficiencies of energy conversion, storage, transportation and use.
c) CO_2 from other than energy.
d) CO_2 recovery and storage: separation system, ocean disposal, subterranean storage, and biological and chemical fixation.
e) CO_2 absorption from atmosphere: afforestation/reforestation, rock weathering, and ocean absorption.
f) Policy and economic options: carbon, energy and environmental taxes, emissions market, etc.

Before describing the detailed evaluation of the individual measures in the order of the conventional classification, first of all, we present several selected evaluation factors, and the evaluation results based on the following discussion are shown as maps in Figures 21.1–21.3.

First of all, we adopted "scale" as the ordinate and "stability" as the abscissa of Figure 21.1. The word "scale" refers to the possible amount of CO_2 (incl. other greenhouse gases; this note is omitted hereafter) reduced when the measures are fully carried out. The word "stability" refers to the possible ratio of CO_2 reduction to the possible CO_2 emission during the performance of the measures. The measures after which the possibility of the re-release of CO_2 is pointed out are also evaluated as non-stable measures. The measures with both large scale and high stability are considered to have a high potential.

Two other evaluation factors are the effects of these measures on other environmental problems and on the resources problems. The abscissa of Figure 21.2 is the contribution of these measures to the conservation of various kinds of non-renewable resources, e.g., fossil fuels and metal resources. For the case when the measure consumes more energy and/or other non-renewable resources than the case without carrying out it, it is evaluated as being negative. For the case when the measure causes a shift from one with less reserves to another with more reserves, it is positive, and vise versa. The other factor is the effects of these measures on other environmental problems, which is shown on the ordinate of Figure 21.2. If the measure is of benefit to other environmental problems, it should be carried out even if it requires more energy and/or other non-renewable resources.

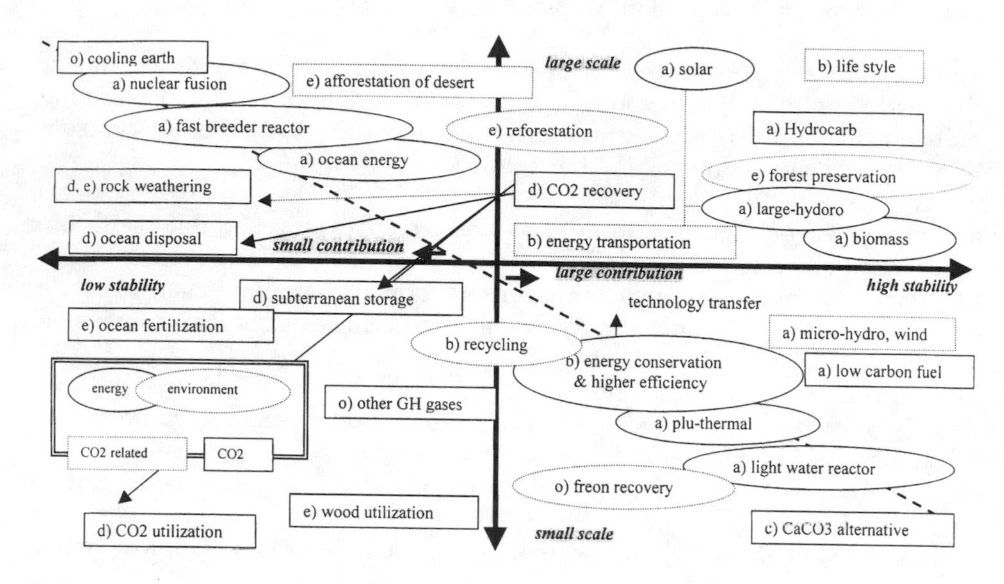

Figure 21.1 Map of measures against CO_2 problem-1; scale and stability (Kojima, 1999).

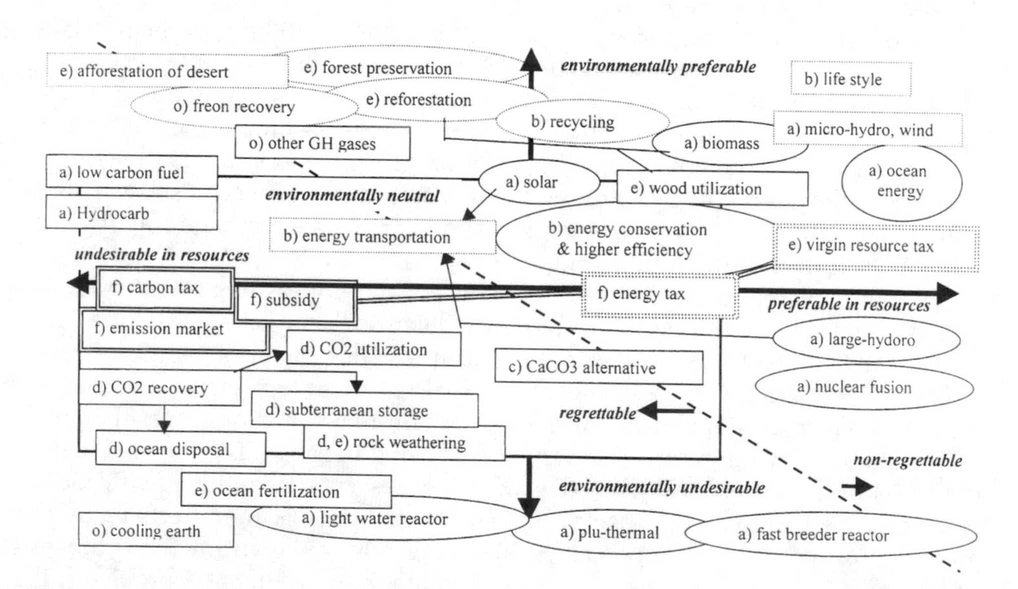

Figure 21.2 Map of measures against CO_2 problem-2; effects on resources and other environmental problems (Kojima, 1999).

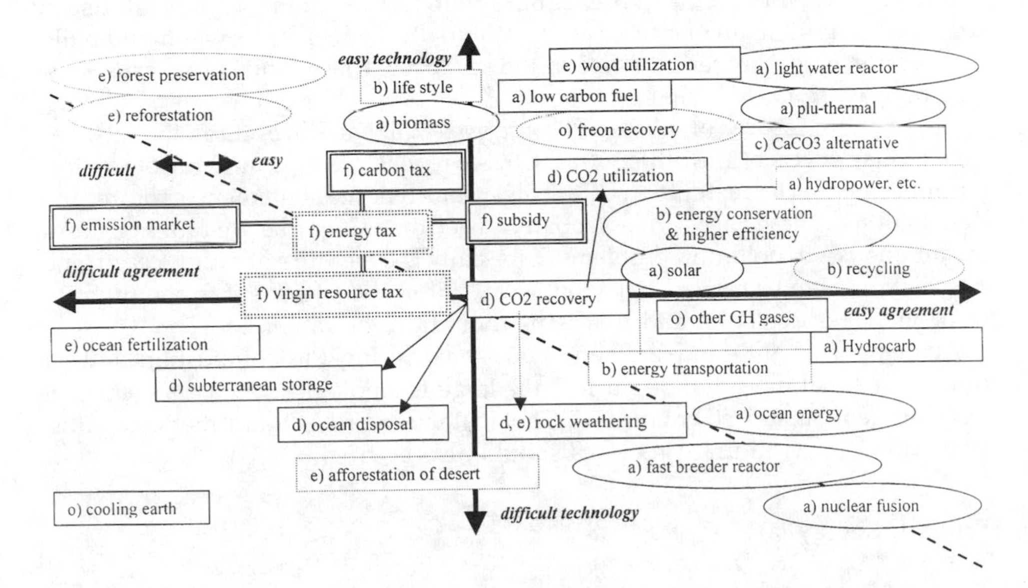

Figure 21.3 Map of measures against CO_2 problem-3; technological and international difficulties (Kojima, 1999). Figures 21.1–21.3 Reproduced from *Multi-Channel Evaluation of CO_2 Mitigation Technologies*, T. Kojima, in Proc. 5th ASME/JSME Joint Thermal Engineering Conference, March 15–19, 1999, San Diego, copyright © 1999 ASME, with permission from ASME.

Finally, we are able to classify the various measures as above into two categories; regrettable and non-regrettable. The measures which give positive effects on the resources and/or other environmental problems are classified into non-regrettable measures: those that will not be regretted even if global warming does not in the end take place, or if its effect is minimal. The boundary of the two categories is shown as the diagonal line in Figure 21.2.

The last two evaluation factors are related to the difficulties of the measures. The difficulties arise from technological and international problems shown as the ordinate and the abscissa, respectively, in Figure 21.3. The technological difficulty also implies economic difficulties, because some economic difficulties are possibly solved by adequate technologies. The international difficulty means the difficulty in attaining international agreement.

Cooling, Symptomatic and Miscellaneous Greenhouse Gases (o)

One possible way to *cool the earth* is to increase the aerosols which interrupt the sunlight and another is to increase the albedo, i.e., reflectivity of the earth surface. But these are thought to be unrealistic. The global temperature might be reduced by the cold water in the deep ocean. But it will just result in a significant rise in sea levels. The ultimate measures are symptomatic. For example, we should build up some levees to prevent the rise in the sea level. These kinds of measures are also omitted from the maps.

As for the measures of *other greenhouse gases*, Global Warming Potentials, i.e., greenhouse effects of the other gases, CFCs (freons), methane and nitrous oxide, per mole are much greater than CO_2. This means that the abatement, capture, and destruction of these gases are essentially easier than CO_2, like we fought against the air and water pollution problems. Especially the measure of *freon recovery* and destruction, is to be carried out soon because it is also of benefit to the problem of ozone layer depletion and its cost is thought to be rather cheap.

Abatement of N_2O and methane emissions during the combustion and/or mining of fossil fuels is thought to have large technological potentials. Further development should be conducted. The abatement of N_2O and methane emissions during agricultural processes should also be tried.

Primary Energy (a)

The total world production of nuclear and *hydro*power amounts to merely a little over 10% of the total energy production, though most of *biomass* energy is not included in the statistics. Only a very small amount of energy is produced from solar cell or solar thermal and other renewable energies other than hydropower. The remainder (nearly 90%) is accounted for by fossil fuels. As long as fossil fuels are burned, carbon dioxide is given off. For various fossil fuels, Table 21.1 shows the typical values of the hydrogen/carbon (H/C) ratio, the higher heating value, the amount of carbon dioxide released, and the size of existing reserves plus their oil equivalents.

Natural gas has the greatest hydrogen content, followed by oil, whereas coal contains the maximum amount of carbon. If the use of coal and oil is completely replaced by the adoption of natural gas, which would be expected to bring about a considerable reduction in carbon dioxide emissions. However, the estimated reserves of natural gas is less than 2% of coal. If coal were no longer used, even if we used all estimated reserves of both natural gas and oil (including the uneconomic ones), supplies would be exhausted in 64 years (see E/T values in Table 21.1). This figure indicates that the conversion to *low-carbon fuels* has no meaning in the term of one hundred years, unless innovative technologies are developed for solar or other natural energy use. These measures of shift from rich to rare resources are regrettable measures. It may technologically possible to burn only the hydrogen without using carbon in the coal (*Hydrocarb process*). But it is also an obviously regrettable measure.

At least, developed countries should not take this option because many developing countries are presently using coals in an inefficient and highly-polluting manner. It is necessary for advanced nations to quickly establish alternative ways of handling renewable energy (including conversion and transportation) while minimizing the use of fossil fuels which will be exhausted within several hundred year even if the CO_2 problem did not exist. The CO_2 reduction in developed countries should not be done just by the conversion to low-carbon fuels or by economical conversion to less energy-intensive industries.

The situation is same for the atomic energy using *light water reactors* from the view point of resources. The use of fuels mixed with plutonium (plu-thermal) will not improve the situation very much. The development of *fast breeder reactors* is essential to effectively use the uranium resources, and from the view points of energy resources and CO_2 problems, though their effects on the environment or their safety should also be considered. The large-scale use of natural energy may also cause other environmental problems, e.g., *large scale hydro-power*. *Nuclear fusion* and *ocean energy* may also arise environmental problems though they will be future technologies.

The output of a *solar cell* is different based on the area where the cell is installed. The *transportation of these renewable energies* is technologically essential if we consider the development of these energies. *Biomass* energy is also essentially renewable, but it is often coupled with deforestation. Furthermore, its conversion efficiency from solar energy is much lower than the solar cell. It should be limited to local use.

The needs or consumption of energy, especially electricity in developing countries will drastically increase in the near future though it is the usual case that they do not have sufficient infrastructures for energy transportation. Based on the above conditions, it may be the best choice to install solar cell, *wind or micro-hydropower* for local electricity supplies in developing countries, with financial and technological support from developed countries.

Secondary Energy System and Energy Conservation (b)

The importance of a *change in our life style* and *energy conservation* goes without saying. From the "global reduction in CO_2" viewpoint, technology transfer is one of the most important issues as well as the development of *high efficient* technologies for energy conversion and use. The *Recycling* of waste comes from the problem of the lack of final sites for waste disposal and more energy may be necessary for some materials to be recycled. A critical evaluation is essential. Also only the energy from non-renewable resources could not be recycled, leading to CO_2 emissions, even though all of the materials could be recycled with energy.

The conservation of energy could be realized with various kinds of efficiency improvements: power generation, home electrical tools, energy conservative houses, and so on. However, the author thinks that the most promising way to reduce CO_2 emission is the improvement of the transportation system. In other words, one must construct the best system around the cities. In addition, the use of unused energy should be effectively utilized, though it usually costs more. It may also be said that future cities should be established at sites near natural energy supplies.

CO_2 from Other than Energy (c)

The amount of CO_2 emitted from cement manufacturers by the calcination of lime is less than several percent of the fossil energy use but their contribution cannot be neglected. If cement is produced from *alternative silicates to lime*, CO_2 emission from cement processes will be drastically reduced (Kojima *et al.*, 1998). Slag from blast furnaces is thought to be the best raw material for cement processes. All of the slug from the blast furnaces should be recycled to the cement manufacturing from the viewpoint of CO_2 reduction.

Recovery and Storage (d)

Separation or *recovery* of CO_2 from flue gas requires energy. As the amount of *CO_2 artificially used* or stocked is extremely small, it is essential to sequester the recovered CO_2. Various sites are nominated such as *subterranean* and *ocean*. However, this needs more energy. Therefore, these measures are essentially regrettable options. Investigation or research work is necessary to evaluate these measures but it is unrealistic that these will be carried out soon. These measures may be conduced only when the CO_2 effects become serious and other more economical or energy-conservative options are found not to be effective for the prompt improvement of the effects.

Various chemical and biological processes have so far been proposed for carbon dioxide fixation; artificially recovered CO_2 is chemically or biologically converted into organic matter that is re-utilized. In these processes, the equivalent of more energy should be input to convert CO_2 to organic matter than that produced when CO_2 is emitted. The source of the energy should be non-fossil fuel energy (denoted as A) to avoid any additional CO_2 emission. In the case that the organic matter is again used to produce electricity (B), CO_2 is again emitted from the power plant, but is again recovered and "fixed". Finally, B is produced from A without the consumption of fossil fuel or CO_2 emission, only with CO_2 recycling at an ideal steady state. This process is essentially one of the conversion processes from A to B. Thus the process should be evaluated and compared with other paths from A to B, such as electricity production using solar cells. Such a process when CO_2 is used as a medium of transportation of a renewable energy, e.g., methanol/CO2 cycle, should be evaluated along with other transportation systems, e.g., liquid hydrogen transportation and the benzene/hexane system.

When the organic matter produced from CO_2 is used as chemical stocks or automobile fuels, the emitted CO_2 cannot be recycled. In this case, the chemical and biological processes are thought to be one option for us to convert a renewable energy (A) into a secondary energy or a chemical stock (C) using CO_2 that is originally produced with electricity (B) production from the fossil fuel (D). This process is to be evaluated by comparing it, with the system that A is used for B production while C is produced from D.

In both cases, CO_2 acts as just a medium for the use of the renewable energy and these are not CO_2 fixation processes. Furthermore, CO_2 is the most stable form of carbonaceous material. It is usually nonsense to use the lowest energy level chemicals as a medium. It is essential to convert from D to A but it is not necessary to use CO_2 as a medium, from the view point of the CO_2 problem. Therefore, the chemical and biological fixation processes of recovered CO_2 are excluded in the maps.

CO_2 Absorption from Atmosphere (e)

Chemical and biological processes which should be separately evaluated from new soft energy paths are only those which essentially require no artificial energy, such as the afforestation of deserts, ocean fertilization (to increase biota in the ocean) and utilization of the rock weathering reaction.

It is said that 60 atm of CO_2 at the beginning of the earth's history has been absorbed by the *weathering* of alkaline silicate and aluminate rocks. The bicarbonates thus produced were converted and fixed as calcium carbonates like coral reefs. Thus CO_2 fixation using inorganic chemicals such as the weathering of silicate rocks is thought to be one of the most reasonable ways to absorb CO_2 from the atmosphere, though its rate is too low (Kojima *et al.*, 1996). How to increase its rate is the key to this technology.

The growing of coral reefs will essentially not contribute to the CO_2 reduction, because ca. 0.6 mole of CO_2 is emitted with 1 mol of $CaCO_3$ formation from 1.6 mol of bicarbonate ions. Therefore, the measure of coral reefs growing is excluded in the maps. However, if a large amount of Ca ion is supplied (e.g., by rock weathering), organic matter is produced using a lower amount of P and N, and the produced organic matter is transported to the deep ocean, it may become one possible measure, though scientific evidence is essential.

As an average, the supply of N and P is short to grow biota in the ocean. Therefore, *fertilization of the ocean* with these nutrients could be effective for increasing the ocean absorption though this kind of measures may produce other environmental problems such as a deficiency of dissolved oxygen. Furthermore, the lack of phosphate resources is pointed out. Fertilization of the ocean with iron may be a promising measure because of the small amount of iron is necessary, though the target area is restricted compared to the ocean fertilization with N and/or P. However, most of these measures should be classified into regrettable ones.

On the other hand, the measure of the afforestation is classified as a non-regrettable one from the viewpoint of the environment, though a small amount of additional energy should be input. It is true that there will not be a net absorption of CO_2 after the maturation of forests, however, we should recognize the present situation that CO_2 from the deforestation at the rate more than ten Mha/y is one to two-fifths of that from fossil fuels. Furthermore, this figure is equivalent to half the CO_2 amount accumulating in the atmosphere. This figure indicates that the accumulation would be stopped if the deforestation were stopped and afforestation were continued at the same rate of ten Mha/y. It should be pointed out that the most important and possible measure is *forest preservation* and the *reforestation* of commercial forests. The price of products from forests should be higher; the extra money should be again invested in reforestation.

Afforestation of others than deserts may cause a battle with agriculture. Furthermore, the desert area is sufficiently large. We should consider that the desertification is now expanding, possibly artificially. The use of young trees for cooking, etc., prevents their growth, evapotranspiration is reduced and run-off is increased, which leads to the reduction in the precipitation of the specific area. The only measure to prevent it is afforestation with the reasonable management of water, which results in carbon fixation.

The improvement in the energy efficiency of this area is of benefit to the global CO_2 problem. The conversion of solid fuel, even if it is coal, into a fluid for its efficient use is to be one of the most important measures, as well as renewable energy use.

Policy and Economic Options (f)

Various policy and economic options have also been proposed. However, we should consider the effects of these options on the energy utilization system. One proposal for restraining the emission of CO_2 is the introduction of a *carbon tax*. For the same amount of energy produced, the tax levied on natural gas would be half of that levied on coal. Thus the regrettable measure of the change from coal to natural gas or nuclear fuel will occure. The introduction of an *emission market* of CO_2 will cause the same effects. The *energy tax* will not cause such a problem, however, the introduction of renewable energies will not be enhanced by the tax. Some additional options such as a *subsidy* should be introduced together with the energy tax. Taking these conditions into account, a *tax on all virgin resources* including all fossil and nuclear fuels and resources from forests without reforestation is thought to be the most suitable, though the critical estimation of reserves and a relative evaluation among the resources are essential. If these options are carried out in a nation, an equivalent amount of taxes should be assessed/reduced to imported/exporting manufacturers based on LCI (life cycle inventory) data.

In Figure 21.2, these measures were classified into two categories: regrettable and non-regrettable. The measures that give positive effects on the resources and/or other environmental problems were classified into non-regrettable. It is obvious that the non-regrettable measures with a larger capacity and reliability, such as reforestation and solar, should be undertaken as early as possible, however, most of these measures have their own difficulties, from technological, economical, and/or international points of view. The social system which promotes these measures should be established.

The present author is certainly not opposed to a carbon tax in itself, but from the above argument, it would seem that the tax should really be an energy tax with subsidies for renewable energies and reforestation, etc. Again a tax on all virgin resources is recommended to comprehensively promote the non-regrettable measures.

FINAL REMARKS

In the present chapter, the history from the pollution to the global environment was first reviewed. It was demonstrated that the role of engineers, especially that of chemical engineers, was significant in the course to combat pollution problems.

The global warming/climate change issue was then focussed on. Various measures against the CO_2 problem were listed, classified and critically evaluated. In the course of the evaluation, it was clarified that the chemical and biological fixation processes using other sources of energy are not intrinsic measures for CO_2 abatement. This conclusion is naturally acceptable if we use the concept of mass/energy balance within a system boundary, which appears during the first stage of a chemical engineering course. We should not draw the boundary between CO_2 emission and energy conversion, but should evaluate the entire process taking the energy source into account.

Yes, the role of chemical engineers is also expected to be significant in the course of the developing individual measures, not only in the conventional areas, e.g., the renewable energy, energy conservation, recovery, recycling and so on, but also in the frontier areas of chemical engineers, e.g., afforestation, utilization of oceans, etc. A group of Japanese chemical engineers is now trying to demonstrate the possibility of the afforestation of deserts as a cure for the CO_2 problem. Of course, we do not have any detail knowledge of specific areas and we should cooperate with soil scientists, meteorologists, foresters, hydrologists and so on. However, our main task is to guide the entire project to the final goal of effective CO_2 fixation combining the knowledge of scientists, like the construction of a chemical plant in cooperation with other engineers, such as electrical and mechanical engineers.

Our recent society is very complicated. Problems are not only environmental ones. Chemical engineers are system engineers who can qualitatively evaluate material and energy flows. I believe chemical engineers can solve complicated environmental problems and contribute to the construction of an ideal future society system.

REFERENCES

Kojima, T. (1998) *The Carbon Dioxide Problem*. London: Gordon and Breach (translated and English edited by Harrison, B. Originally published in Japanese as *Nisankatansomondai Uso to Honto* by Agune Shofu Sha, Tokyo, 1994)

Kojima, T. (1999) *Multi-Channel Evaluation of CO_2 Mitigation Technologies*. In *Proc. 5th ASME/JSME Joint Thermal Engineering Conference*, March 15–19, 1999, San Diego

Kojima, T. and Saito, S. (2000) Refer to Chap 20 *Recovery and Recycling of Resources.*, of this book.

Kojima, T., Nagamine, A., Ueno, N. and Uemiya, S., (1996), "Absorption and Fixation of Carbon Dioxide by Rock Weathering." *Energy Convers. & Mgmt.*, **38**, S461–S466

Meadows, D. H., Meadows, D. L., Randers, J. and Behrens III, W. W. (1972) *The Limits to Growth*. New York: University Press

INDEX